Autonomous Marine Vehicle Operations

Editors

Xiao Liang
Rubo Zhang
Xingru Qu

MDPI

Basel • Beijing • Wuhan • Barcelona • Belgrade • Novi Sad • Cluj • Manchester

Editors

Xiao Liang	Rubo Zhang	Xingru Qu
Dalian Maritime University	Dalian Minzu University	Dalian Minzu University
Dalian	Dalian	Dalian
China	China	China

Editorial Office
MDPI AG
Grosspeteranlage 5
4052 Basel, Switzerland

This is a reprint of articles from the Special Issue published online in the open access journal *Journal of Marine Science and Engineering* (ISSN 2077-1312) (available at: https://www.mdpi.com/journal/jmse/special_issues/598X93147M).

For citation purposes, cite each article independently as indicated on the article page online and as indicated below:

Lastname, A.A.; Lastname, B.B. Article Title. *Journal Name* **Year**, *Volume Number*, Page Range.

ISBN 978-3-7258-2397-0 (Hbk)
ISBN 978-3-7258-2398-7 (PDF)
doi.org/10.3390/books978-3-7258-2398-7

Contents

About the Editors

Xiao Liang

Xiao Liang received the B.S. degree in Naval Architecture and Ocean Engineering from Harbin Engineering University, China, in 2003, the M.S. degree in Fluid Mechanics from Harbin Engineering University, in 2006, and the Ph.D. degree in Design and Construction of Naval Architecture and Ocean Structure, in 2009. He is currently a Professor with the School of Naval Architecture and Ocean Engineering, Dalian Maritime University, China. He has authored and co-authored more than 80 papers in scientific journals. His current research interests include decision making and advanced control, unmanned technology, and swarm intelligence in maritime applications.

Rubo Zhang

Rubo Zhang received the B.S. degree in Automatic Control of Aircraft and the M.S. degree in Navigation Control of Aircraft from Harbin Shipbuilding Engineering Institute, Harbin, China in 1984 and in 1987, respectively. He received the Ph.D. degree in Control Theory and Control Engineering from Harbin Engineering University, Harbin, China in 1999. He is currently a Professor with the School of Mechanical and Electronic Engineering, Dalian Minzu University, Dalian, China. He has published over 100 papers in journals and conferences. His research interest includes the intelligent robot hardware and software architecture, multi-robot technology, and autonomous decision making technology in complex environments.

Xingru Qu

Xingru Qu received the B.S. degree in Naval Architecture and Ocean Engineering from Shandong Jiaotong University, Jinan, China, in 2015, and the M.S. and Ph.D. degrees in Design and Construction of Naval Architecture and Ocean Structure from Dalian Maritime University, Dalian, China, in 2018 and in 2022, respectively. He is currently a Lecturer with the School of Mechanical and Electronic Engineering, Dalian Minzu University, Dalian, China. He has published over 20 papers in journals and conferences. His research interest includes guidance and control as well as game of autonomous marine vehicles.

Journal of
Marine Science and Engineering

MDPI

Editorial

Autonomous Marine Vehicle Operations

Xiao Liang [1,*], Rubo Zhang [2,*] and Xingru Qu [2,*]

1 School of Naval Architecture and Ocean Engineering, Dalian Maritime University, Dalian 116026, China
2 School of Mechanical and Electronic Engineering, Dalian Minzu University, Dalian 116600, China
* Correspondence: liangxiao@dlmu.edu.cn (X.L.); zhangrubo@dlnu.edu.cn (R.Z.); qxr@dlnu.edu.cn (X.Q.)

Citation: Liang, X.; Zhang, R.; Qu, X. Autonomous Marine Vehicle Operations. *J. Mar. Sci. Eng.* **2024**, *12*, 355. https://doi.org/10.3390/jmse12020355

Received: 22 January 2024
Accepted: 28 January 2024
Published: 19 February 2024

1. Introduction

The world has witnessed the rapid development of autonomous marine vehicles, such as surface vehicles and underwater vehicles, which have created fruitful innovative approaches to previously unsolvable problems in marine and ocean engineering [1,2]. In practice applications, single-vehicle operations inevitably face many difficulties and challenges, such as non-linearity, strong coupling, multi-inputs/multi-outputs, uncertainties and multi-constraints. In addition, the information interaction-based swarm system has complicated dynamics, such as high state dimensions and complicated connection topology, thus leading to great challenges for autonomous operations. At present, the main subjects for autonomous marine vehicle operations can be divided into perception, decision making and control. In this context, this Special Issue "Autonomous Marine Vehicle Operations" has been launched, including 11 selected hot topics.

Perception is the essential precondition for autonomous and intelligent operations [3,4]. Environment information with high precision is difficult to obtain by a single sensor. In this context, multi-sensor fusion-based perception technology receives more attention. All kinds of sensors are used to realize multi-level and multi-spatial information complementation and optimal combination, thus generating the consistency explanation for the observation environment. With the challenges of complicated marine environments, such as ocean disturbances, uneven illumination and non-significant targets, the vehicles need to realize information fusion of the appointed target based on the typical attributes of each sensor. The system navigation unit and electronic chart are utilized to build a multi-dimensional tridimensional situation picture of the task environment, contributing to the tasks of target tracking, target detection, target recognition and cognition. For example, in [4], a space-scale attention-based context-aware detection network model is proposed to realize marine multi-scale target detection and recognition. In [5], considering the distributed multi-platform data fusion, by utilizing the Dempster–Shafer evidence theory to analyze multi-source information, the confidence conflict of different information sources during information fusion is addressed successfully.

Decision making is at the core of autonomous and intelligent operations [6,7]. Under complicated marine environments, obtaining effective information and generating feasible control commands face great challenges. According to the differences in control objectives and spatial constraints, the main results of single-vehicle decision making include target tracking, path following and trajectory tracking, where light-of-sight guidance, backstepping and artificial potential field are widely utilized [8,9]. The main results of swarm decision making include cooperative target tracking, cooperative path following and cooperative trajectory tracking, where leader following and graph theory are widely employed [10]. Considering operation environments and operation constraints, marine vehicles are allocated one or more tasks in order during decision making. And the objective is to optimize the system task efficiency in terms of task types, spatial constraints and system performance [11]. In this context, the allocation model and the allocation algorithm are available, where the former transforms the allocation problem to the multi-traveling salesman or mixed integer linear programming, and the latter is responsible for obtaining a solution based on the Hungarian algorithm, ant colony algorithm, game theory, Bayesian theory, auction algorithm, etc.

Control is utilized to stabilize and control the system attitude, forcing vehicles to accomplish the desired command generated by decision making [12]. According to the differences in solving disturbances, the main results include passive anti-disturbance control and active anti-disturbance control. The former depends on the inherent robustness, such as adaptive control, sliding mode control and model predictive control, etc. And the latter is also called disturbance observation-based control, where the unknown disturbance is approximated by the approximator/observer and then applied to the feedback controller as the feedforward compensation. The main types of observation-based control include neural network approximators, fuzzy logic systems and normal model-based observers. For example, in [13], the radial basis function neural network-based finite-time controller is designed. In [14], the fixed-time disturbance observation-based terminal synovial controller is proposed, reducing the time of disturbance estimation efficiently. In [15,16], based on the form of differential inclusion, a finite time observer is designed to compensate system unknowns and time-varying disturbances exactly. In [17], the active anti-disturbance controller is developed by combining a fuzzy logic system and speed error-based prediction, guaranteeing that the desired command can be tracked accurately despite the existence of unknown dynamics.

2. An Overview of Published Articles

This Special Issue focuses on autonomous marine vehicle operations under complex ocean environments and contains 12 published articles. The main contributions are as follows.

In contribution 1, considering the influence of situational static loads and varying hydrodynamic forces during high-speed movement, a novel S-plane controller is developed for an underwater vehicle by using a sliding mode variable structure. Prototype experiments including cruise control and path-following control are successfully carried out, thereby demonstrating the superiority of the S-plane controller and enriching our motion control technology for marine vehicles.

In contribution 2, to enhance the identification accuracy of underwater objectives under the complex environment, an accurate identification and detection method is employed using the sizeable convolutional-network-based You-Only-Look-Once (YOLO) algorithm. With the aid of the improved YOLO network, the problem of low image quality and dense objectives is well alleviated. The proposed network model can provide effective guidance for the intelligent aquaculture of fishes.

In contribution 3, two key problems including path planning and tracking control of USVs are considered. Within the planning module, a safe and optimal path can be generated using rapidly exploring randomized trees. Within the control module, an active anti-disturbance controller is designed for unmanned vehicles with unknown dynamics. In addition, considering the potential occurrence of thruster failures, different fault-tolerant schemes are developed based on the neural-network-based model predictive control. The proposed path planning and control methods can guide obstacle avoidance for marine vehicles in faulty conditions.

In contribution 4, an automatic alignment approach of an underwater charging system is employed with the aid of monocular vision recognition, where the vehicle number can be identified, guiding the charging pile to accurately insert into the charging port of the vehicle. To enhance the accuracy and robustness of decoding, this research proposes a redundant information-based encoding and ArUco code reconstruction approach. In addition, the target position can be determined, thereby overcoming the difficulty of an underwater two-dimensional location and meeting the accuracy requirements for alignment.

In contribution 5, considering the complexities of the maritime environment and the non-holonomic characteristic of the operation system, a novel path planning method for surface vehicles is creatively established, consisting of an optimization model, a meta-heuristic solver, and a Clothoid-based path connector. By virtue of the proposed path planner, a path with both optimally safe and quick convergence can be generated, enhancing its adaptability to time-varying environments. Compared with the existing results, this study guarantees the continuity of paths and the consistency of planners and controllers, solving the non-holonomic limitations.

In contribution 6, an underwater object detection model is developed for underwater unmanned vehicles by using a mobile vision transformer and YOLOX, providing a good balance between accuracy and memory. In addition, the double coordinate attention strategy with fewer parameters is built to strengthen the ability of extracting data from difficult objectives. This study is beneficial for underwater vehicles to identify small and difficult targets in water.

In contribution 7, to enhance the hydrodynamic performance of the underwater vehicle, the steady-stream (SS) active flow control (AFC) method is employed via numerical calculation. The mechanism of the SS-AFC influencing the lift-to-drag performance is revealed in terms of the flow field, contributing to operations of underwater vehicles with large angles of attack. This proposed method improves the hydrodynamic performance of the vehicle, thus guaranteeing vehicle stability, maneuverability and safety.

In contribution 8, a swarm key node identification method is developed using network structure entropy to address the critical identification of underwater vehicles. The network structure of multiple vehicles is built using the motion similarity model and the information entropy of swarm nodes is solved by the aid of the weighted network structure entropy method. Simulation and lake experiments are carried out, where the time-varying trajectory of the swarm and the importance ranking of the swarm nodes can be successfully calculated. This research provides a valuable reference for underwater cluster countermeasures.

In contribution 9, a dynamic data-driven operation system is employed, effectively overcoming the difficulties of high-resolution and accurate flow field forecast in ocean environments. The neural network structure is developed using information extracted from historic flow data, thereby enhancing the flow forecast performance. In addition, the Kalman filter is applied to assimilate spatially correlated flow-sensing data from vehicles, thereby enabling efficient learning and accurate flow forecasts. This study offers a feasible solution to high-resolution and accurate flow field forecasts in practice engineering.

In contribution 10, an accurate path-following controller including kinematics and dynamics is developed for an unmanned surface vehicle suffering from system uncertainties and wind, wave and current disturbances. At the kinematics level, the desired guidance signals can be generated by the proposed surge-heading joint guidance method. At the dynamics level, the deep-reinforcement-learning-based surge and heading control laws are designed using the error feedback between the actual and desired signals. In addition, actor networks and critic networks are established by utilizing the long-short time memory network, helping the vehicle to take advantage of historical data. This study provides an optional plan for path-following operations of surface vehicles.

In contribution 11, the hydrodynamic interaction between the underwater vehicle and the submarine is studied using computational fluid dynamics, where the two systems are defined as relatively static states. Simulation tests show that, in the recovery phase, the submarine appreciably affects the velocities and relative attitudes of underwater vehicles. This research offers a valuable reference for submarine recovery operations involving underwater vehicles in terms of stability, safety and efficiency.

In contribution 12, aiming at small target detection on water surfaces using cameras in complex environments, a novel millimeter-wave radar-based visual detection technology is developed, achieving robust coordination of radar data and images in the presence of inaccurate extrinsic parameters. Note that the developed technology has lower computational complexity and has been successfully applied in practical engineering. And the results show that the proposed target detection technology has obvious advantages compared with the existing work.

3. Conclusions

This Special Issue delves into the forefront of autonomous marine vehicle operations in terms of perception, decision making and control. Within the perception community of surface and underwater vehicles, an improved YOLO mode for identification and a mobile vision transformer-based object detection model are developed under the complex environment, alleviating the situation of low-quality underwater images and dense objectives. Within the decision-making community of vehicles, with the aid of geometric optimization, several path planning methods have been successfully developed, guaranteeing path continuity and reliable coordination. Within the control community, improved S-plane and reinforcement-learning-based dynamic control is proposed for underactuated systems suffering from model uncertainties and environment disturbances, achieving accurate path following and motion control. Additionally, this Special Issue explores the hydrodynamic interaction of the submarine and the underwater vehicle by using computational fluid dynamics, providing a valuable reference for submarine recovery operations. The current studies presented in this Special Issue should not only be considered as the results of an investigation accomplished by the respective scholars, but as a starting point, encouraging readers to continue with new studies.

Acknowledgments: As Guest Editors of the Special Issue "Autonomous Marine Vehicle Operations", we would like to express our gratitude towards all the authors whose valuable works were published.

Conflicts of Interest: The authors declare no conflicts of interest.

List of Contributions:

1. Jiang, C.; Lv, J.; Wan, L.; Wang, J.; He, B.; Wu, G. An Improved S-Plane Controller for High-Speed Multi-Purpose AUVs with Situational Static Loads. *J. Mar. Sci. Eng.* **2023**, *11*, 646. https://doi.org/10.3390/jmse11030646.
2. Zhou, S.; Cai, K.; Feng, Y.; Tang, X.; Pang, H.; He, J.; Shi, X. An Accurate Detection Model of *Takifugu rubripes* Using an Improved YOLO-V7 Network. *J. Mar. Sci. Eng.* **2023**, *11*, 1051. https://doi.org/10.3390/jmse11051051.
3. Song, Y.; Chen, Y.; Gao, J.; Wang, Y.; Pan, G. Collision Avoidance Strategy for Unmanned Surface Vessel Considering Actuator Faults Using Kinodynamic Rapidly Exploring Random Tree-Smart and Radial Basis Function Neural Network-Based Model Predictive Control. *J. Mar. Sci. Eng.* **2023**, *11*, 1107. https://doi.org/10.3390/jmse11061107.
4. Yu, A.; Wang, Y.; Li, H.; Qiu, B. Automatic Alignment Method of Underwater Charging Platform Based on Monocular Vision Recognition. *J. Mar. Sci. Eng.* **2023**, *11*, 1140. https://doi.org/10.3390/jmse11061140.
5. Wang, F.; Bai, Y.; Zhao, L. Physical Consistent Path Planning for Unmanned Surface Vehicles under Complex Marine Environment. *J. Mar. Sci. Eng.* **2023**, *11*, 1164. https://doi.org/10.3390/jmse11061164.
6. Sun, Y.; Zheng, W.; Du, X.; Yan, Z. Underwater Small Target Detection Based on YOLOX Combined with MobileViT and Double Coordinate Attention. *J. Mar. Sci. Eng.* **2023**, *11*, 1178. https://doi.org/10.3390/jmse11061178.
7. Du, X.; Liu, X.; Song, Y. Analysis of the Steady-Stream Active Flow Control for the Blended-Winged-Body Underwater Glider. *J. Mar. Sci. Eng.* **2023**, *11*, 1344. https://doi.org/10.3390/jmse11071344.
8. Chen, Y.; Liu, L.; Zhang, X.; Qiao, W.; Ren, R.; Zhu, B.; Zhang, L.; Pan, G.; Yu, Y. Critical Node Identification of Multi-UUV Formation Based on Network Structure Entropy. *J. Mar. Sci. Eng.* **2023**, *11*, 1538. https://doi.org/10.3390/jmse11081538.
9. Jin, Q.; Tian, Y.; Zhan, W.; Sang, Q.; Yu, J.; Wang, X. Dynamic Data-Driven Application System for Flow Field Prediction with Autonomous Marine Vehicles. *J. Mar. Sci. Eng.* **2023**, *11*, 1617. https://doi.org/10.3390/jmse11081617.
10. Qu, X.; Jiang, Y.; Zhang, R.; Long, F. A Deep Reinforcement Learning-Based Path-Following Control Scheme for an Uncertain Under-Actuated Autonomous Marine Vehicle. *J. Mar. Sci. Eng.* **2023**, *11*, 1762. https://doi.org/10.3390/jmse11091762.

11. Luo, W.; Ma, C.; Jiang, D.; Zhang, T.; Wu, T. The Hydrodynamic Interaction between an AUV and Submarine during the Recovery Process. *J. Mar. Sci. Eng.* **2023**, *11*, 1789. https://doi.org/10.3390/jmse11091789.
12. Zhu, J.; Yang, Y.; Cheng, Y. A Millimeter-Wave Radar-Aided Vision Detection Method for Water Surface Small Object Detection. *J. Mar. Sci. Eng.* **2023**, *11*, 1794. https://doi.org/10.3390/jmse11091794.

References

1. Dong, K.; Liu, T.; Zheng, Y.; Shi, Z.; Du, H.; Wang, X. Visual detection algorithm for enhanced environment perception of unmanned surface vehicles in complex marine environments. *J. Intell. Robot. Syst.* **2024**, *110*, 1. [CrossRef]
2. Li, J.; Zhang, G.; Jiang, C.; Zhang, W. A survey of marine unmanned search system: Theory, applications and future directions. *Ocean. Eng.* **2023**, *285*, 115359. [CrossRef]
3. Huy, D.; Sadjoli, N.; Azam, A.; Elhadidi, B.; Cai, Y.; Seet, G. Object perception in underwater environments: A survey on sensors and sensing methodologies. *Ocean. Eng.* **2023**, *267*, 113202. [CrossRef]
4. Zhang, G.; Lu, S.; Zhang, W. CAD-Net: A context-aware detection network for objects in remote sensing imagery. *IEEE Trans. Geosci. Remote Sens.* **2019**, *57*, 12. [CrossRef]
5. Xiao, F. Multi-sensor data fusion based on the belief divergence measure of evidences and the belief entropy. *Inf. Fusion* **2019**, *46*, 23–32. [CrossRef]
6. Ntakolia, C.; Kladis, G.; Lyridis, D. A fuzzy logic approach of pareto optimality for multi-objective path planning in case of unmanned surface vehicle. *J. Intell. Robot. Syst.* **2023**, *109*, 1. [CrossRef]
7. Qu, X.; Gan, W.; Song, D.; Zhou, L. Pursuit-evasion game strategy of USV based on deep reinforcement learning in complex multi-obstacle environment. *Ocean. Eng.* **2023**, *273*, 114016. [CrossRef]
8. Gu, N.; Wang, D.; Peng, Z.; Wang, J.; Han, Q. Advances in line-of-sight guidance for path following of autonomous marine vehicles: An overview. *IEEE Trans. Syst. Man Cybern.-Syst.* **2023**, *53*, 1. [CrossRef]
9. Wei, H.; Shi, Y. MPC-based motion planning and control enables smarter and safer autonomous marine vehicles: Perspectives and a tutorial survey. *IEEE/CAA J. Autom. Sin.* **2023**, *10*, 1. [CrossRef]
10. Peng, Z.; Wang, J.; Wang, D.; Han, Q. An overview of recent advances in coordinated control of multiple autonomous surface vehicles. *IEEE Trans. Ind. Inform.* **2021**, *17*, 2. [CrossRef]
11. Abbasi, A.; MahmoudZadeh, S.; Yazdani, A. A cooperative dynamic task assignment framework for COTSBOT AUVs. *IEEE Trans. Autom. Sci. Eng.* **2022**, *19*, 2. [CrossRef]
12. Degorre, L.; Delaleau, E.; Chocron, O. A survey on model-based control and guidance principles for autonomous marine vehicles. *J. Mar. Sci. Eng.* **2023**, *11*, 430. [CrossRef]
13. Huang, C.; Zhang, X.; Zhang, G. Adaptive neural finite-time formation control for multiple underactuated vessels with actuator faults. *Ocean. Eng.* **2021**, *222*, 108556. [CrossRef]
14. Gao, Z.; Guo, G. Fixed-time sliding mode formation control of AUVs based on a disturbance observer. *IEEE/CAA J. Autom. Sin.* **2020**, *7*, 2. [CrossRef]
15. Wang, N.; Su, S. Finite-time unknown observer-based interactive trajectory tracking control of asymmetric underactuated surface vehicles. *IEEE Trans. Control Syst. Technol.* **2021**, *29*, 2. [CrossRef]
16. Qu, X.; Liang, X.; Hou, Y.; Li, Y.; Zhang, R. Finite-time sideslip observer-based synchronized path-following control of multiple unmanned underwater vehicles. *Ocean. Eng.* **2020**, *217*, 107941. [CrossRef]
17. Qu, X.; Liang, X.; Hou, Y. Fuzzy State Observer Based Cooperative Path-Following Control of Autonomous Underwater Vehicles with Unknown Dynamics and Ocean Disturbances. *Int. J. Fuzzy Syst.* **2021**, *23*, 6. [CrossRef]

Journal of
Marine Science and Engineering

MDPI

Article

A Millimeter-Wave Radar-Aided Vision Detection Method for Water Surface Small Object Detection

Jiannan Zhu [1,2], Yixin Yang [1,*] and Yuwei Cheng [2,3]

1 School of Marine Science and Technology, Northwestern Polytechnical University, Xi'an 710068, China; jacknyzhu@orca-tech.com.cn
2 ORCA-Uboat, Xi'an 710004, China
3 Department of Electronic Engineering, Tsinghua University, Beijing 100080, China; chengyw@orca-tech.com.cn
* Correspondence: yxyang@nwpu.edu.cn

Abstract: Unmanned surface vehicles (USVs) have wide applications in marine inspection and monitoring, terrain mapping, and water surface cleaning. Accurate and robust environment perception ability is essential for achieving autonomy in USVs. Small object detection on water surfaces is an important environment perception task, typically achieved by visual detection using cameras. However, existing vision-based small object detection methods suffer from performance degradation in complex water surface environments. Therefore, in this paper, we propose a millimeter-wave (mmWave) radar-aided vision detection method that enables automatic data association and fusion between mmWave radar point clouds and images. Through testing on real-world data, the proposed method demonstrates significant performance improvement over vision-based object detection methods without introducing more computational costs, making it suitable for real-time application on USVs. Furthermore, the image–radar data association model in the proposed method can serve as a plug-and-play module for other object detection methods.

Keywords: unmanned surface vehicle; object detection; visual–radar fusion

Citation: Zhu, J.; Yang, Y.; Cheng, Y. A Millimeter-Wave Radar-Aided Vision Detection Method for Water Surface Small Object Detection. *J. Mar. Sci. Eng.* **2023**, *11*, 1794. https://doi.org/10.3390/jmse11091794

Academic Editor: Mohamed Benbouzid

Received: 23 August 2023
Accepted: 27 August 2023
Published: 14 September 2023

1. Introduction

In recent years, unmanned surface vehicles (USVs) have been gradually used in various fields, such as autonomous surface transportation [1], water quality testing [2], autonomous surface cleaning [3], etc. To ensure that USVs complete their tasks safely and intelligently, an excellent and robust perception system is essential. Among all the perception tasks, object detection plays an important role in both safe navigation and special task completion, and small object detection causes the most challenges, for example, the small reefs and other small obstacles that may affect USVs or small floating wastes that a cleaning USV needs to collect.

Recent development in computer vision makes vision-based object detection one of the most cost-effective solutions for the detection system of USVs. However, for vision-based small object detection on water surfaces, many can be missed and falsely detected due to the water surface environments. For vision-based small object detection on water surfaces, on the one hand, as the sky and water surfaces occupy the most area of the image, the reflection of sunlight may cause overexposure. The small objects can be shaded by the halo or fused with the background, which can cause miss detection. Besides, the reflection of objects in the surrounding environments also disturbs the detection system and causes false detection. In addition to the camera, LiDAR is also widely used for object detection as it can provide precise location and shape information of the objects [4]. However, for the small object detection on water surfaces, for LiDAR with a low number of beams, the possibility of LiDAR beams falling on small objects is low and the objects might be unstable in sequential frames. In addition, dense fog is easy to appear on the water surface, which can disturb the propagation of LiDAR and lead to more clutter points [5].

With the development of integrated circuits, the low-cost single-chip 77 GHz millimeter-wave (mmWave) radar is gradually used in autonomous vehicles and mobile robots recently. The mmWave radar can provide measurements of the range, azimuth, and Doppler velocity of the objects. Besides, benefiting from the inherent propagation characteristics of 77 GHz electromagnetic wave, the mmWave radar shows better robustness to harsh weather conditions and lighting conditions compared to camera and LiDAR [6] and can be used during all types of weather and all day. Despite this, there are still some challenges in using mmWave radar for small object detection on water surfaces. The angular resolution of mmWave radar point clouds is relatively low and the points of the objects are usually more sparse [7]. Furthermore, the semantic information of mmWave radar point clouds is often insufficient, making it difficult to accurately discern the types of targets.

Therefore, for small object detection on water surfaces, vision and mmWave radar data complement each other effectively, and fusion of vision and radar can improve the detection performance. Compared to other levels of fusion, decision-level fusion has greater robustness and adaptability, and the fused results are also more interpretable. However, there are two challenges in the decision-level fusion of camera and radar in USVs scenes:

- Extrinsic Calibration. To perform decision-level fusion, the spatial relationship between the mmWave radar and camera needs to be found, which is referred to as extrinsic calibration. Due to the characteristics of glittery and sparsity of mmWave radar point clouds, extrinsic calibration between mmWave radar and cameras typically requires specific markers, and the calibration process is usually complex. Current extrinsic calibration is mainly conducted offline with human assistance. However, the positions of sensors on the platform may change due to vibrations, shocks, or structural deformations of USVs, leading to some degree of variation in the extrinsic parameter between the mmWave radar and the camera.
- Data association. Traditional methods tend to manually craft various distance metrics to represent the similarities between vision and mmWave radar data. However, these manually crafted metrics are not adaptable when the data from different sensors degrade, and setting the parameters is also challenging.

In this paper, we propose a water surfaces small object detection method based on the decision-level fusion of vision and mmWave radar data. Compared to traditional methods, the proposed method has the following advantages: (1) With an initial offline calibrated extrinsic parameter, the proposed method is adapted to changes in extrinsic parameters to some degree during USVs' online operation; (2) The method has lower computational complexity and can run in real time on embedded systems; (3) The method achieves a higher detection accuracy in the water surface small object detection task.

The contribution of this paper mainly lies in the following aspects:

- We propose a new mmWave radar-aided visual small object detection method.
- We propose a new image–radar association model based on the metric learning model, which can achieve a robust association of mmWave radar data and images with inaccurate extrinsic parameters to some degree.
- We test the proposed method on real-world data, and the results show that our method achieves significantly better performance than current vision detection methods.

The detailed composition of this paper is listed as follows. In Section 2, we discuss the related works, including object detection on water surfaces and the visual–radar fusion-based detection method. In Section 3, we introduce the proposed mmWave radar-aided visual small object detection method in detail. Section 4 gives the results of experiments based on real-world data. Finally, Section 5 concludes this paper.

2. Related Works

2.1. Object Detection on Water Surfaces

Attention from researchers has been paid to object detection on water surfaces. Hammedi et al. [8] proposed a relevant dataset for inland water navigation that contains categories of riverside, vessel, person, etc. Moosbauer et al. [9] proposed a benchmark for object detection in maritime environments based on the Singapore Marine Dataset [10] to support relevant research. Vision-based detection methods are the ones that are mainly used for water surface object detection. For example, the method proposed in [11] is based on MobileNet for feature extraction and SSD for fast multi-scale detection to achieve real-time marine object detection of high-speed USVs. Zhang et al. [12] proposed a method for marine object detection and tracking based on improved YOLOv3 and used their method on a real USV experiment platform. The authors of [13] fused DenseNet in YOLOv3 for robust detection of marine objects under various weather conditions.

The vision-based methods for object detection on water surfaces are easily disturbed by weather and lighting conditions. Besides, the methods mainly aim at detection and cannot provide relative location information of the object. Therefore, methods based on the fusion of LiDAR data and images are proposed to improve detection accuracy and support object localization. Wu et al. [14] proposed a 3D object detection method based on the fusion of image and LiDAR point cloud for USVs in marine environments. They used a two-stage network which contains the proposal generation network and the deep fusion detection network. Cardillo et al. [15] analyzed the detection performance of radars with different frequency bands for USVs obstacle avoidance tasks, providing a valuable reference for the perception applications of mmWave radar in USVs. Im et al. [16] conducted object detection and tracking in USVs using frequency-modulated continuous wave (FMCW) radar with improved density-based spatial clustering of applications with noise (DBSCAN). Ha et al. [17] achieved autonomous obstacle avoidance tasks of USVs based the on marine radar. Stanislas et al. [18] utilized the fusion of LiDAR point clouds, camera, and 2D sparse radar point clouds for robust detection and classification in marine environments. The fusion-based methods can provide location information of the object in addition to object detection.

Current water surface object detection research mainly aims at maritime object detection. The objects are mostly vessels and other objects which are relatively big. However, for USVs, there are many other small objects that may cause dangers, such as small fountain nozzles, or are the searching targets of USVs, such as floating wastes. Besides, the Lidar that can be applied to complex water body environments is relatively expensive.

2.2. Visual–Radar Fusion Detection

Using solely visual information for object detection is susceptible to the influence of factors such as weather conditions, lighting, and object motion, which can result in detection errors and unreliability. In contrast, mmWave radar offers robust localization and velocity information for objects even in adverse weather conditions. Consequently, the fusion of visual and radar modalities, known as camera–radar fusion detection, has garnered increasing attention in the field of computer vision in recent years. Various fusion methods have been proposed to combine the strengths of camera and radar modalities and achieve improved detection performance in diverse scenarios. Based on the fusion stage within the network, the fusion methods of camera and radar can be broadly categorized as data-level fusion, feature-level fusion, and decision-level fusion. Data-level fusion [19–21] integrates raw or preprocessed data from radar and camera sensors at the early stages of deep learning models. Such methods necessitate addressing the correspondence between the camera and mmWave radar data, often requiring object matching or association operations. Long et al. [19] introduced Radar-Camera Pixel Depth Association (RC-PDA), which enhances and densifies radar images by associating radar point clouds with nearby image pixels. This approach resolves the challenge of associating radar point clouds with image pixels. Nobis et al. [20] input cascaded camera and radar point clouds into a network and extract features from the combined data using VGG [22]. However, data-level fusion methods typically impose high computational complexity and real-time requirements due to the potential disparate update rates between the camera and mmWave radar data.

Feature-level fusion [23–27] combines features extracted from radar data and camera images at the intermediate stages of deep learning-based fusion networks. Leveraging the distinct characteristics and advantages of these two sensor types, fusing their features provides a more comprehensive description of target objects. Chadwick et al. [24] proposed generating image and radar features separately using ResNet [28] and subsequently fusing them through concatenation and addition operations. Li et al. [25] introduced a feature pyramid layer attention module that integrates radar information, extending the feature pyramid module through the input interface of radar-projected images and attention modules. Nevertheless, feature-level fusion methods face challenges in striking a balance between fusion and aligning different sensor features.

Decision-level fusion [29] entails conducting separate object detection using camera and radar, followed by combining their results through weighted averaging or voting to obtain a comprehensive outcome. By amalgamating detection results from multiple sensors, the reliability of object detection experiences significant improvement. Jha et al. [29] employed YOLOv3 [30] as the image detector, projecting radar-detection results onto the image plane using transformation matrices, and subsequently aligning independently detected objects from the two sensors. Compared to the first two fusion methods, decision-level fusion exhibits greater robustness and adaptability, facilitating adaptive adjustments based on real-world scenarios and requirements. However, decision-level fusion encounters challenges associated with data inconsistency.

The existing methods primarily focus on road scenes, where visual information plays a dominant role and radar information serves as a supplementary source. However, the water surface environment is considerably more complex, characterized by water reflections and a prevalence of small objects. Relying predominantly on visual information in such scenarios can lead to a higher rate of false detections. Currently, there are limited camera–radar fusion methods specifically designed for water surface detection. Only RISFNet [23] has been proposed, which maps radar point clouds onto the image plane. It incorporates global attention and self-attention mechanisms to achieve deep multi-scale feature fusion between the two sensors, demonstrating robustness in detecting small objects on the water surface. Nevertheless, feature-level fusion alone fails to address the issue of unreliable camera sensors, and RISFNet heavily relies on accurate extrinsic parameters between radar and camera.

3. Our Method

For the task of small object detection on water surfaces, vision-based detection methods always generate false detection due to the sunlight reflection and surrounding scene reflection. The mmWave radar is robust to different lighting conditions but contains limited semantic information compared to the RGB image, which makes it difficult to distinguish objects of similar sizes using the radar-based detection method. Besides, the radar-based detection method may generate false detection on water surfaces due to the water clutter. Therefore, to improve the accuracy and robustness of small object detection on water surfaces, we propose a radar-aided visual small object detection method on water surfaces.

3.1. Network Overview

Due to the inherent shortcomings of camera and radar sensors, in the water surface small object detection task, both vision-based and radar-based detection methods have false detection. However, the reasons that the two sensors generate false detection are different, and the statistical probabilities of error occurrence in detection methods based on the two sensors are also independent. Hence, we adopt a detection method based on the decision-level fusion of vision and radar data. The visual object detection results are gained first, and then the detection results are associated with radar data to reduce false detection.

However, for the decision-level fusion method, the spatial position correlation of different sensors is of vital importance and acquires accurate extrinsic parameters. Due to the sparse and glittery characteristics of mmWave radar point clouds, corner reflectors or LiDAR are usually needed as the auxiliary in the extrinsic calibration between radar and camera, which involves complex calibration procedures [31]. For the applications of USVs, there can be certain variations in the extrinsic parameters between the radar and camera due to vibrations, shocks, or structural deformations of USVs during operations. In this case, we propose a new image–radar association model based on the metric learning model. By training the model using data based on the provided initial extrinsic parameters, the model is adaptable to variations in extrinsic parameters in practical application.

As shown in Figure 1, there are two main stages in the proposed radar-aided visual small object detection method: the detection stage and the association stage. Next, we will introduce more details about the two stages.

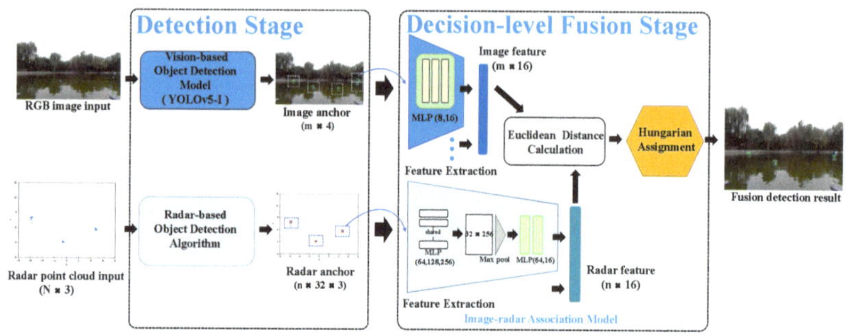

Figure 1. The architecture of the proposed method.

3.2. Detection Stage

The detection stage includes a vision-based detection model and a radar-based detection algorithm. We adopt YOLOv5-l [32] as the vision-based model. YOLOv5-l shows good performance in visual object detection tasks and it is a lightweight model which can carry out real-time inference in an embedded system.

3.2.1. Vision-Based Detection

To make the object detection model specialize on our fusion algorithm, we modify the original YOLOv5-l [32] as the vision-based model. As our fusion detection algorithm can remove the false positive detection results efficiently through the radar-detection results and vision detection results, we need to generate more detection results to improve the recall rate of the vision-based model. The framework of enhanced YOLOv5-l is illustrated in Figure 2. We adjust the prediction head of YOLOv5-l using a double prediction head and transformer decoder module, then we will introduce the architecture of the prediction head in detail.

(1) Double prediction heads. YOLOv5 object detector uses a single prediction head to predict the location and classification of the detected bounding box at the same time. In our vision-based model, we design a double prediction head including a classification head and location regression head to predict, respectively, the location and classification of objects. Independent double prediction heads will benefit from searching both the location and classification of objects. While we utilize the full connection (FC) layer to obtain more semantic information about objects in the classification head, we obtain the position of detection objects in the location regression head.

(2) Transformer decoder module. Inspired by the vision transformer [33], we use a transformer decoder module to replace the convolution blocks in the prediction head. Compared with convolution operation, the transformer decoder module can capture global information and abundant contextual information. Each transformer decoders contain a

multi-head attention layer and a fully-connected layer. Furthermore, there are residual connections between each sublayer. As the prediction head is at the end of the network and the feature map has low resolution, applying a transformer decoder module in a low-resolution feature map explores the feature representation potential with a self-attention mechanism and enlarges the receptive field of the prediction head with low computation and memory cost.

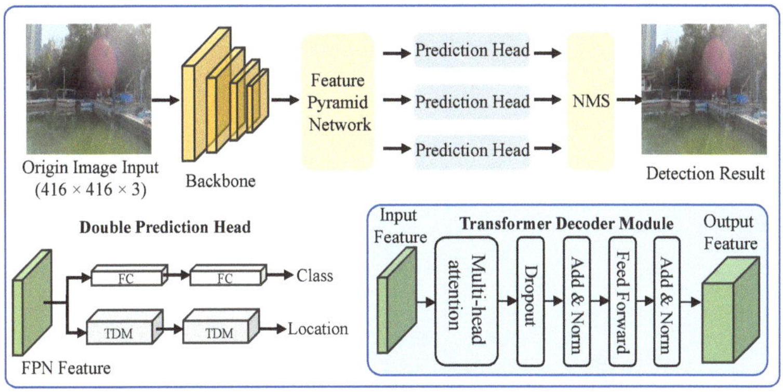

Figure 2. The framework of enhanced YOLOv5-l.

After applying the vision-based detection model to an RGB image, the image anchors, expressed as $\boldsymbol{B}_1, \boldsymbol{B}_2, \ldots, \boldsymbol{B}_m$, where m denotes the number of image anchors, are extracted. Each image anchor $\boldsymbol{B}$ contains four parameters, including the u-axis position, v-axis position, box width, and box height in the u-v image coordinate system. Therefore, for each image, the output size in the detection stage is $m \times 4$.

3.2.2. Radar-Based Detection

A mmWave radar system senses its surroundings by transmitting and receiving FMCW signals. The transmitted and reflected signals are mixed using a frequency mixer to obtain beat signals. Then, 1D (range) Fast Fourier Transformation (FFT) and 2D (velocity) FFT are applied to the sampled beat signals along the fast time and slow time, respectively, resulting in the well-known range–Doppler matrix (RDM). The cells with strong energy in the RDM are detected as targets. The most commonly employed detector for FMCW signal processing is the constant false alarm rate (CFAR) detector, which adaptively estimates the noise level based on nearby cells relative to the cell under test. After detection, the direction of arrival (DOA) is estimated for each detected target using signals from multiple antennas. Consequently, we obtain what is referred to as 4D radar point clouds, representing various detected targets with distinct 3D positions and Doppler velocities. The illustration of the radar signal processing chain is shown in Figure 3.

For radar-based detection, we use the spatial information of mmWave point clouds and the size of the input radar point cloud is $N \times 3$, where N denotes the number of radar points in the current frame and each point contains three coordinates information x,y,z. The radar point clouds are clustered into groups and the discrete radar clutter points are also removed using DBSCAN [34]. The point clouds are divided into n clusters $\boldsymbol{C}_1, \boldsymbol{C}_2, \ldots, \boldsymbol{C}_n$, where n denotes the number of radar point clusters. Then, we use the farthest point sampling (FPS) [35] to sample the point clouds of each group $\boldsymbol{C}_i$ into a fixed number 32. Therefore, the final size of outputs of radar-based detection is $n \times 32 \times 3$.

Through the detection stage, the vision-based and radar-based detection results are gained. Then, the detection results are sent to the fusion association stage to generate fusion detection results.

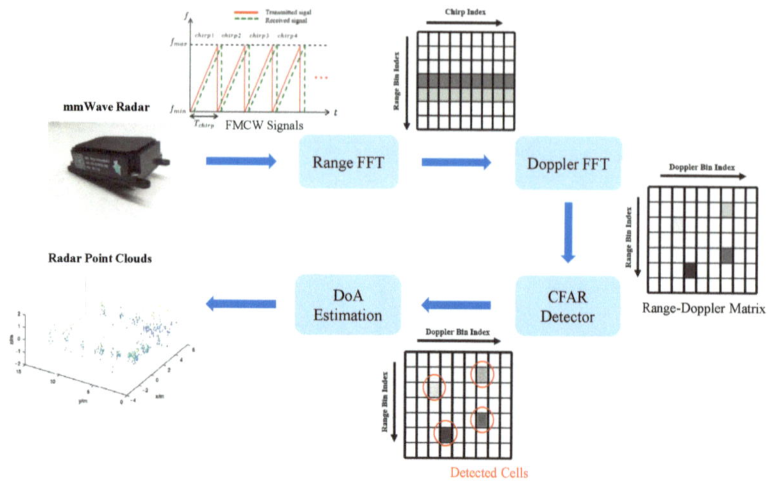

Figure 3. Illustration of the radar signal processing chain.

3.3. *Fusion Association Stage*

The fusion association stage extracts image feature vectors from object detection bounding boxes in the image plane and extracts radar feature vectors from radar point clouds. The image features can represent the position and size of detection-bounding boxes of corresponding objects in the image plane. The radar feature contains the spatial information of objects in the radar coordinate system as well as the shape information of objects. Therefore, by measuring the similarity between image and radar features, the association of vision and radar-detection results can be achieved. The Hungarian algorithm [36] is used for matching image and radar data according to the L2 distance between two feature vectors. Thus, an end-to-end spatial correlation between image and radar data can be achieved without the extrinsic parameter calibration procedure.

Next, we will introduce the radar and vision feature extraction model in detail. For a frame of RGB image, m bounding boxes are generated from the detection stage and the size of each bounding box is 1×4. We use the multi-layer perception (MLP) to extract the image feature $\boldsymbol{F}_{img}$ whose size is $m \times 16$ from each vision detection result $\boldsymbol{B}_i$:

$$\boldsymbol{F}_{img} = \{\boldsymbol{F}_{img_i} | \boldsymbol{F}_{img_i} = \mathrm{MLP}(\boldsymbol{B}_i), i = 1, 2, ..., m\}, \tag{1}$$

where $\boldsymbol{F}_{img_i}$ is the feature tensor of ith vision detection result $\boldsymbol{B}_i$.

A frame of radar data contains n point cloud groups and the output size of the radar-detection result is $n \times 32 \times 3$, where 32×3 denotes each group consisting of 32 points with each point containing x, y, z coordinates. For radar feature extraction, we adopt the mini-PointNet [37] architecture, which is a famous method to extract point cloud features. Through the shared weighted MLPs, the max-pooling, and another MLP, each point cluster generates a feature of size 1×16. The $n \times 16$ radar feature vector of a whole frame is generated by combining the n cluster features. The radar feature extraction can be represented as follows:

$$\boldsymbol{F}_r = \{\boldsymbol{F}_{r_j} | \boldsymbol{F}_{r_j} = \mathrm{MLP}(\mathrm{maxpool}(\mathrm{MLP}(\boldsymbol{C}_j))), j = 1, 2, ..., n\}, \tag{2}$$

where $\mathrm{F}_{\mathrm{r_j}}$ denotes the radar point cloud feature of the ith cluster.

After obtaining a frame of image feature $\boldsymbol{F}_{img}$ and corresponding radar feature $\boldsymbol{F}_r$, we compute the L2 distance between each object's image feature $\boldsymbol{F}_{img_i}$ and each object's radar feature $\boldsymbol{F}_{r_j}$ and obtain a cost matrix of size $m \times n$. Based on the cost matrix, within the minimum distance threshold, the matching results are gained using the Hungarian assignment algorithm.

Through the fusion association stage, the final fusion detection results which contain the vision-based detection box, object classification result, and the range and azimuth of the objects can be gained.

3.4. Loss Function

In our method, the detection model and the image–radar association model are trained separately. The training loss function of vision-based detection model $\mathcal{L}_{\text{vis}}$ is the same as the YOLOv5 object detection model, which is computed as:

$$\mathcal{L}_{\text{vis}} = \alpha_b * \mathcal{L}_{\text{box}} + \alpha_o * \mathcal{L}_{\text{obj}} + \alpha_c * \mathcal{L}_{\text{cla}}, \tag{3}$$

where $\mathcal{L}_{\text{box}}$ denotes the location loss, $\mathcal{L}_{\text{obj}}$ denotes the confidence loss, and $\mathcal{L}_{\text{cla}}$ denotes the classification loss. The three loss weights α_b, α_o, and α_c are constants. For the training of the image–radar association model, we choose the triplet loss [38], which is commonly used as the training loss function in the metric learning. Each training data pair for triplet loss contains three samples: a vision-based detection bounding box $\boldsymbol{B}_i$ as a base anchor, a positive radar sample $\boldsymbol{C}_j$, which is the radar-detection cluster corresponding with $\boldsymbol{B}_i$, and a negative radar sample, which is randomly selected from rest of the radar-detection clusters. The image and radar features are extracted from the training data pair, and the triplet loss is used to minimize the L2 distance d_{pos} between the image feature and positive radar feature while maximizing the L2 distance d_{neg} between the image feature and negative radar feature using:

$$\mathcal{L}_{\text{triplet}} = \max(d_{\text{pos}} + d_{\text{neg}} + \beta, 0), \tag{4}$$

where β is a constant to express the minimum distance loss.

4. Experiment and Evaluation

4.1. Dataset, Evaluation Metric, and Baseline

To evaluate the performance of our method, we use the public FloW RI dataset [39], which contains synchronized images and mmWave radar data of floating bottles on water surfaces. Besides, to test the model's generalization performance on a broader range of data, we supplement a new dataset for water surface small object detection using a USV platform equipped with an RGB camera and a Texas Instruments 77 GHz single-chip mmWave radar AWR1843. The dataset includes 1600 frames of synchronized RGB images and mmWave radar point cloud data. The newly added data are shown in Figure 4. Finally, we use 4400 frames of data as the training set and 1200 frames of data as the test set.

To quantitatively evaluate the performance of our method, we use the mean of average precision (mAP), which is widely used in object detection as the evaluation metric, and compare the performance of our method with some famous baseline methods in object detection. For vision-based methods, the YOLO [40] series object detection methods are widely used in mobile robots due to the high inference speed. Therefore, we choose the newest YOLOv5-l as one baseline method. Compared to the single-stage object detection methods, the two-stage methods are usually slower but can achieve a higher detection accuracy. Therefore, Fast R-CNN [41] and Cascade R-CNN [42] are also selected as baselines in the experiment. In addition to methods based on the convolutional neural network, in recent years, methods based on transformers also achieve SOTA performance in some tasks. Therefore, we also choose the Swin Transformer [43] as one baseline method. For the mmWave radar-based object detection method, we choose the VoteNet [44] and the method of Danzer et al. [45] as baselines. In addition, we also compare our method with other visual–radar fusion-based methods including feature-level fusion method RISFNet [23], CRF-Net [20], the method in Li et al. [25], and a decision-level fusion method [29].

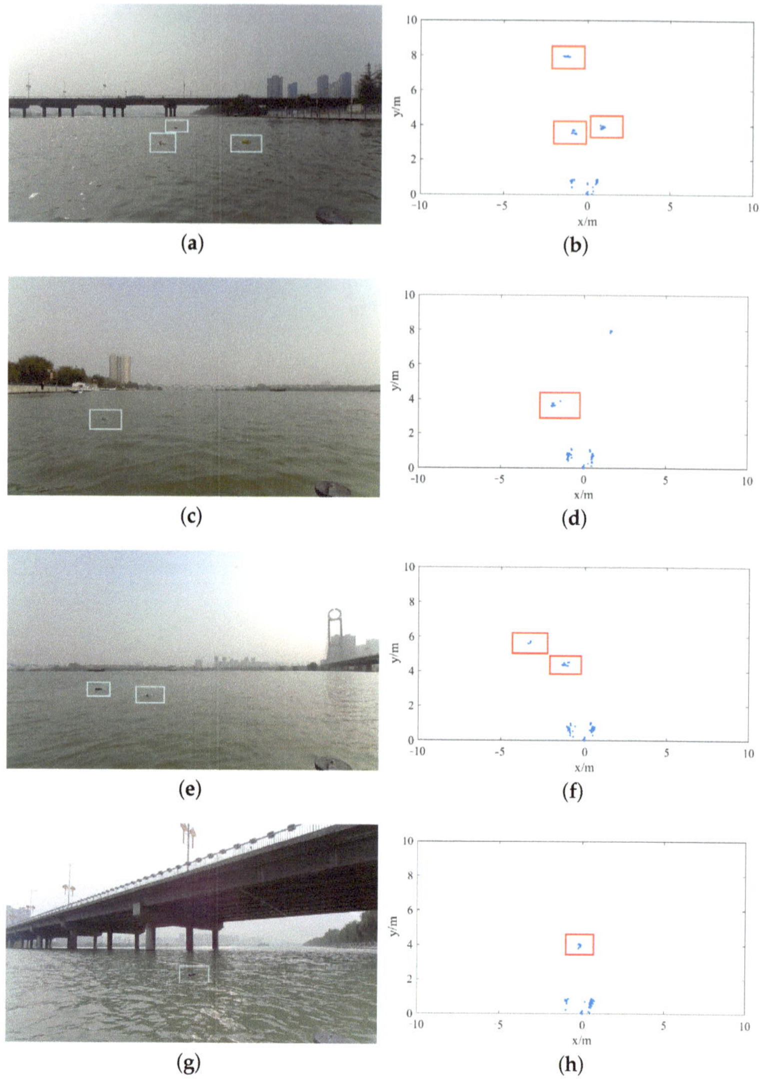

Figure 4. Examples of the supplementary data. The blue and red boxes indicate the targets in the image and radar point cloud respectively. (**a**,**c**,**e**,**g**) Images; (**b**,**d**,**f**,**h**) Corresponding radar point clouds.

4.2. Training Details

In our method, the detection model and the image–radar association model are trained separately. In the training of the vision-based detection model, our model based on PyTorch implements is pre-trained from the COCO dataset [46] and is trained on an Nvidia GTX 3090 with an initial learning rate set to 10^{-3} and the momentum of 0.937. The network is trained for 100 epochs using the SGD optimizer with a batch size of 8 and the mini-batch StepLR descent algorithm with step-size = 1, gamma = 0.94. Besides, in order to enhance the performance of the detection model, multiscale data augmentation methods such as image resizing image placing, color adjustment, and image left-right flipping are used for the training images. During the training of the image–radar association model, using the extrinsic parameters provided in the dataset, we generate 4600 pairs of objects' radar point clouds and RGB image bounding boxes to train the metric learning model. The training implementation of our association network is based on Pytorch and is trained with a batch size set to 16 and an initial learning rate set to 10^{-4}. We train the image–radar association model for 200 epochs using an ADAM optimizer with weight decay of 5×10^{-4}.

4.3. Quantitative Evaluation

To evaluate the performance of our method, we compared our method with other baseline methods. All the baselines and our model are trained on the same training set. As different model training parameters influence the final results, in the experiment, the training parameters for the baseline models are primarily set by following its recommended default values with only minor adjustments.

The result in Table 1 shows that, compared with other vision-based baseline methods, the proposed method achieves better detection accuracy while keeping a relatively low computation cost. The result in Table 2 shows that our mmWave radar-aided vision detection method outperforms other radar-based and most visual–radar fusion-based methods in detection accuracy. Although the RISFNet achieves higher detection accuracy, it has a higher computation cost and requires accurate extrinsic parameters between the radar and camera. When the extrinsic parameter is inaccurate, the performance of RISFNet decreases significantly.

In addition, we also combine our image–radar association model with other vision-based object detection methods. As shown in Table 3, by directly applying our image–radar association model, other vision-based methods all achieve obvious improvement in detection accuracy. As our image–radar association model has low computational complexity and the inference speed of the image–radar association model is extremely fast with about 280 FPS, the image–radar association model can also be seen as an independent plug-and-play model to improve the detection accuracy of the vision-based detection method.

Table 1. Comparison of the detection accuracy using vision-based baseline methods and our method on the dataset.

Method	mAP (IoU = 0.35, %)	FPS
YOLOv5-l [32]	74.66	29
Cascade-RCNN [42]	78.36	17
Faster-RCNN [41]	74.34	19
Swin-Transformer [43]	77.33	15
Ours	**81.41**	**29**

Table 2. Comparison of the detection accuracy using radar-based and fusion-based methods and our method on the dataset.

Modality	Method	mAP (IoU = 0.35, %)
Radar	VoteNet [44]	45.24
	Danzer et al. [45]	32.65
Vision + Radar	CRF-Net [20]	74.35
	Li et al. [25]	77.23
	Jha et al. [29]	77.98
	RISFNet [23]	83.25
	Ours	**81.41**

Table 3. The results of combining our image–radar association model with other vision-based object detection methods.

Method	mAP (IoU = 0.35, %) (with Image–Radar Association Model)	FPS
YOLOv5-l [32]	81.41 (**+6.75**)	29
Cascade-RCNN [42]	83.62 (**+5.26**)	15
Faster-RCNN [41]	79.53 (**+5.19**)	17
Swin-Transformer [43]	82.42 (**+5.09**)	19

4.4. Robustness Analysis

In order to validate the effect of camera–radar extrinsic parameter changes on the performance of the proposed model, based on the known extrinsic parameter, we artificially add a rotation and translation bias to the overall radar point clouds, to simulate variations in the camera–radar extrinsic parameters. As shown in Table 4, when the extrinsic parameters change slightly (with ±5° rotation bias, ±1 m translation bias), the model's performance is nearly not affected, indicating the model's adaptability to small changes in the extrinsic parameters. However, when the extrinsic parameters change a lot (with ±20° rotation bias, ±4 m translation bias), there is a significant decrease in the model's performance. Nevertheless, in the practical application of USVs, extrinsic parameters are unlikely to have large variations, and the model can adapt well to most scenes.

Table 4. Comparison of results of the fusion model with different extrinsic parameter variations.

Method	mAP (IoU = 0.35, %)
Using origin radar data	81.41
Using radar data with slight bias	80.83
Using radar data with large bias	56.29

4.5. Ablation Analysis

To verify the contributions of the proposed modifications to the YOLOv5 model's prediction head, we conduct an ablation analysis by replacing it with the original prediction head. Furthermore, to test the effectiveness of the newly proposed image–radar association model for data fusion, we compare it with the traditional manual configuration fusion method. The method directly projects the mmWave radar point cloud onto the RGB image plane based on the initial extrinsic parameters. Then, data association is performed based on the spatial relationships between radar point cloud clusters and 2D image boxes in the image plane with a predefined distance threshold. The results are shown in Table 5, indicating that the proposed improved double prediction heads effectively enhance the object detection accuracy. Besides, the proposed metric learning-based image–radar association model achieves better fusion results compared to the traditional manual association method.

Table 5. Results of the ablation analysis.

Method	mAP (IoU = 0.35, %)
Without double prediction heads	79.85
Without image–radar association model	78.17
Our method	81.41

4.6. Discussion

The visualization of the detection results of our fusion detection method compared with the vision-based YOLOv5-l is shown in Figure 5. As can be seen, our method achieves a lower false object detection rate in various surrounding scenes.

Figure 5. Visualization of results of the proposed method and vision-based YOLOv5-l. The red and green boxes in the figures represent correct and false detection results, respectively.(**a**,**c**,**e**,**g**) Ours; (**b**,**d**,**f**,**h**) YOLOv5-l.

To enhance the detection performance of the YOLOv5-l visual detector, we integrate a transformer decoder module to replace the conventional convolutional blocks within its prediction head. The transformer decoder uses multi-head attention to enhance the low-resolution feature representation capability. We visualize the input and output feature maps of the transformer decoder module in Figure 6. In Figure 6, each row represents a frame, where the first column displays the original image with the final detection results (highlighted in red bounding boxes), the second column shows the output features of the backbone network, and the third column show cases the features strengthened through the transformer decoder module. The highlighted regions denote areas of high response, indicating a higher probability of object presence in those regions. As can be seen, the results demonstrate that small objects in the enhanced features of this module are more distinguishable, so that our vision-based detector can locate all objects more accurately.

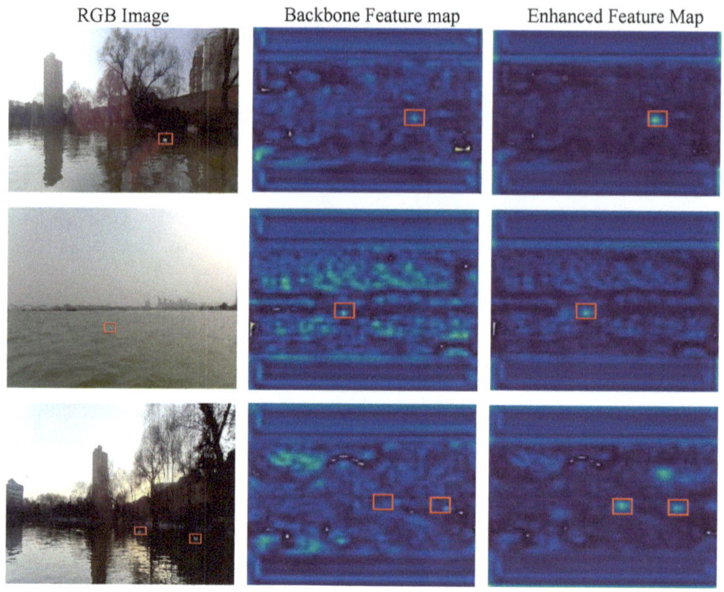

Figure 6. Feature maps of the transformer decoder module in enhanced YOLOv5-l.

5. Conclusions

In this paper, we propose a new mmWave radar-aided visual water surfaces small object detection method. The method associates mmWave radar data and images through the metric learning model, and is adapted to changes of extrinsic parameters to some degree. Through the detection stage and the fusion association stage, the proposed method outputs the final fusion detection results. Finally, we conduct experiments on the real-world dataset to test the proposed method. The results show that our method outperforms other visual detection methods on water surface small object detection.

Author Contributions: Conceptualization, J.Z., Y.Y. and Y.C.; methodology, J.Z. and Y.Y.; validation, J.Z.; data curation, J.Z.; coding and experiments, Y.C.; writing—original draft preparation, J.Z. and Y.C.; writing—review and editing, Y.Y. All authors have read and agreed to the published version of the manuscript.

Funding: This work was funded by the National Natural Science Foundation of China under Grant Number 11974286.

Institutional Review Board Statement: Not applicable.

Informed Consent Statement: Not applicable.

Data Availability Statement: Not applicable.

Conflicts of Interest: The authors declare no conflict of interest.

Abbreviations

The following abbreviations are used in this manuscript:

CFAR	Constant false alarm rate
DBSCAN	Density-based spatial clustering of applications with noise
DOA	Direction of arrival
FC	Full connect
FFT	Fast Fourier Transformation
FMCW	Frequency-modulated continuous wave
FPS	Farthest point sampling
GNSS	Global navigation satellite system
IMU	Inertial measurement unit
mAP	Mean of average precision
MLP	Multi-layer perception
mmWave	Millimeter wave radar
RDM	Range–Doppler matrix
USV	Unmanned surface vehicle

References

1. Wang, W.; Gheneti, B.; Mateos, L.A.; Duarte, F.; Ratti, C.; Rus, D. Roboat: An autonomous surface vehicle for urban waterways. In Proceedings of the 2019 IEEE/RSJ International Conference on Intelligent Robots and Systems (IROS), Venetian Macao, Macau, 4–8 November 2019; pp. 6340–6347.
2. Chang, H.C.; Hsu, Y.L.; Hung, S.S.; Ou, G.R.; Wu, J.R.; Hsu, C. Autonomous Water Quality Monitoring and Water Surface Cleaning for Unmanned Surface Vehicle. *Sensors* **2021**, *21*, 1102. [CrossRef]
3. Zhu, J.; Yang, Y.; Cheng, Y. SMURF: A Fully Autonomous Water Surface Cleaning Robot with A Novel Coverage Path Planning Method. *J. Mar. Sci. Eng.* **2022**, *10*, 1620. [CrossRef]
4. Wu, Y.; Wang, Y.; Zhang, S.; Ogai, H. Deep 3D object detection networks using LiDAR data: A review. *IEEE Sens. J.* **2020**, *21*, 1152–1171. [CrossRef]
5. Carballo, A.; Lambert, J.; Monrroy, A.; Wong, D.; Narksri, P.; Kitsukawa, Y.; Takeuchi, E.; Kato, S.; Takeda, K. LIBRE: The multiple 3D lidar dataset. In Proceedings of the 2020 IEEE Intelligent Vehicles Symposium (IV), Las Vegas, NV, USA, 19 October–13 November 2020; pp. 1094–1101.
6. Patole, S.M.; Torlak, M.; Wang, D.; Ali, M. Automotive Radars: A Review of Signal Processing Techniques. *IEEE Signal Process. Mag.* **2017**, *34*, 22–35. [CrossRef]
7. Brodeski, D., Bilik, I.; Giryes, R. Deep radar detector. In Proceedings of the 2019 IEEE Radar Conference (RadarConf), Boston, MA, USA, 22–26 April 2019; pp. 1–6.
8. Hammedi, W.; Ramirez-Martinez, M.; Brunet, P.; Senouci, S.M.; Messous, M.A. Deep learning-based real-time object detection in inland navigation. In Proceedings of the 2019 IEEE Global Communications Conference (GLOBECOM), Waikoloa, HI, USA, 9–13 December 2019; pp. 1–6.
9. Moosbauer, S.; Konig, D.; Jakel, J.; Teutsch, M. A benchmark for deep learning based object detection in maritime environments. In Proceedings of the IEEE/CVF Conference on Computer Vision and Pattern Recognition Workshops, Long Beach, CA, USA, 16–17 June 2019.
10. Prasad, D.K.; Rajan, D.; Rachmawati, L.; Rajabally, E.; Quek, C. Video processing from electro-optical sensors for object detection and tracking in a maritime environment: A survey. *IEEE Trans. Intell. Transp. Syst.* **2017**, *18*, 1993–2016. [CrossRef]
11. Zhou, Z.; Yu, S.; Liu, K. A Real-time Algorithm for Visual Detection of High-speed Unmanned Surface Vehicle Based on Deep Learning. In Proceedings of the 2019 IEEE International Conference on Signal, Information and Data Processing (ICSIDP), Chongqing, China, 11–13 December 2019; pp. 1–5.
12. Zhang, W.; Gao, X.z.; Yang, C.f.; Jiang, F.; Chen, Z.y. A object detection and tracking method for security in intelligence of unmanned surface vehicles. *J. Ambient. Intell. Humaniz. Comput.* **2020**, *13*, 1279–1291. [CrossRef]
13. Li, Y.; Guo, J.; Guo, X.; Liu, K.; Zhao, W.; Luo, Y.; Wang, Z. A novel target detection method of the unmanned surface vehicle under all-weather conditions with an improved YOLOV3. *Sensors* **2020**, *20*, 4885. [CrossRef] [PubMed]
14. Wu, Y.; Qin, H.; Liu, T.; Liu, H.; Wei, Z. A 3D object detection based on multi-modality sensors of USV. *Appl. Sci.* **2019**, *9*, 535. [CrossRef]
15. Cardillo, E.; Ferro, L. Multi-frequency analysis of microwave and millimeter-wave radars for ship collision avoidance. In Proceedings of the 2022 Microwave Mediterranean Symposium (MMS), Pizzo Calabro, Italy, 9–13 May 2022; pp. 1–4.
16. Im, S.; Kim, D.; Cheon, H.; Ryu, J. Object Detection and Tracking System with Improved DBSCAN Clustering using Radar on Unmanned Surface Vehicle. In Proceedings of the 2021 21st International Conference on Control, Automation and Systems (ICCAS), Jeju, Republic of Korea, 12–15 October 2021; pp. 868–872.

17. Ha, J.S.; Im, S.R.; Lee, W.K.; Kim, D.H.; Ryu, J.K. Radar based Obstacle Detection System for Autonomous Unmanned Surface Vehicles. In Proceedings of the 2021 21st International Conference on Control, Automation and Systems (ICCAS), Jeju, Republic of Korea, 12–15 October 2021; pp. 863–867.
18. Stanislas, L.; Dunbabin, M. Multimodal sensor fusion for robust obstacle detection and classification in the maritime RobotX challenge. *IEEE J. Ocean. Eng.* **2018**, *44*, 343–351. [CrossRef]
19. Long, Y.; Morris, D.; Liu, X.; Castro, M.; Chakravarty, P.; Narayanan, P. Radar-camera pixel depth association for depth completion. In Proceedings of the IEEE/CVF Conference on Computer Vision and Pattern Recognition, Nashville, TN, USA, 20–25 June 2021; pp. 12507–12516.
20. Nobis, F.; Geisslinger, M.; Weber, M.; Betz, J.; Lienkamp, M. A deep learning-based radar and camera sensor fusion architecture for object detection. In Proceedings of the 2019 Sensor Data Fusion: Trends, Solutions, Applications (SDF), Bonn, Germany, 15–17 October 2019; pp. 1–7.
21. Nabati, R.; Qi, H. Rrpn: Radar region proposal network for object detection in autonomous vehicles. In Proceedings of the 2019 IEEE International Conference on Image Processing (ICIP), Taipei, Taiwan, 22–25 September 2019; pp. 3093–3097.
22. Simonyan, K.; Zisserman, A. Very deep convolutional networks for large-scale image recognition. *arXiv* **2014**, arXiv:1409.1556.
23. Cheng, Y.; Xu, H.; Liu, Y. Robust Small Object Detection on the Water Surface Through Fusion of Camera and Millimeter Wave Radar. In Proceedings of the IEEE/CVF International Conference on Computer Vision, Montreal, BC, Canada, 10–17 October 2021; pp. 15263–15272.
24. Chadwick, S.; Maddern, W.; Newman, P. Distant vehicle detection using radar and vision. In Proceedings of the 2019 International Conference on Robotics and Automation (ICRA), Paris, France, 31 May–4 June 2019; pp. 8311–8317.
25. Li, L.q.; Xie, Y.l. A feature pyramid fusion detection algorithm based on radar and camera sensor. In Proceedings of the 2020 15th IEEE International Conference on Signal Processing (ICSP), Beijing, China, 6–9 December 2020; Volume 1, pp. 366–370.
26. Nabati, R.; Qi, H. Centerfusion: Center-based radar and camera fusion for 3D object detection. In Proceedings of the IEEE/CVF Winter Conference on Applications of Computer Vision, Online, 5–9 January 2021; pp. 1527–1536.
27. Chang, S.; Zhang, Y.; Zhang, F.; Zhao, X.; Huang, S.; Feng, Z.; Wei, Z. Spatial attention fusion for obstacle detection using mmwave radar and vision sensor. *Sensors* **2020**, *20*, 956. [CrossRef] [PubMed]
28. He, K.; Zhang, X.; Ren, S.; Sun, J. Deep residual learning for image recognition. In Proceedings of the IEEE Conference on Computer Vision and Pattern Recognition, Las Vegas, NV, USA, 27–30 June 2016; pp. 770–778.
29. Jha, H.; Lodhi, V.; Chakravarty, D. Object detection and identification using vision and radar data fusion system for ground-based navigation. In Proceedings of the 2019 6th International Conference on Signal Processing and Integrated Networks (SPIN), Noida, India, 7–8 March 2019; pp. 590–593.
30. Redmon, J.; Farhadi, A. Yolov3: An incremental improvement. *arXiv* **2018**, arXiv:1804.02767.
31. Domhof, J.; Kooij, J.F.; Gavrila, D.M. An extrinsic calibration tool for radar, camera and lidar. In Proceedings of the 2019 International Conference on Robotics and Automation (ICRA), Montreal, QC, Canada, 20–24 May 2019; pp. 8107–8113.
32. Ultralytics. YOLO-v5. 2020. Available online: https://github.com/ultralytics/yolov5 (accessed on 20 August 2023).
33. Dosovitskiy, A.; Beyer, L.; Kolesnikov, A.; Weissenborn, D.; Zhai, X.; Unterthiner, T.; Dehghani, M.; Minderer, M.; Heigold, G.; Gelly, S.; et al. An image is worth 16 × 16 words: Transformers for image recognition at scale. *arXiv* **2020**, arXiv:2010.11929.
34. Bäcklund, H.; Hedblom, A.; Neijman, N. A density-based spatial clustering of application with noise. *Data Min. TNM033* **2011**, *33*, 11–30.
35. Qi, C.R.; Yi, L.; Su, H.; Guibas, L.J. Pointnet++: Deep hierarchical feature learning on point sets in a metric space. *Adv. Neural Inf. Process. Syst.* **2017**, *30*, 5105–5114.
36. Kuhn, H.W. The Hungarian method for the assignment problem. *Nav. Res. Logist. Q.* **1955**, *2*, 83–97. [CrossRef]
37. Qi, C.R.; Su, H.; Mo, K.; Guibas, L.J. Pointnet: Deep learning on point sets for 3d classification and segmentation. In Proceedings of the IEEE Conference on Computer Vision and Pattern Recognition, Honolulu, HI, USA, 21–26 July 2017; pp. 652–660.
38. Schroff, F.; Kalenichenko, D.; Philbin, J. Facenet: A unified embedding for face recognition and clustering. In Proceedings of the IEEE Conference on Computer Vision and Pattern Recognition, Boston, MA, USA, 7–12 June 2015; pp. 815–823.
39. Cheng, Y.; Zhu, J.; Jiang, M.; Fu, J.; Pang, C.; Wang, P.; Sankaran, K.; Onabola, O.; Liu, Y.; Liu, D.; et al. FloW: A Dataset and Benchmark for Floating Waste Detection in Inland Waters. In Proceedings of the IEEE/CVF International Conference on Computer Vision, Montreal, BC, Canada, 11–17 October 2021; pp. 10953–10962.
40. Redmon, J.; Divvala, S.; Girshick, R.; Farhadi, A. You only look once: Unified, real-time object detection. In Proceedings of the IEEE Conference on Computer Vision and Pattern Recognition, Las Vegas, NV, USA, 27–30 June 2016; pp. 779–788.
41. Girshick, R. Fast r-cnn. In Proceedings of the IEEE International Conference on Computer Vision, Washington, DC, USA, 7–13 December 2015, pp. 1440–1448.
42. Cai, Z.; Vasconcelos, N. Cascade r-cnn: Delving into high quality object detection. In Proceedings of the IEEE Conference on Computer Vision and Pattern Recognition, Salt Lake City, UT, USA, 18–23 June 2018; pp. 6154–6162.
43. Liu, Z.; Lin, Y.; Cao, Y.; Hu, H.; Wei, Y.; Zhang, Z.; Lin, S.; Guo, B. Swin transformer: Hierarchical vision transformer using shifted windows. In Proceedings of the IEEE/CVF International Conference on Computer Vision, Montreal, BC, Canada, 11–17 October 2021; pp. 10012–10022.
44. Qi, C.R.; Litany, O.; He, K.; Guibas, L.J. Deep hough voting for 3d object detection in point clouds. In Proceedings of the IEEE/CVF International Conference on Computer Vision, Seoul, Republic of Korea, 27 October–2 November 2019; pp. 9277–9286.

45. Danzer, A.; Griebel, T.; Bach, M.; Dietmayer, K. 2d car detection in radar data with pointnets. In Proceedings of the 2019 IEEE Intelligent Transportation Systems Conference (ITSC), Auckland, New Zealand, 17–30 October 2019; pp. 61–66.
46. Lin, T.Y.; Maire, M.; Belongie, S.; Hays, J.; Perona, P.; Ramanan, D.; Dollár, P.; Zitnick, C.L. Microsoft coco: Common objects in context. In Proceedings of the European Conference on Computer Vision, Zurich, Switzerland, 6–12 September 2014; Springer: Cham, Switzerland, 2014; pp. 740–755.

Journal of
Marine Science
and Engineering

Article

The Hydrodynamic Interaction between an AUV and Submarine during the Recovery Process

Wanzhen Luo [1,2], Caipeng Ma [1,2], Dapeng Jiang [1,2,*], Tiedong Zhang [1,2] and Tiecheng Wu [1,2]

1 School of Ocean Engineering and Technology, Sun Yat-sen University, Guangzhou 510275, China; luowzh5@mail.sysu.edu.cn (W.L.); macp3@mail2.sysu.edu.cn (C.M.); zhangtd5@mail.sysu.edu.cn (T.Z.); wutch7@mail.sysu.edu.cn (T.W.)

2 Southern Marine Science and Engineering Guangdong Laboratory (Zhuhai), Zhuhai 519000, China

* Correspondence: jiangdp5@mail.sysu.edu.cn

Abstract: The hydrodynamic interaction between an AUV (Autonomous Underwater Vehicle) and a recovery device, such as a real-scale submarine, is a crucial factor affecting the safe recovery of the AUV. This paper employs the CFD (Computational Fluid Dynamics) method to investigate the hydrodynamic interaction of the AUV and the submarine during the recovery process. Both the AUV and the submarine are considered to be relatively stationary. The results indicate that the submarine has a significant impact on the AUV during the recovery process, with sailing speed and relative positions identified as key influential factors. Due to the influence of the submarine, it can be difficult for the AUV to approach the submarine and be recovered safely. This study provides valuable insights into the hydrodynamic interaction between the AUV and the recovery device, and offers guidance for future submarine recovery operations involving AUVs. By considering the influence of the submarine's position and motion, as well as other relevant factors, it may be possible to improve the stability, safety, and efficiency of AUV recovery operations.

Keywords: hydrodynamic interaction; AUV; submarine; AUV recovery

Citation: Luo, W.; Ma, C.; Jiang, D.; Zhang, T.; Wu, T. The Hydrodynamic Interaction between an AUV and Submarine during the Recovery Process. *J. Mar. Sci. Eng.* **2023**, *11*, 1789. https://doi.org/10.3390/jmse11091789

Academic Editor: Constantine Michailides

Received: 5 August 2023
Revised: 6 September 2023
Accepted: 7 September 2023
Published: 13 September 2023

1. Introduction

Autonomous Underwater Vehicles (AUVs) have gained widespread applications in various fields, including bathymetry, environmental monitoring, underwater operations, and antimine activities, owing to their remarkable benefits [1–3]. However, with the increasing demand for AUV utilization, it is crucial to enhance their performance by increasing their underwater operation time, work efficiency, and endurance [4].

The recovery of an AUV after completing its mission is a critical aspect of AUV operation, as it requires energy replenishment, data processing, and maintenance [5,6]. Over the years, researchers have proposed various methods for recovering AUVs, which can be broadly classified into two categories: surface mother-ship recovery and underwater docking recovery [7–9]. Surface mother-ship recovery typically involves the AUV returning to a predetermined area and approaching the mother ship for recovery, where it is lifted using cranes and other equipment onboard the surface vessel [10–12]. More recently, unmanned surface vehicles (USVs) have been used in the recovery process, providing a flexible and efficient solution for AUV recovery [13]. Underwater docking recovery involves various methods and equipment, including capture rod docking [14], guided docking [15,16], seated docking [17], and docking recovery through the torpedo launch tube of submarines [18–20].

The hydrodynamic interaction between the AUV and recovery device is a crucial factor that must be taken into consideration during the recovery process [21,22]. The hydrodynamic interaction can have varying degrees of impact on the recovery process, including issues such as long recovery time, failure of recovery, and even collisions between the AUV and recovery device. As such, it is essential to study the hydrodynamic interaction between the AUV and recovery device to ensure a successful and safe recovery process. This requires taking into account several factors, such as the size and shape of the AUV, the hydrodynamic characteristics of the recovery device, and the surrounding water currents.

The hydrodynamic interaction during the AUV recovery process has indeed garnered the interest of numerous researchers, and significant work has been devoted to understanding this issue. One notable contribution to this field was made by Molland [23,24], who conducted an extensive investigation into the viscous interaction between a pair of ellipsoids in close proximity. This included low-speed wind tunnel experiments and numerical simulations, which considered the effects of various factors, such as the Reynolds number and the transverse distance between the ellipsoids on the drag and side force of a single ellipsoid. The results of this work have enhanced our knowledge of viscous interactions between two objects and provided valuable reference points for basic CFD algorithms. Other researchers, such as Husaini [25], Zhang [26], and Rattanasiri [27], have focused on studying the hydrodynamic interaction between AUV formations. Zhang's research analyzed a range of factors that might impact the hydrodynamic interaction between AUVs, such as spacing, drift angle, appendages, and formation structure. Rattanasiri's study, on the other hand, focused on the detailed analysis of the influence of the relative positioning between AUVs on the hydrodynamic interaction, dividing the dominant space around the AUV into seven areas based on drag.

Research on the hydrodynamic issues related to AUV docking has also received significant attention from researchers [28,29]. Wu et al. [30] investigated the docking of AUVs to a cone-shaped dock, considering various factors, such as different velocities, accelerations, dock shapes, gliding modes, cross currents from different directions, and rudder angles, in their simulations. Their findings indicated that an unclosed dock tends to decrease the drag of the AUV, and decelerating at a constant speed is beneficial for successful docking. Additionally, Meng et al. [31,32] conducted extensive research on the underwater docking of AUVs. This included exploring the influence of various docking methods and different structural forms of the docking device on the recovery process, as well as studying the hydrodynamic interaction between the AUV and the docking device when stationary and in motion. Their work has yielded important results, including the observation that installing the docking device in the middle of the submarine provides the best stability and that the AUV's speed plays a vital role in the recovery process.

The hydrodynamic interaction between submarines and Unmanned Underwater Vehicles (UUVs) has been a focus of research for several years, and notable contributions have been made in this field. Fedor [33] studied the hydrodynamic interaction between a submarine and a much smaller UUV, attempting to identify an area around the submarine where the effect of hydrodynamic interaction on UUV launch and recovery is relatively small. His analysis considered the static situation of the submarine, and his results lacked experimental verification. Leong and Randeni [34–36] used CFD methods and captive model experiments to carry out steady-state analysis to estimate the hydrodynamic interaction effects on an AUV operating in proximity to a submarine. They considered the longitudinal and lateral positions of the AUV relative to the submarine and the diameter ratio between the AUV and the submarine. Their results indicated that the force/moment coefficients of the AUV due to hydrodynamic interaction were independent of the Reynolds number at test speeds within the fully turbulent regime, and the interaction effects are minimal around amidships of the submarine. Furthermore, Du [37] analyzed the variation of hydrodynamic coefficients of the AUV moving around a submarine by solving the Reynolds-Averaged Navier–Stokes (RANS) equation. His numerical results showed that the attack angles and sideslip angles of the AUV had a significant influence on hydrodynamic coefficients, while the effect of Reynolds numbers could be ignored.

Past studies on the hydrodynamic interaction between an AUV and submarines have primarily focused on the model size. However, in practical applications, the size of the submarine is significantly larger than that of the AUV. Thus, this paper seeks to address the hydrodynamic interaction between a full-sized submarine and an AUV. The commercial

CFD software Siemens PLM STAR-CCM+16.02 was used to carry out this research. The submarine employed in this study was utilized as a recovery device and was magnified by 20 times from the full appendage SUBOFF model. This approach enabled us to capture the effects of a real-world-sized submarine on the hydrodynamic interaction with the AUV. Moreover, to enhance the realism of the simulation, the AUV shape used in this study is flat, unlike the traditional revolving body often used in previous research.

To provide a more comprehensive overview of the research conducted, the remainder of this paper is organized as follows. Section 2 provides an introduction to both the geometric and numerical models utilized in this study. In Section 3, we verify the numerical method employed in our research. Next, in Section 4, we analyze the hydrodynamic interaction between the AUV and submarine when they are relatively stationary, with consideration given to various relative positions and speeds. Finally, we draw conclusions based on our findings in Section 5.

2. Geometric Models and Numerical Methodology

2.1. Geometric Models

The prototype of the submarine in this paper is the SUBOFF AFF-8 developed by DARPA (Defense Advanced Research Projects Agency) [38]. In order to approximate the size of the actual submarine, SUBOFF AFF-8 has been enlarged by 20 times. In addition, the AUV used in this paper is also different from the traditional AUV in the form of revolution. The geometry and main dimensions of SUBOFF and AUV are shown in Figure 1 and Table 1, respectively. For convenience, the length of the SUBOFF AFF-8 and submarine are defined as L_{SUBOFF} and $L_{submarine}$, respectively. The length, width, and height of the AUV are defined as L_{AUV}, W_{AUV}, and H_{AUV}, respectively.

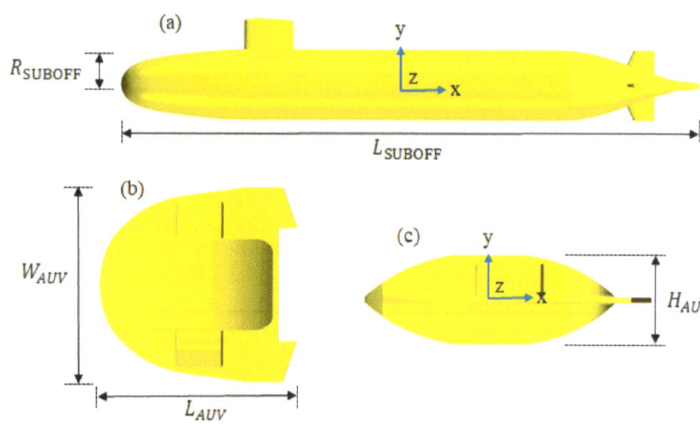

Figure 1. Schematic of simulation mode; (**a**) SUBOFF AFF-8; (**b**) AUV (Top view); (**c**) AUV (Side view).

Table 1. The main parameters for the SUBOFF model, the submarine, and AUV (m).

-	Length (m)	Radius/Width (m)	Height (m)
SUBOFF AFF-8	4.356 (L_{SUBOFF})	0.254 (R_{SUBOFF})	-
submarine	87.120 ($L_{submarine}$)	5.080 ($R_{submarine}$)	-
AUV	1.080 (L_{AUV})	0.923 (W_{AUV})	0.307 (H_{AUV})

2.2. Numerical Methodology

2.2.1. Governing Equations and Numerical Setting

The CFD software utilized in this study is Siemens PLM STAR-CCM+16.02; the Reynolds-Averaged Navier–Stokes (RANS) equations discretized by the finite volume method (FVM) are applied to carry out the numerical simulation. The governing equations for unsteady, three-dimensional, incompressible flow are described below [39,40]:

$$\frac{\partial u_i}{\partial x_i} = 0 \tag{1}$$

$$\frac{\partial u_i}{\partial t} + \frac{\partial (u_i u_j)}{\partial x_j} = -\frac{\partial P}{\partial x_i} + \frac{\partial}{\partial x_j}\left[v\left(\frac{\partial u_i}{\partial x_j} + \frac{\partial u_j}{\partial x_i}\right)\right] - \frac{\partial \overline{u_i' u_j'}}{\partial x_j} \tag{2}$$

where u_i and x_i denote the velocity component and coordinate component in i direction, respectively; I = (1,2,3). P, v, and $-\overline{u_i' u_j'}$ are the static pressure, kinematic viscosity, and the Reynolds stress, respectively. δ_{ij} is the Kronecker delta, and the Reynolds stress is expressed as below:

$$-\overline{u_i' u_j'} = v_t\left(\frac{\partial u_i}{\partial x_j} + \frac{\partial u_j}{\partial x_i}\right) - \frac{2}{3}\delta_{ij} k \tag{3}$$

Here, $v_t = C_\mu k^2/\varepsilon$ is the turbulent viscosity, where C_μ is an empirical number ($C_\mu = 0.09$). k is the turbulent kinetic energy, and ε is the turbulent dissipation rate.

The k-ε turbulent model is applied to close the momentum equation; the two transport equations for k-ε are defined as:

$$\frac{\partial k}{\partial t} + \frac{\partial (k u_j)}{\partial x_j} = \frac{\partial}{\partial x_j}\left[\left(v + \frac{v_t}{\sigma_k}\right)\frac{\partial k}{\partial x_j}\right] + P_k - \varepsilon \tag{4}$$

$$\frac{\partial \varepsilon}{\partial t} + \frac{\partial (k u_j)}{\partial x_j} = \frac{\partial}{\partial x_j}\left[\left(v + \frac{v_t}{\varepsilon_k}\right)\frac{\partial \varepsilon}{\partial x_j}\right] + C_{\varepsilon 1} P_k \frac{\varepsilon}{k} - C_{\varepsilon 2}\frac{\varepsilon^2}{k} \tag{5}$$

In this study, spatial discretization employs second-order schemes for increased numerical accuracy. For temporal discretization, the first-order implicit Euler scheme is adopted, striking a balance between computational efficiency and accuracy in capturing time-dependent flow behavior. To solve the flow equations, a segregated flow solver based on the SIMPLE algorithm is implemented. This algorithm iteratively couples the pressure and velocity components to reach a converged solution. In each simulation, the time step size is determined using the Courant–Friedrichs–Lewy (CFL) condition, determined by CFL = $U\Delta t/\Delta x$, where U represents the characteristic velocity, Δt is the time step size, and Δx is the characteristic length scale. The Courant number is calculated for each cell and is typically kept below or equal to 1 for numerical stability. Finally, convergence is considered achieved when the forces acting on the submarine and AUV, the flow field, and the residual values reach a stable state. This indicates that the calculations have reached a reliable and consistent solution.

2.2.2. Fluid Domain and Boundary Conditions

Figure 2 illustrates the schematic of the fluid domain used in our simulation, where the AUV is positioned directly beneath the submarine. As the AUV is much smaller than the submarine, we magnified it ten times for increased clarity. The length of the fluid domain is 5.0 L, with 1.0 L located before the submarine and 3.0 L behind it. The width and height of the fluid domain are both 1.0 L. Our simulation required a sufficiently large fluid domain to prevent any potential backflow from affecting the numerical results. The boundary conditions are shown in Figure 2.

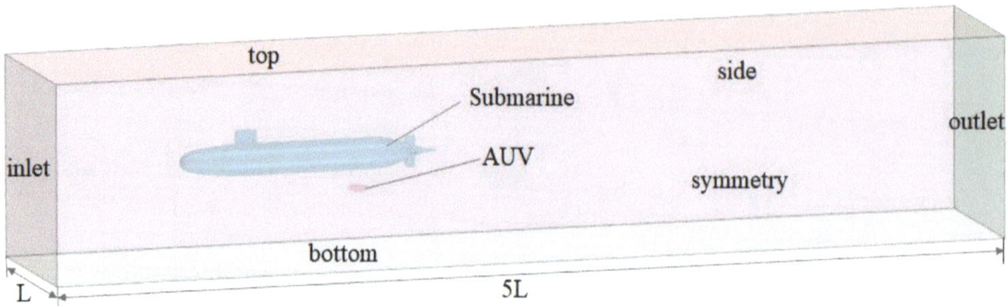

Figure 2. Schematic of fluid domain.

The inlet boundary of the region is specified with a velocity inlet condition, while the outlet boundary is set with a pressure outlet condition. The bottom, side, and top boundaries are treated as velocity inlets, consistent with the velocity inlet boundary. The symmetry plane serves as a symmetric boundary condition. The submarine and AUV are treated as impermeable walls with a no-slip condition. For detailed boundary conditions, please refer to Table 2.

Table 2. Boundary parameters.

Boundary	Boundary Parameter
Inlet	Velocity Inlet, flow speed = (Uo,0,0) m/s, turbulence intensity 0.01
Outlet	Pressure Outlet, hydrostatic pressure
Bottom/side/top	Same as inlet
Submarine	Impermeability wall with no-slip condition
AUV	Impermeability wall with no-slip condition
Symmetry plane	Symmetry plane

2.2.3. Meshing

In this study, to ensure accurate calculations of the flow field and resistance, the mesh set must be able to accurately depict the model's shape, while also capturing the surrounding flow field and the force and vortex structure of the boundary layer of the hull. Mesh generation in this study was performed using the advanced automatic meshing capabilities of STAR-CCM+, which employs the highly efficient Cartesian cut-cell method. To accurately capture the boundary layer near the surfaces of the submarine and AUV, an orthogonal prism layer was generated. This layer consisted of prismatic cells that gradually increased in thickness by a constant scale factor of 1.1, ensuring proper resolution of the flow behavior. The first layer of the prism layer mesh was specifically placed within the logarithmic region of the boundary layer, targeting the range of $30 < Y+ < 100$. This approach helped to accurately model the near-wall flow phenomena. In addition, the trimmer mesh, consisting mostly of hexahedral elements, discretized the majority of the computational domain. The trimmer mesh was chosen for its ability to efficiently represent the complex geometry of the simulation. The total number of meshes employed in the calculations amounted to 6,136,820. Around the AUV, the minimum grid step size was determined to be 0.018 m ($L_{AUV}/242$), whereas around the submarine, it was 0.4 m ($L_{submarine}/217.8$). Figure 3 shows the meshes of the computational domain, Submarine, and AUV.

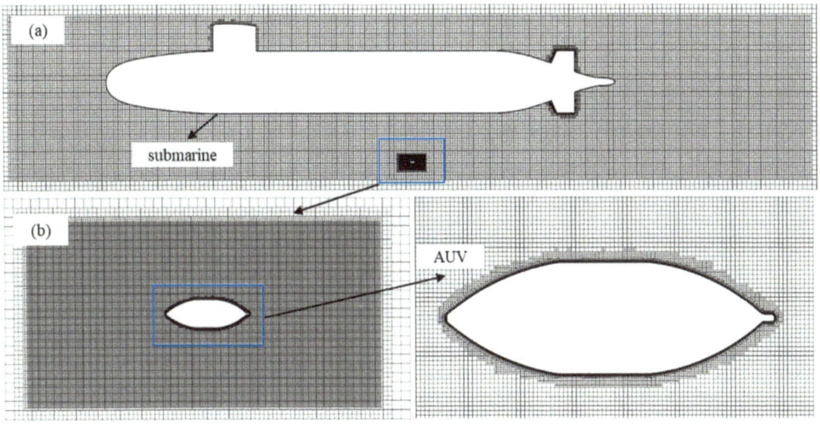

Figure 3. Mesh refinement of the computational domain; (**a**) Overall meshes; (**b**) Local mesh refinement of the AUV.

3. Reliability of Numerical Models

Before applying any numerical method to carry out a simulation, it is essential to first verify its reliability. In this section, we present two typical experimental results to demonstrate the credibility of our numerical method. The first result involves the pressure distribution on the surface of the SUBOFF in a uniform incoming flow. We compared the simulated results against the corresponding experimental data and found that they were in good agreement, validating the accuracy of our numerical method. The second result involved a pair of similar ellipsoids in proximity at separation-length ratios.

3.1. The Pressure Distribution on the Longitudinal Section and Resistance of the SUBOFF

To verify the mesh independence of our research, we utilized SUBOFF as the research object under the sailing condition of infinite depth with a speed of 5.144 m/s. We then adjusted the mesh base size and defined three distinct mesh sizes: coarse mesh, medium mesh, and fine mesh. The number of meshes under the same base size were 2.582 million, 3.846 million, and 6.823 million, respectively. Based on our calculations, the resistance values under these three mesh conditions were 275.6 N, 280.25 N, and 281.4 N, respectively. Notably, the experimental value is 283.8 N [41]. However, the errors between the calculated results and experimental values under various mesh conditions were all within 5%. Considering both calculation accuracy and efficiency, we determined that the medium mesh was the most suitable for our research. See Table 3.

Table 3. Resistance results for different mesh numbers.

-	Coarse Mesh	Medium Mesh	Fine Mesh	Experimental Results
Mesh number (Million)	2.582	3.846	6.823	-
Resistance (N)	275.6	280.25	281.4	283.8
Error (%)	2.89	1.25	0.85	-

Huang et al. [42] have conducted a lot of work on the DARPA SUBOFF model under various Reynolds numbers, whether drag experiments or flow field measurements. Here, the pressure distributions on the SUBOFF AFF-8 surface when the Reynolds number is 1.2×10^6 are used to verify the reliability of the numerical method. The dimensionless pressure coefficient is defined as

$$C_p = \frac{P - P_\infty}{\rho U^2/2} \tag{6}$$

where P_∞ and U are the free-stream static pressure and uniform flow velocity in the velocity inlet, respectively. The C_p on the middle longitudinal section of the SUBOFF AFF-8 model obtained by the numerical simulation is compared with the experimental results [41]. The comparison between the numerical results and experimental results is shown in Figure 4. It can be seen that the results obtained from the numerical simulation have good agreement with the experiment results, which proves the reliability of the numerical methods used in this paper.

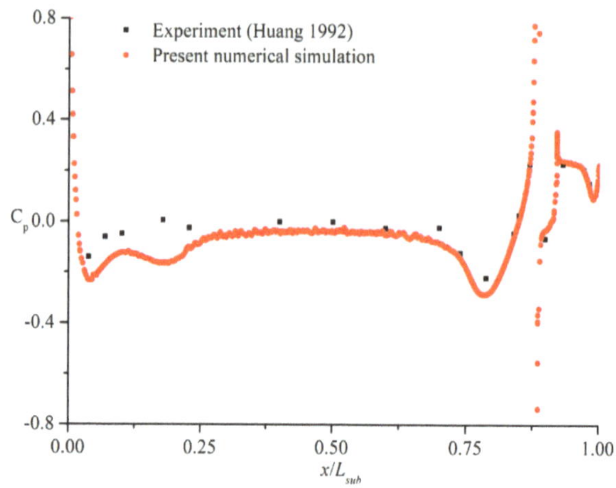

Figure 4. Pressure coefficients on the middle longitudinal section of the SUBOFF AFF-8 model [42].

3.2. A Pair of Similar Ellipsoids in Proximity at Separation-Length Ratios

Molland [24,25] has conducted a lot of work on the hydrodynamic interaction between simple structures, such as the resistance change and pressure distribution when two ellipsoids are very close. The experimental photographs and 2D (two-dimensional) schematic of the two similar ellipsoids are shown in Figure 5. As shown in the figure, the model ellipsoids were constructed from glass-reinforced plastic (GRP). They have an overall length of 1200 mm, a maximum diameter of 200 mm, an eccentricity of 0, and a surface area of 0.601 m^2. The two ellipsoids are exactly the same, the ratio of their long and short axes is 6, and s is defined as the separation between the center lines of the two ellipsoids. Here, the pressure coefficient on the longitudinal section of the above ellipsoid is used to verify the numerical method, where the Reynolds number Is 3.2×10^6 and s/L_e = 0.27 and 0.37.

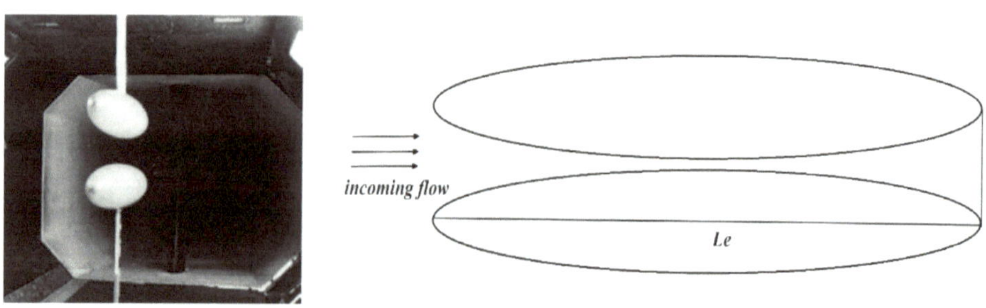

Figure 5. The experimental photographs (**left**) and 2D schematic (**right**) of a pair of ellipsoids in proximity.

The results of the numerical simulation and the experiment are shown in Figure 6. It can be seen that there is almost no difference between the two, which further proves the accuracy of the numerical method in this paper.

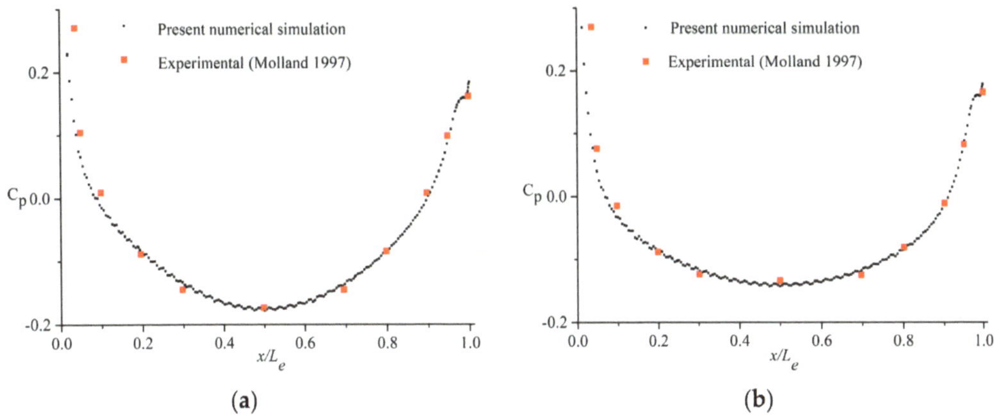

Figure 6. The pressure coefficient on the longitudinal section of the above ellipsoid: (**a**) $s/L_e = 0.27$, (**b**) $s/L_e = 0.37$ [23].

4. Results Analysis

This section aims to analyze the hydrodynamic interaction when the submarine and AUV remain relatively stationary. Two factors are considered in this analysis: the speed and position of the AUV relative to the submarine. The main focus is on the forces acting on the AUV, with additional important flow field information also presented. As the size of the submarine is significantly larger than that of the AUV, the hydrodynamic interaction has minimal effect on the submarine. Accordingly, this paper chooses to disregard the influence of the submarine on the simulation and analysis.

Figure 7 illustrates the relative position between the AUV and the submarine, with the AUV magnified ten times for better visualization, as was done in Figure 2. The AUV is positioned directly beneath the submarine, as shown. To accurately represent the position of the AUV relative to the submarine, we introduce two additional parameters: s and l. The length of the submarine (L) is 87.12 m, as previously defined. The parameter s represents the vertical distance between the center line of the submarine and the AUV. The parameter l represents the longitudinal distance from the geometric center of the AUV to the forefront of the submarine.

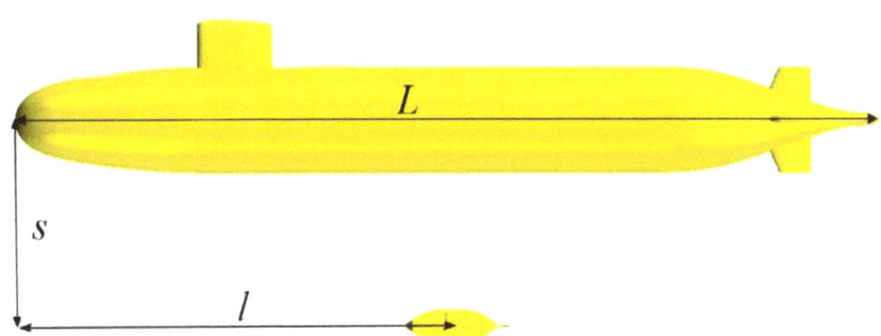

Figure 7. The schematic of the relative position of the AUV and submarine.

In the following simulation, the position of the AUV relative to the submarine will be dimensionless through the three parameters L, s, and l. In order to fully investigate the hydrodynamic interaction of the AUV at different positions, l/L is taken as 0, 0.1, 0.2, 0.4, 0.6,

0.8, 0.9, and 1.0. And, s/L is taken as 0.20, 0.15, 0.12, 0.10, 0.08, and 0.062. In addition, the influence of velocity has also been considered. The simulations are investigated with five speeds at each position, which are 2.0 kn (1.029 m/s), 3.0 kn (1.543 m/s), 4.0 kn (2.058 m/s), 5.0 kn (2.572 m/s), and 6.0 kn (3.087 m/s).

4.1. Results for Drag and Lift

In this study, we selected eight longitudinal positions along the length of the submarine from bow to stern, which are denoted by l/L values of 0, 0.1, 0.2, 0.4, 0.6, 0.8, 0.9, and 1.0, respectively. Additionally, we selected six vertical positions from far to near, which are denoted by s/L values of 0.20, 0.15, 0.12, 0.10, 0.08, and 0.062. As a result, there are 48 positions of the AUV relative to the submarine that are considered in this analysis. The primary objective of this section is to examine the changes in the drag and lift of the AUV with respect to speed at these 48 positions and then compare the influence of the submarine on the AUV at different positions. All of the drag and lift values presented in this section are dimensionless by $0.5\rho U^2 L_a^2$; ρ is the density of water, which is 997.561 kg/m^3. The length of the AUV (L_a) is 1.08 m, as shown in Table 1. Finally, U represents the speed of the AUV.

Figure 8 depicts the changes in drag with respect to speed at different vertical positions of the AUV relative to the submarine, while keeping the longitudinal position fixed. As evident from Figure 8a–h, the drag increases with an increase in speed. However, as the vertical distance between the AUV and the submarine decreases gradually, the drag does not change significantly. This observation suggests that when the AUV and the submarine are relatively stationary, the drag of the AUV is primarily influenced by its speed rather than the presence of the submarine. It is worth noting that when the AUV is positioned in the middle of the submarine, i.e., l/L values of 0.4, 0.6, and 0.8, the drag curves almost fully coincide. On the other hand, when the AUV is located towards the bow or stern of the submarine, the drag curves differ slightly. This difference can be attributed to the significant changes in the shape of the submarine towards its bow and stern resulting in a larger disturbance to the flow field. As the AUV moves closer to the submarine, this disturbance is stronger, thereby altering the drag. However, when the AUV is positioned in the middle of the submarine, where the geometric shape of the submarine remains relatively constant, the flow field is not disturbed, and the drag of the AUV remains nearly unchanged.

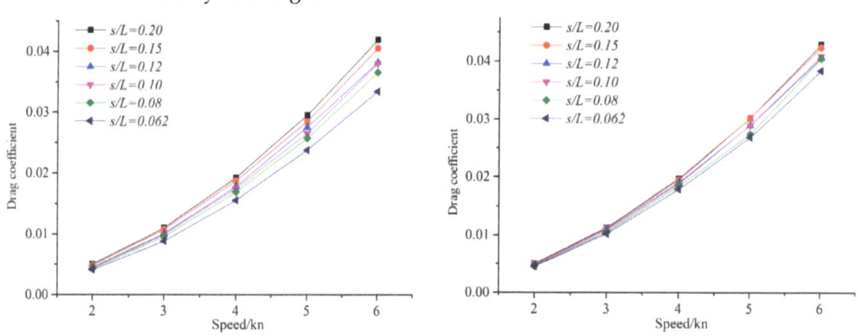

(a) (b)

Figure 8. *Cont.*

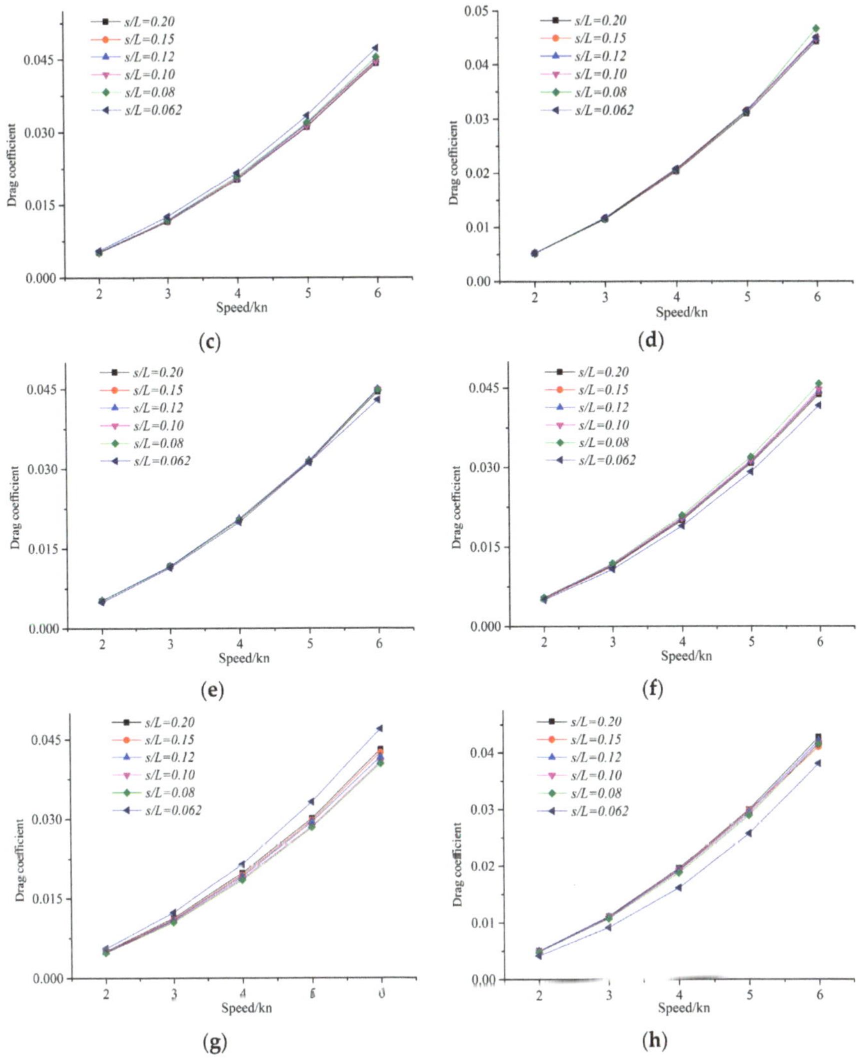

Figure 8. The dimensionless drag of the AUV in different vertical positions: (**a**) $l/L = 0$, (**b**) $l/L = 0.1$, (**c**) $l/L = 0.2$, (**d**) $l/L = 0.4$, (**e**) $l/L = 0.6$, (**f**) $l/L = 0.8$, (**g**) $l/L = 0.9$, (**h**) $l/L = 1.0$.

The lift, as shown in Figure 9, increases with an increase in speed, similar to drag. However, the lift also exhibits a dependence on the vertical distance between the AUV and the submarine, unlike drag. As the vertical distance decreases, the lift increases, with this trend becoming more apparent as the AUV moves closer to the submarine. Another crucial difference is that the lift properties differ significantly based on the position of the AUV relative to the submarine. When the AUV is located at longitudinal positions $l/L = 0$, 0.1, 0.2, 0.4, 0.6, and 1.0, the lift is negative, indicating repulsive hydrodynamic interaction between the AUV and the submarine. Moreover, when the AUV is closer to the submarine, the repulsive force becomes stronger. However, at l/L values of 0.8 and 0.9, if the AUV is positioned far away from the submarine, the lift remains negative. But, as the AUV approaches the submarine, the lift becomes positive, indicating attractive hydrodynamic interaction. This attractive force increases rapidly with the increase in speed as the AUV moves closer to the submarine.

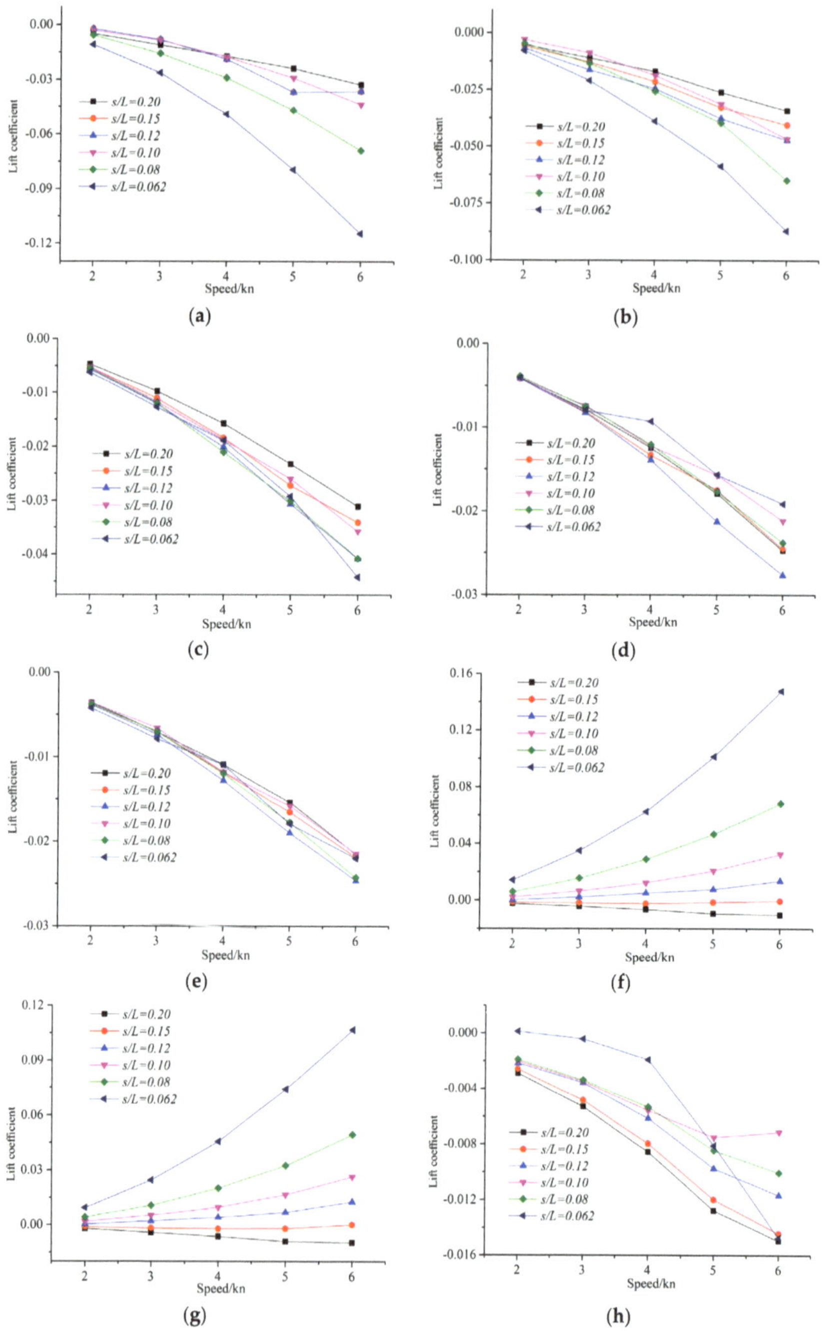

Figure 9. The dimensionless lift of the AUV in different vertical positions: (**a**) $l/L = 0$, (**b**) $l/L = 0.1$, (**c**) $l/L = 0.2$, (**d**) $l/L = 0.4$, (**e**) $l/L = 0.6$, (**f**) $l/L = 0.8$, (**g**) $l/L = 0.9$, (**h**) $l/L = 1.0$.

Figure 10 presents a comparison of the drag of the AUV at various longitudinal positions relative to the submarine. It is noticeable that when the vertical distance between the AUV and the submarine is significant, i.e., $s/L = 0.20$, the hydrodynamic interaction has little impact on the drag of the AUV. Consequently, the drag remains practically constant at different longitudinal positions. However, as the AUV moves closer to the submarine, the differences in drag become more pronounced across various positions. It can be observed that the drag of the AUV is highest when it is located at the middle of the submarine, while the drag is lowest when it is located towards the bow. These results indicate that the shape of the submarine has a critical impact on the flow field around the AUV. The changes in the shape of the submarine towards its bow and stern result in a larger disturbance to the flow field, thereby modifying the drag coefficient. However, when the AUV is placed in the middle of the submarine, where the geometric shape of the vessel remains relatively constant, the flow field is not disturbed, resulting in higher drag values.

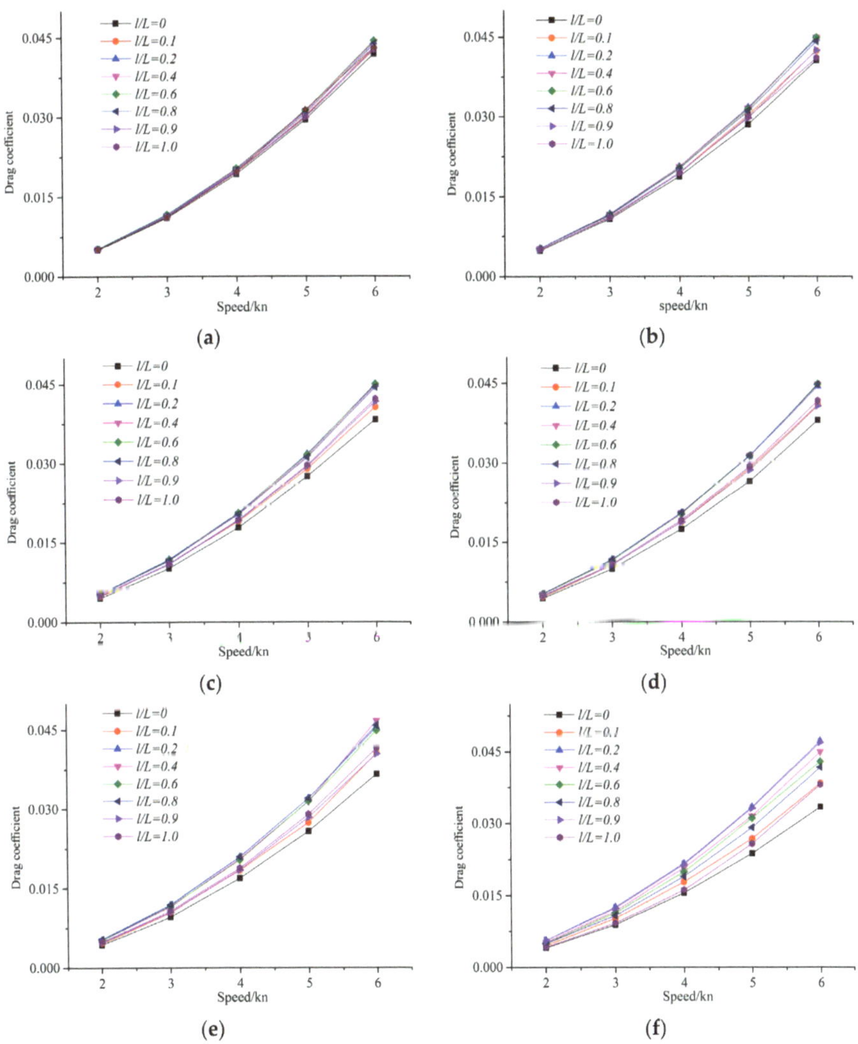

Figure 10. The dimensionless drag of the AUV in different longitudinal positions: (**a**) $s/L = 0.2$, (**b**) $s/L = 0.15$, (**c**) $s/L = 0.12$, (**d**) $s/L = 0.10$, (**e**) $s/L = 0.08$, (**f**) $s/L = 0.062$.

Figure 11 illustrates a comparison of the lift of the AUV at different longitudinal positions relative to the submarine. It is apparent that the AUV's lift varies significantly, depending on its position in relation to the submarine. When the AUV is positioned towards the bow and middle of the submarine, i.e., l/L = 0, 0.1, 0.2, 0.4, and 0.6, the lift exhibits a similar trend. In these positions, the lift is always repulsive. Moreover, the magnitude of the lift at the bow of the submarine is larger than that at the middle for the same speed. This indicates that the geometry of the submarine has a critical impact on the lift properties of the AUV. The flow field around the AUV is more disturbed by the presence of the submarine towards its bow than in the middle region, resulting in higher lift values.

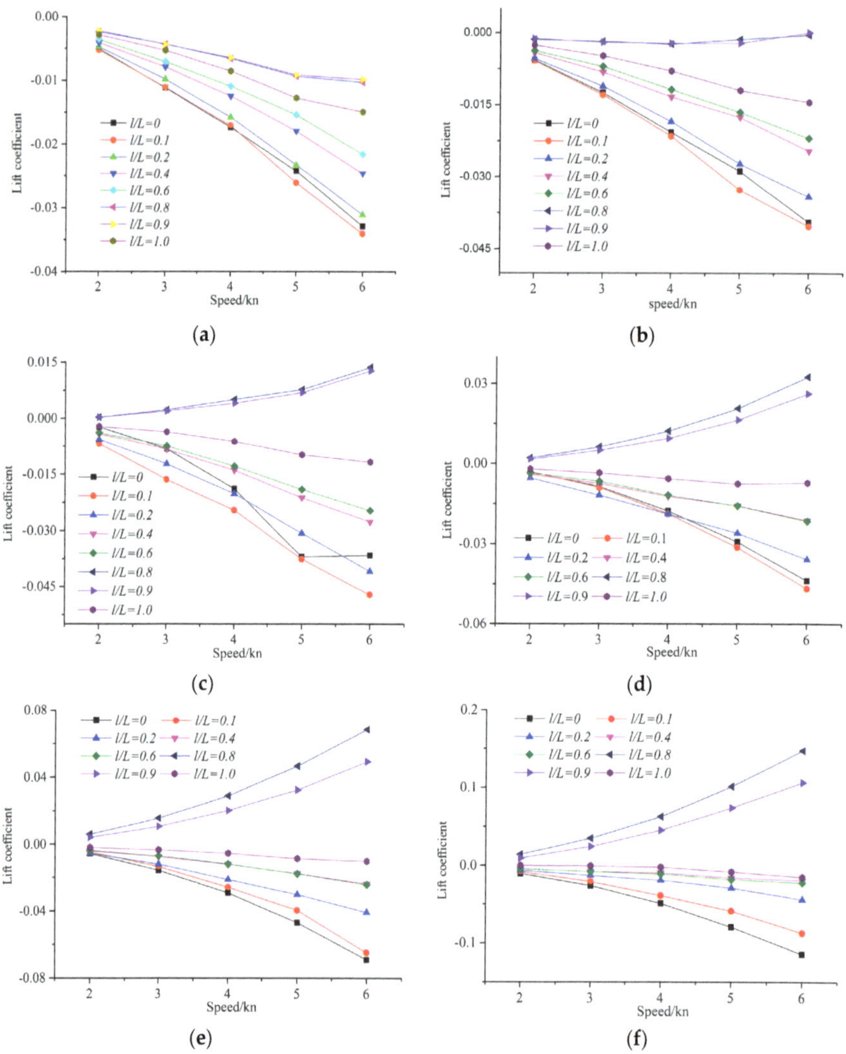

Figure 11. The dimensionless lift of the AUV in different longitudinal positions: (**a**) s/L = 0.2, (**b**) s/L = 0.15, (**c**) s/L = 0.12, (**d**) s/L = 0.10, (**e**) s/L = 0.08, (**f**) s/L = 0.062.

When the AUV is positioned towards the stern of the submarine, the situation is different. For l/L = 0.8 and 0.9, the lift exhibits a similar trend, where an increase in speed results in a rapid increase in lift. Additionally, when the AUV is far from the submarine, the lift force is repulsive, while the lift force becomes attractive when the AUV is close to the submarine. However, it can be observed that the lift is always a repulsive force for $l/L = 1.0$. It is worth noting that the impact of speed on lift is relatively small, and the increase in lift with speed is negligible. Furthermore, the influence of speed decreases as the AUV moves closer to the submarine. These observations highlight the complexity of the flow field around the AUV and submarine system. The intricate geometrical features of the submarine result in an intricate flow pattern, affecting the lift characteristics of the AUV.

4.2. Velocity Field

4.2.1. $l/L = 0$, $s/L = 0.08$

In this study, the AUV is much smaller than the submarine, making it more susceptible to flow disruption. Therefore, only the influence of the AUV's velocity field is analyzed in this section. The distribution of AUV's velocity field under different conditions is analyzed. Figure 12 illustrates the velocity field distribution when the vertical distance between the AUV and submarine is $l/L = 0$ and the longitudinal distance $s/L = 0.08$, with speeds of 2 kn, 4 kn, 5 kn, and 6 kn. The X-axis speed U_x is dimensionless and denoted as U_x/U_0, where U_0 represents the initial sailing speed. The AUV navigates in the flow field of the submarine's bow area, where the surrounding flow field is disturbed by the submarine. Due to the suction effect created by the submarine, the velocity field on the upper surface of the AUV exhibits asymmetric distribution, while the lower surface is less disturbed, showing a relatively symmetrical velocity field distribution. Low-speed areas appear in the bow and stern regions of the AUV, with high-speed areas present on both the upper and lower surfaces. At different speeds, the low-speed recirculation zone at the stern of the AUV slightly decreases with an increase in speed.

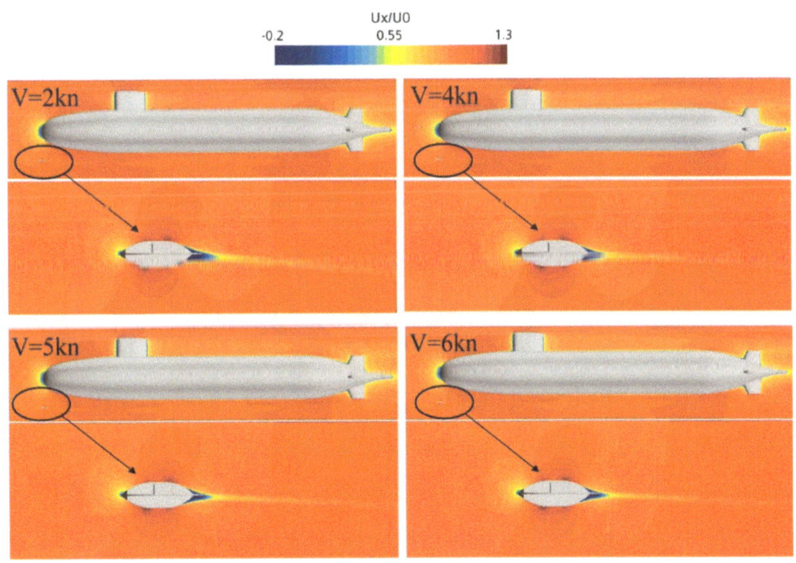

Figure 12. Velocity field distribution with different speeds at $l/L = 0$, $s/L = 0.08$.

Figures 13 and 14 present the velocity field distribution at different speeds for $l/L = 0.6$, $s/L = 0.062$ and $l/L = 1.0$, $s/L = 0.08$, respectively. As shown in Figure 13, when the AUV is sailing in the middle of the submarine, the flow field on the upper surface is significantly disrupted by the presence of the submarine. Moreover, as the speed increases, the high-speed flow field area increases. On the other hand, the lower surface of the AUV exhibits relatively low disturbance, showing a relatively symmetrical velocity field distribution. In Figure 14, the AUV is sailing at the stern of the submarine. Here, due to the relatively far position from the wake area of the submarine, the influence on the AUV's flow field at different speeds is minimal, with negligible changes in the velocity field distribution observed.

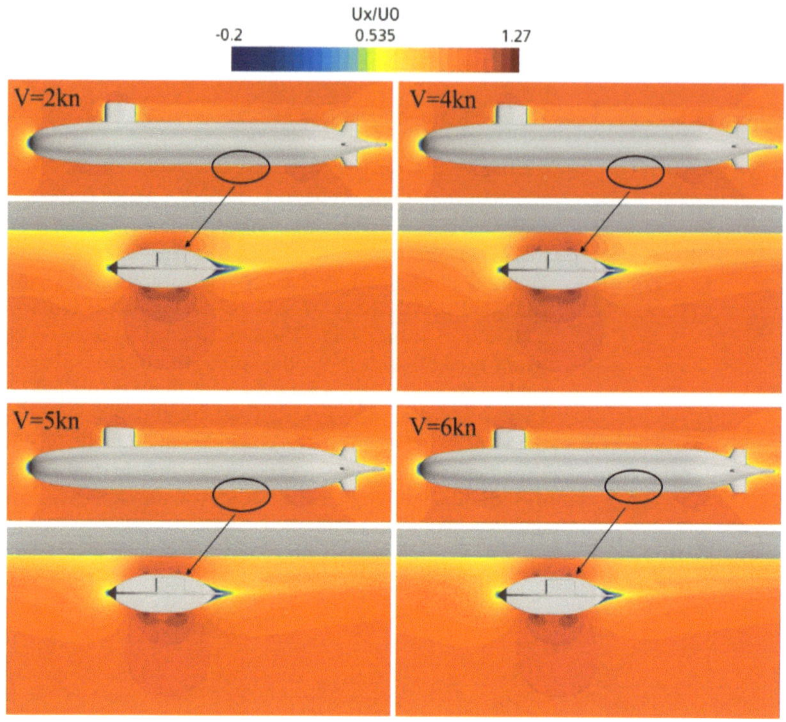

Figure 13. Velocity field distribution with different speeds at $l/L = 0.6$, $s/L = 0.062$.

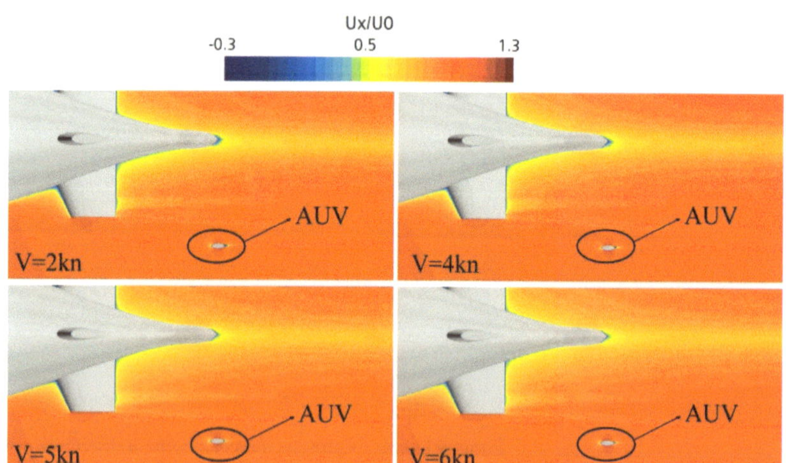

Figure 14. Velocity field distribution with different speeds at $l/L = 1.0$, $s/L = 0.08$.

4.2.2. $s/L = 0.08$, V = 2 kn

This study analyzes the velocity field distribution at different longitudinal distances. Specifically, the longitudinal distance between the submarine and AUV is set at $s/L = 0.08$, while the speed is 2 kn. Figure 15 shows the longitudinal position diagram of the AUV. As the AUV is much smaller than the submarine, an enlarged view of the velocity field distribution is presented in Figure 16 to provide a clearer visualization of the flow field.

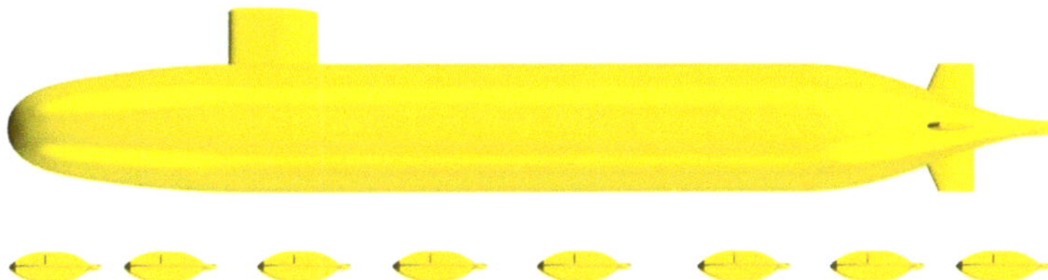

Figure 15. Longitudinal position diagram of the AUV.

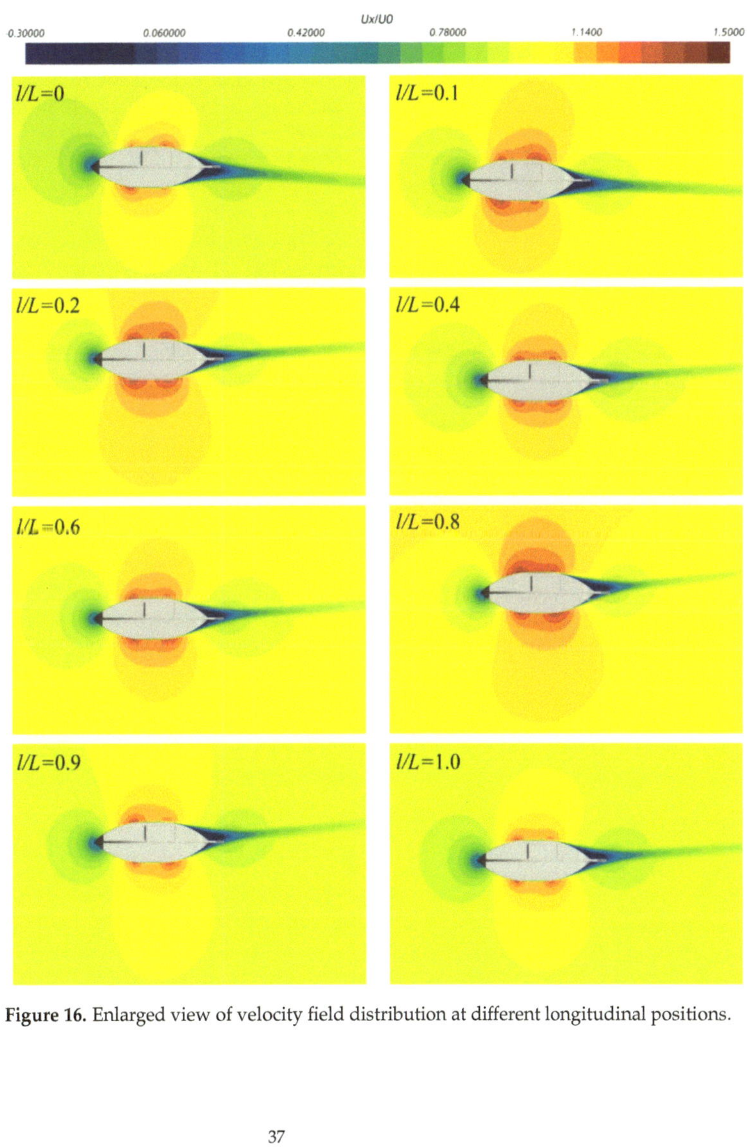

Figure 16. Enlarged view of velocity field distribution at different longitudinal positions.

The velocity field surrounding the AUV is significantly influenced by the presence of the submarine, causing disturbances in the flow field. The distribution of velocity fields under different conditions reveals that low-speed reflux zones appear at the bow and stern regions of the AUV, with high-speed zones appearing on both the upper and lower surfaces. The velocity field distribution on the upper surface of the AUV appears asymmetric when $l/L = 0$, 0.1, and 0.2, while the distribution on the lower surface is relatively symmetrical. In contrast, the velocity field distribution on both the upper and lower surfaces of the AUV is symmetric when $l/L = 0.4$ and 0.6. At $l/L = 0.8$, the velocity field distribution on the upper surface of the AUV is symmetrical, while it becomes asymmetric on the lower surface. In addition, when the AUV approaches the stern area of the submarine ($l/L = 0.9$), the low-speed areas on both the upper and lower surfaces increase due to the interference of the stern rudder. When $l/L = 1.0$, since the AUV is further away from the submarine, it is less affected, and the distribution of velocity fields on both the upper and lower surfaces is basically symmetrical.

4.2.3. $l/L = 0.6$, V = 2 kn

In this study, the velocity field distribution at different vertical distances has been analyzed. To provide a clearer visualization of the flow patterns under different conditions, Figure 17 presents an enlarged view of the velocity field distribution at different vertical positions. As shown in Figure 16, all conditions have been presented in one diagram for ease of comparison. When $s/L = 0.062$, it can be observed that the flow field on the upper surface and bow and stern areas of the AUV is significantly disturbed due to the adsorption effect of the submarine, resulting in a low-speed zone. This effect is caused by the submarine's interference with the surrounding water flow, leading to an uneven velocity field distribution. However, when the longitudinal distance between the submarine and AUV is further increased to $s/L = 0.08$, 0.10, and 0.12, the AUV is far away from the submarine, and the influence of the submarine on the surrounding flow field is negligible. Consequently, the flow field surrounding the AUV is basically unaffected by the presence of the submarine, resulting in a more stable velocity field distribution.

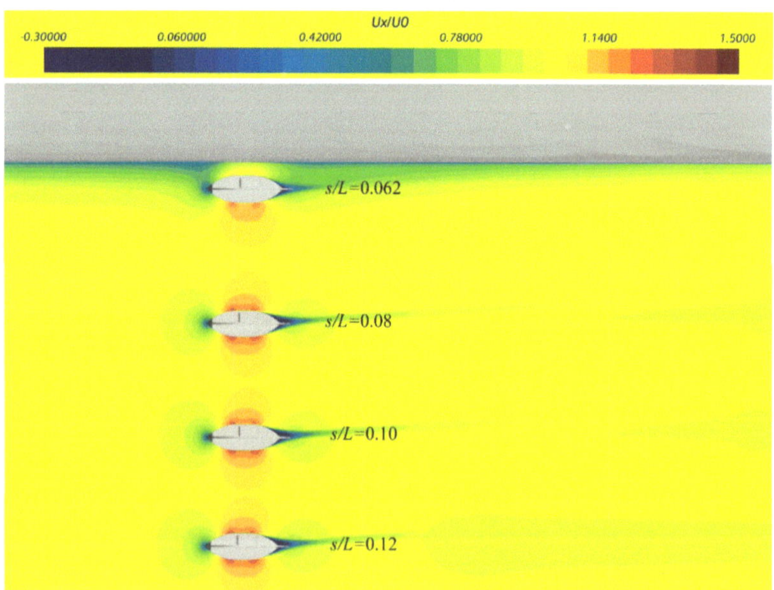

Figure 17. Enlarged view of velocity field distribution at different vertical positions.

Figure 18 depicts the velocity vectors of the AUV at different vertical positions when $l/L = 0.6$ and $V = 2$ kn. The analysis indicates that when $s/L = 0.062$, the adsorption effect of the submarine causes the low-speed reflux area at the stern of the AUV to be smaller compared to other conditions. Conversely, when $s/L = 0.08$, 0.10, 0.12, 0.15, and 0.20, the AUV is less affected by the submarine, leading to a relatively large low-speed reflux zone. To visualize the vortex structure of the velocity vector affected by the submarine more clearly, Figure 19 presents an enlarged view of the aft region of the AUV under the $s/L = 0.062$ condition. As shown in Figure 18, a reflux zone appears at the stern area of the AUV, accompanied by the formation of vortex structures. This observation further confirms the effect of the submarine's presence on the flow field surrounding the AUV.

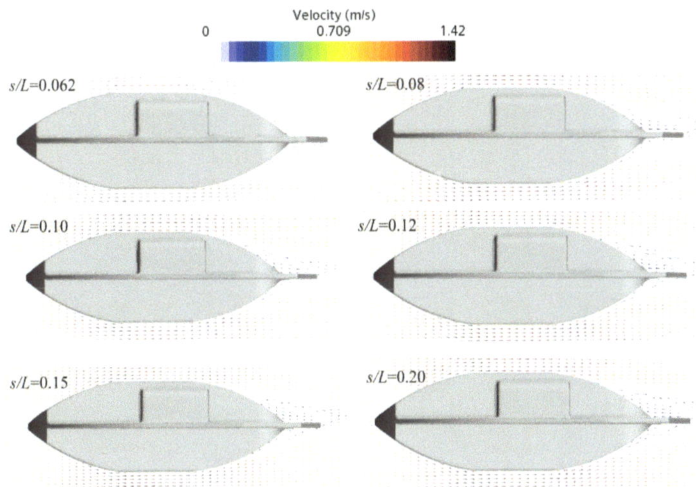

Figure 18. Velocity vectors of AUV at different vertical positions.

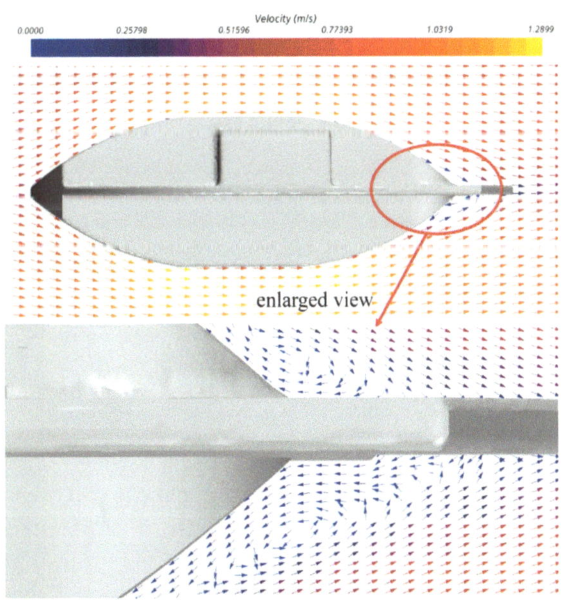

Figure 19. Enlarged view of velocity vectors of AUV at $l/L = 0.6$, $s/L = 0.062$, and $V = 2$ kn.

4.3. *Pressure s/L = 0.08, V = 2 kn*

Figure 20 illustrates the pressure distribution of the AUV under different vertical distances, with $s/L = 0.08$ and $V = 2$ kn. Observations indicate that the high-pressure region of the AUV primarily appears in the bow area and the shape transition region of the upper surface. The AUV is significantly disrupted when it enters the bow flow field area of the submarine. Therefore, as l/L increases from 0 to 0.8, the AUV gradually moves away from the bow area, and the range of the high-pressure area gradually decreases. Furthermore, at $l/L = 0.9$ and 1.0, the high-pressure range increases near the stern rudder area due to the interference of the submarine's stern rudder. The AUV experiences increasing pressure levels in this area, which can affect its overall stability and maneuverability.

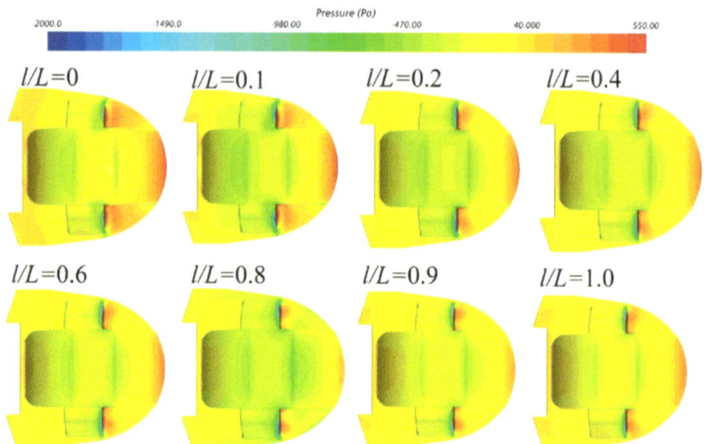

Figure 20. Pressure distribution.

5. Conclusions

The research background of this paper is based on the interference of the AUV by the mother ship during the recycling process. The AUV remains relatively stationary with the mother ship during the recovery process. The AUV sails at different vertical positions and longitudinal positions relative to the mother ship. And, the optimal position for the AUV to be least affected by the mother ship during the recovery process was obtained. The details are as follows:

(1) When the vertical distance between the AUV and submarine changes, the resistance has little change, but when the speed changes, the resistance changes significantly.
(2) When approaching the bow and stern area of the submarine, the AUV is more susceptible to interference, while in the parallel middle section, it is less susceptible to interference because the shape of the bow and stern of the submarine changes greatly, and the flow field disturbance is obvious.
(3) The lift increases as the vertical distance between the AUV and the submarine decreases, and the closer it approaches the submarine, the more significant the change in lift. And, when the AUV is in different longitudinal positions on the submarine, when the lift is negative, there is a repulsive effect between the submarine and the AUV. When the lift is positive, the submarine exhibits an upward suction effect on the AUV.
(4) The velocity and pressure fields of the AUV are affected by the submarine when sailing at different positions. At the mid-longitudinal position of the submarine, the impact is minimal, and it is the best position for recovery.
(5) This paper focuses on the interference effects of the AUV and submarine when they remain relatively stationary. Further research is needed to investigate the interference effects of oblique motion recovery and non-relative static recovery.

In conclusion, the AUV moving through the water generates a resistance force due to the viscosity of the fluid. The submarine must, therefore, accelerate slowly during the recovery process. The movement of the submarine can cause disturbances in the surrounding fluid flow field. These disturbances may affect the motion of the AUV, causing it to move erratically or become unstable. Therefore, careful consideration must be given to the position and motion of the submarine during the recovery process.

Author Contributions: W.L.: writing—original draft, conceptualization, methodology, software, funding acquisition, resources; C.M.: writing—original draft, resources; D.J.: writing—review and editing, software; T.Z.: visualization, data curation; T.W.: funding acquisition, resources. All authors have read and agreed to the published version of the manuscript.

Funding: This work was financially supported by the National Key Research and Development Program of China (Grant No. 2022YFC2806300); the National Natural Science Foundation of China (Grant No. 52101379); and the Natural Science Foundation of Guangdong Province, China (2021A1515012134).

Institutional Review Board Statement: Not applicable.

Informed Consent Statement: Not applicable.

Data Availability Statement: The data that support the findings of this study are available within the article.

Conflicts of Interest: The authors declare no conflict of interest.

References

1. Nicholson, J.W.; Healey, A.J. The present state of autonomous underwater vehicle (AUV) applications and technologies. *Mar. Technol. Soc. J.* **2008**, *42*, 44–51. [CrossRef]
2. Wang, H.J.; Long, X.; Li, J.; Zhou, H.N. Design, construction of a small unmanned underwater vehicle. In Proceedings of the 2013 MTS/IEEE OCEANS-Bergen, Bergen, Norway, 10–14 June 2013; pp. 1–6.
3. Nouri, N.M.; Zeinali, M.; Jahangardy, Y. AUV hull shape design based on desired pressure distribution. *J. Mar. Sci. Technol.* **2016**, *21*, 203–215. [CrossRef]
4. Yan, K.; Wu, L. A survey on the key technologies for underwater AUV docking. *Robot* **2007**, *29*, 267–273.
5. Zhang, W.; Wu, W.; Teng, Y.; Li, Z.; Yan, Z. An underwater docking system based on UUV and recovery mother ship: Design and experiment. *Ocean Eng.* **2023**, *281*, 114767. [CrossRef]
6. Szczotka, M. AUV launch & recovery handling simulation on a rough sea. *Ocean Eng.* **2022**, *246*, 110509.
7. Fan, S.; Liu, C.; Li, B.; Xu, Y.; Xu, W. UUV docking based on USBL navigation and vision guidance. *J. Mar. Sci. Technol.* **2019**, *24*, 673–685. [CrossRef]
8. Page, B.R.; Mahmoudian, N. Simulation-driven optimization of underwater docking station design. *IEEE J. Ocean. Eng.* **2019**, *45*, 404–413. [CrossRef]
9. Zhang, W.; Jia, G.; Wu, P.; Yang, S.; Huang, B.; Wu, D. Study on hydrodynamic characteristics of AUV launch process from a launch tube. *Ocean Eng.* **2021**, *232*, 109171. [CrossRef]
10. Brizzolara, S.; Chryssostomidis, C. Design of an Unconventional ASV for Underwater Vehicles Recovery: Simulation of the motions for operations in rough seas. In Proceedings of the ASNE International Conference on Launch and Recovery, Linthicum, MD, USA, 14–15 November 2012.
11. Sarda, E.I. *Automated Launch and Recovery of an Autonomous Underwater Vehicle from an Unmanned Surface Vessel*; Florida Atlantic University: Boca Raton, FL, USA, 2016.
12. Bai, G.; Gu, H.; Zhang, H.; Meng, L.; Tang, D. V-shaped wing design and hydrodynamic analysis based on moving base for recovery AUV. In Proceedings of the 2018 WRC Symposium on Advanced Robotics and Automation (WRC SARA), Beijing, China, 16 August 2018; pp. 320–325.
13. Palomeras, N.; Vallicrosa, G.; Mallios, A.; Bosch, J.; Vidal, E.; Hurtos, N.; Carreras, M.; Ridao, P. AUV homing and docking for remote operations. *Ocean Eng.* **2018**, *154*, 106–120. [CrossRef]
14. Meng, L.; Lin, Y.; Gu, H.; Bai, G.; Su, T.C. Study on dynamic characteristics analysis of underwater dynamic docking device. *Ocean Eng.* **2019**, *180*, 1–9. [CrossRef]
15. Li, Y.; Jiang, Y.; Cao, J.; Wang, B.; Li, Y. AUV docking experiments based on vision positioning using two cameras. *Ocean Eng.* **2015**, *110*, 163–173. [CrossRef]
16. Lin, M.; Yang, C. Auv docking method in a confined reservoir with good visibility. *J. Intell. Robot. Syst.* **2020**, *100*, 349–361. [CrossRef]
17. Vu, M.T.; Choi, H.S.; Nhat, T.Q.M.; Nguyen, N.D.; Lee, S.D.; Le, T.H.; Sur, J. Docking assessment algorithm for autonomous underwater vehicles. *Appl. Ocean Res.* **2020**, *100*, 102180. [CrossRef]

18. Hardy, T.; Barlow, G. Unmanned Underwater Vehicle (UUV) deployment and retrieval considerations for submarines. In Proceedings of the International Naval Engineering Conference and Exhibition 2008, Hamburg, Germany, 13 April 2008.
19. Mawby, A.; Wilson, P.; Bole, M.; Fiddes, S.P.; Duncan, J. *Manoeuvring Simulation of Multiple Underwater Vehicles in Close Proximity*; SEA (Group) Ltd.: Frome, UK, 2010.
20. Kim, J.; Lee, G. A study on the UUV docking system by using torpedo tubes. In Proceedings of the 2011 8th International Conference on Ubiquitous Robots and Ambient Intelligence (URAI), Incheon, Republic of Korea, 23–26 November 2011; pp. 842–844.
21. Meng, L.; Lin, Y.; Gu, H.; Su, T.C. Study on dynamic docking process and collision problems of captured-rod docking method. *Ocean Eng.* **2019**, *193*, 106624. [CrossRef]
22. Yang, Q.; Liu, H.; Yu, X.; Zhang, W.; Chen, J. Attitude constraint-based recovery for under-actuated AUVs under vertical plane control during the capture stage. *Ocean Eng.* **2023**, *281*, 115012. [CrossRef]
23. Molland, A.F.; Utama, I.K.A.P. *Wind Tunnel Investigation of a Pair of Ellipsoids in Close Proximity*; University of Southampton: Southampton, UK, 1997.
24. Molland, A.F.; Utama, I.K.A.P. Experimental and numerical investigations into the drag characteristics of a pair of ellipsoids in close proximity. *Proc. Inst. Mech. Eng. Part M J. Eng. Marit. Environ.* **2002**, *216*, 107–115. [CrossRef]
25. Husaini, M.; Samad, Z.; Arshad, M.R. CFD simulation of cooperative AUV motion. *Indian J. Mar. Sci.* **2009**, *38*, 346–351.
26. Zhang, D.; Chao, L.; Pan, G. Analysis of hydrodynamic interaction impacts on a two-AUV system. *Sh. Offshore Struct.* **2019**, *14*, 23–34. [CrossRef]
27. Rattanasiri, P.; Wilson, P.A.; Phillips, A.B. Numerical investigation of a fleet of towed AUVs. *Ocean Eng.* **2014**, *80*, 25–35. [CrossRef]
28. Mitra, A.; Panda, J.P.; Warrior, H.V. Experimental and numerical investigation of the hydrodynamic characteristics of Autonomous Underwater Vehicles over sea-beds with complex topography. *Ocean Eng.* **2020**, *198*, 106978. [CrossRef]
29. Zhang, W.; Zeng, J.; Yan, Z.; Wei, S.; Tian, W. Leader-following consensus of discrete-time multi-AUV recovery system with time-varying delay. *Ocean Eng.* **2021**, *219*, 108258. [CrossRef]
30. Wu, L.; Li, Y.; Su, S.; Yan, P.; Qin, Y. Hydrodynamic analysis of AUV underwater docking with a cone-shaped dock under ocean currents. *Ocean Eng.* **2014**, *85*, 110–126. [CrossRef]
31. Meng, L.; Lin, Y.; Gu, H.; Su, T.C. Study on the mechanics characteristics of an underwater towing system for recycling an Autonomous Underwater Vehicle (AUV). *Appl. Ocean Res.* **2018**, *79*, 123–133. [CrossRef]
32. Meng, L.S.; Lin, Y.; Gu, H.T.; Su, T.C. Study of the Dynamic Characteristics of a Cone-Shaped Recovery System on Submarines for Recovering Autonomous Underwater Vehicle. *China Ocean Eng.* **2020**, *34*, 387–399. [CrossRef]
33. Fedor, R. *Simulation of a Launch and Recovery of an UUV to an Submarine*; KTH Royal Institute of Technology: Stockholm, Sweden, 2009.
34. Leong, Z.Q.; Saad, K.; Ranmuthugala, S.D.; Duffy, J.T. Investigation into the hydrodynamic interaction effects on an AUV operating close to a submarine. In Proceedings of the Pacific 2013 International Maritime Conference, Sydney, Australia, 7–9 October 2013; pp. 1–11.
35. Leong, Z.Q.; Ranmuthugala, D.; Penesis, I.; Nguyen, H. Quasi-static analysis of the hydrodynamic interaction effects on an autonomous underwater vehicle operating in proximity to a moving submarine. *Ocean Eng.* **2015**, *106*, 175–188. [CrossRef]
36. Leong, Z.Q.; Ranmuthugala, D.; Forrest, A.L.; Duffy, J. Numerical investigation of the hydrodynamic interaction between two underwater bodies in relative motion. *Appl. Ocean Res.* **2015**, *51*, 14–24.
37. Du, X.; Zheng, Z.; Guan, S. Numerical Calculation of Hydrodynamic Interactions of Submarine Flow on AUV. In Proceedings of the OCEANS-MTS/IEEE Kobe Techno-Oceans (OTO), Kobe, Japan, 28–31 May 2018; pp. 1–5.
38. Groves, N.C.; Huang, T.T.; Chang, M.S. *Geometric Characteristics of DARPA SUBOFF Models (DTRC Model Nos 5470 and 5471)*; David Taylor Research Center: Bethesda, MD, USA, 1989.
39. Menter, F.R. Two-equation eddy-viscosity turbulence models for engineering applications. *AIAA J.* **1994**, *32*, 1598–1605. [CrossRef]
40. Larsson, L.; Raven, H. *Ship Resistance and Flow*; Society of Naval Architects and Marine Engineering: New York, NY, USA, 2010.
41. Liu, H.; Huang, T.T. *Summary of DARPA SUBOFF Experimental Program Data*; Naval Surface Warfare Center Carderock Div: Bethesda, MD, USA, 1998.
42. Huang, T.; Liu, H.L.; Groves, N.; Forlini, T.; Blanton, J.; Gowing, S. Measurements of flows over an axisymmetric body with various appendages in a wind tunnel: The DARPA SUBOFF experimental program. In Proceedings of the 19th Symposium on Naval Hydrodynamics, Seoul, Republic of Korea, 23–28 August 1992; pp. 312–346.

Journal of
Marine Science and Engineering

MDPI

Article

A Deep Reinforcement Learning-Based Path-Following Control Scheme for an Uncertain Under-Actuated Autonomous Marine Vehicle

Xingru Qu, Yuze Jiang, Rubo Zhang * and Feifei Long

School of Mechanical and Electrical Engineering, Dalian Minzu University, Dalian 116600, China; qxr@dlnu.edu.cn (X.Q.); jiangyuze@dlnu.edu.cn (Y.J.); longfeifei@dlnu.edu.cn (F.L.)
* Correspondence: zhangrubo@dlnu.edu.cn

Abstract: In this article, a deep reinforcement learning-based path-following control scheme is established for an under-actuated autonomous marine vehicle (AMV) in the presence of model uncertainties and unknown marine environment disturbances is presented. By virtue of light-of-sight guidance, a surge-heading joint guidance method is developed within the kinematic level, thereby enabling the AMV to follow the desired path accurately. Within the dynamic level, model uncertainties and time-varying environment disturbances are taken into account, and the reinforcement learning control method using the twin-delay deep deterministic policy gradient (TD3) is developed for the under-actuated vehicle, where path-following actions are generated via the state space and hybrid rewards. Additionally, actor-critic networks are developed using the long-short time memory (LSTM) network, and the vehicle can successfully make a decision by the aid of historical states, thus enhancing the convergence rate of dynamic controllers. Simulation results and comprehensive comparisons on a prototype AMV demonstrate the remarkable effectiveness and superiority of the proposed LSTM-TD3-based path-following control scheme.

Keywords: autonomous marine vehicle; path-following control; surge-heading joint guidance; twin-delay deep deterministic policy gradient; long-short time memory network

Citation: Qu, X., Jiang, Y., Zhang, R., Long, F. A Deep Reinforcement Learning-Based Path-Following Control Scheme for an Uncertain Under-Actuated Autonomous Marine Vehicle. *J. Mar. Sci. Eng.* **2023**, *11*, 1762. https://doi.org/10.3390/jmse11091762

Academic Editor: Kamal Djidjeli

Received: 1 August 2023
Revised: 2 September 2023
Accepted: 6 September 2023
Published: 9 September 2023

1. Introduction

The Autonomous marine vehicle (AMV) is a marine intelligent platform that performs tasks autonomously or semi-autonomously [1], which is widely applied in military and civilian fields due to its small volume, strong concealment, good flexibility and other advantages [2]. In different missions, path following of the AMV plays a crucial role for realized autonomous operation. Considering that, in practice, the AMV inevitably suffers from marine environment disturbances, the path-following control method with high precision and efficiency is crucial to the success of an operation, where a parameterized path is expected to be tracked as accurately as possible [3].

In generally, the path-following control of an AMV consists of two critical parts: kinematics guidance and dynamics control. In the part of guidance research, by calculating the desired heading angle, path-following errors can converge to zero, and in the part of control research, control inputs including surge force and yaw torque are solved using the desired guidance signals, thus contributing to the path following performance [4]. For the former, the light-of-sight (LOS) guidance was widely applied because of its high precision and simplicity [5–8]. For the latter, fruitful methods were proposed and applied to controller design, such as PID control [9,10], fuzzy control [11–13], adaptive control [14,15], active disturbances rejection control [16,17], sliding mode control [18,19], and backstepping control [20–22]. In [23], considering the path-following control under unknown environment disturbances, the modified integral LOS guidance law and the adaptive sliding mode control law are developed, realizing the desired path following. In [24], to solve the path-following control of an under-actuated autonomous underwater vehicle subject to environment disturbances, an adaptive robust control method is proposed using fuzzy logic, backstepping and sliding mode technology, where the fuzzy logic system is utilized to approximate the unknown uncertainties. In [25], a novel, adaptive, robust path-following scheme is proposed by combining with the trajectory linearization control and the finite-time disturbance observer. In [26], a fuzzy unknown observer-based, robust, adaptive path-following control scheme is proposed, where the fuzzy observer is designed to estimate lumped unknowns and the observer-based, robust, adaptive tracking control laws are synthesized, thus ensuring that the guided signals are globally asymptotically tracked. However, the above control method depends on a system model with high accuracy, and the derivation process is complex.

With increasingly rapid development of machine learning, deep reinforcement learning (DRL) algorithms are widely applied to the relative studies of unmanned system control [27]. The DRL is a combination of deep learning and reinforcement learning, which has strong decision-making ability and anti-disturbance ability of reinforcement learning and strong representation ability of deep neural network, thus effectively reducing the complexity and difficulty of the controller design. At present, the popular DRL algorithms include the soft actor-critic (SAC), the proximal policy optimization (PPO), and the deep deterministic policy gradient (DDPG) [28–30]. In [31], the advantage of actor-critic (A2C) is proposed to solve path-following control for a fish-like robot, where the desired path is a randomly generated curve. In [32], a DRL controller is designed using the DDPG for path following, and simulation shows that the proposed method is better than the PID in terms of transient characteristics. In [33], a distributed DRL method is proposed to solve the path-following control of an under-actuated AMV, where the DDPG-based controller is designed and the radial basis neural network is utilized to approximate the unknown disturbances. In [34], an improved DDPG control method was proposed for path following based on an optimized sampling pool and average motion evaluation network, and the simulation results show that the proposed method effectively improves the utilization rate of samples and avoids falling into a local optimum in the training process. In [35], a linear active disturbances rejection controller based on the SAC was proposed to solve path-following control under wind and wave environments. In [36], the path-following control laws were designed using the twin-delayed deep deterministic policy gradient algorithm (TD3), where desired velocities were generated by the LOS guidance.

Considering the path following of an under-actuated AMV under unknown model parameters and environment disturbances, this article establishes the motion model for an AMV, and proposes a path-following control method by combining with the long short-term memory network (LSTM) and the TD3 algorithm. The main contributions are as follows: (1) within the kinematic level, the surge-heading joint guidance law is developed based on the LOS, where the desired velocity signals are generated to guide the vehicle along the desired path; (2) within the dynamic level, the TD3-based surge-heading controller is developed for the vehicle, where states, actions and reward functions are defined; and (3) to enhance the convergence rate of controller networks, the LSTM layer, using the historical states, is added into the TD3.

The remainder of this article is organized as follows: preliminaries and problem statement are described in Section 2; Section 3 presents the kinematic guidance law and the DRL-based dynamic controller of an AMV; simulation results and analysis are presented in Section 4; and Section 5 contains the conclusion.

2. Preliminaries and Problem Statement

2.1. Reinforcement Learning

Reinforcement learning is based on the Markov decision process. Four basic elements are defined as $\{S, A, P, R\}$, where S is the set of all states, A is the set of all actions, P is the state transition probability, and R is the reward function [37]. The decay sum of all rewards from a certain state to the final state can be calculated by

$$R_t = r_{t+1} + \gamma r_{t+2} + \gamma^2 r_{t+3} + \ldots = \sum_{k=0}^{\infty} \gamma^k r_{t+k+1} \tag{1}$$

where γ is the discount factor, satisfying $\gamma \in [0, 1]$, and r_{t+i} $(i = 1, \cdots, k+1)$ is the reward at the current time.

Additionally, the value functions under the policy η include action value function $Q^\mu(s_t, a_t)$ and state value function $V^\mu(s_t)$, where s_t and a_t are the state and action at the current time. The value functions are described as

$$\begin{cases} Q^\mu(s_t, a_t) = E_\mu[R_t|s_t, a_t] = E_\mu[\sum\limits_{k=0}^{\infty} \gamma^k r_{t+k+1}|s_t, a_t] \\ V^\mu(s_t) = E_\mu[R_t|s_t] = E_\mu[\sum\limits_{k=0}^{\infty} \gamma^k r_{t+k+1}|s_t] \end{cases} \tag{2}$$

where E expresses the expectation, and the optimal policy μ^* can be achieved by maximizing the optimal state-action value functions [38].

$$\mu^* = \text{argmax} V^\mu(s_t) = \text{argmax} Q^\mu(s_t, a_t) \tag{3}$$

2.2. LSTM Network

The LSTM network has better memory ability, where important data is retained and irrelevant noise is deleted, thereby relieving the gradient disappearance of the existing recurrent neural network and the memory burden of networks [39]. The neuronal structure is shown in Figure 1, where x_t is the input; h_t is the output; c_t is the state value of the memory cell at the current time; h_{t-1} and c_{t-1} are the input signals at the previous time; f_t is the forgetting gate; i_t is the input gate; o_t is the output gate; and σ is the sigmoid function.

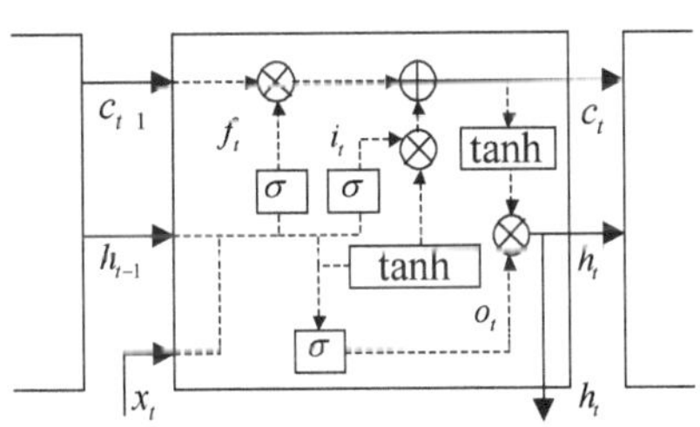

Figure 1. Neuronal structure of the LSTM.

As shown in Figure 1, when the information inputs to the neuron, it firstly goes through the forgetting gate and input gate; then, it goes through the output gate, and the state value of the memory cell c_t are calculated based on the f_t and i_t. Finally, the outputs are calculated based on o_t and c_t. The renewal process can be described by

$$\begin{cases} f_t = \sigma(W_1 x_t + W_2 h_{t-1} + b_1) \\ i_t = \sigma(W_3 x_t + W_4 h_{t-1} + b_2) \\ o_t = \sigma(W_7 x_t + W_8 h_{t-1} + b_4) \\ c_t = c_{t-1} \times f_t + \tanh(W_5 x_t + W_6 h_{t-1} + b_3) \times i_t \\ h_t = o_t \times \tanh(c_t) \end{cases} \tag{4}$$

where W_i is the weight coefficient with $i = 1, 2, \cdots, 8$; b_h is the bias with $h = 1, \cdots, 4$, and tanh is the activation function [40].

2.3. Under-Actuated AMV Model

As described in [41], the under-actuated AMV model of three degrees of freedom in the horizontal plane is written as

$$\begin{cases} \eta = R(\eta)\nu \\ M\dot{\nu} + C(\nu)\nu + D(\nu)\nu = \tau_d + \tau \end{cases} \tag{5}$$

where $\eta = [x, y, \psi]^T$ are the positions and heading angle of AMV in the earth-fixed frame, and $\nu = [u, v, r]^T$ are the surge, sway and yaw velocities in the body-fixed frame. $\tau = [\tau_u, 0, \tau_r]^T$ are the control inputs of path-following, and $\tau_d = [\tau_{ud}, \tau_{vd}, \tau_{rd}]^T$ are the time-varying marine environment disturbances. $R(\eta)$ is the rotation matrix from the body-fixed frame to the earth-fixed frame, which is defined as

$$R(\eta) = \begin{bmatrix} \cos\psi & -\sin\psi & 0 \\ \sin\psi & \cos\psi & 0 \\ 0 & 0 & 1 \end{bmatrix} \tag{6}$$

M is the inertial matrix and satisfies $M = M^T > 0$, which is written as

$$M = \begin{bmatrix} m_{11} & 0 & 0 \\ 0 & m_{22} & m_{23} \\ 0 & m_{32} & m_{33} \end{bmatrix} \tag{7}$$

$C(\nu)$ is the coriolis-centripetal matrix and satisfies $C(\nu) = -C(\nu)^T$, which is written as

$$C(\nu) = \begin{bmatrix} 0 & 0 & c_{13}(\nu) \\ 0 & 0 & c_{23}(\nu) \\ -c_{13}(\nu) & -c_{23}(\nu) & 0 \end{bmatrix} \tag{8}$$

and $D(\nu)$ is the damping matrix, which is written as

$$D(\nu) = \begin{bmatrix} d_{11}(\nu) & 0 & 0 \\ 0 & d_{22}(\nu) & d_{23}(\nu) \\ 0 & d_{32}(\nu) & d_{33}(\nu) \end{bmatrix} \tag{9}$$

with $m_{11} = m - X_{\dot{u}}$, $m_{22} = m - Y_{\dot{v}}$, $m_{33} = I_z - N_{\dot{r}}$, $m_{23} = mx_g - Y_{\dot{r}}$, $m_{32} = mx_g - N_{\dot{v}}$, $c_{13}(\nu) = -m_{11}v - m_{23}r$, $c_{23}(\nu) = -m_{11}u$, $d_{11}(\nu) = -X_u - X_{|u|u}|u| - X_{uuu}|u|^2$, $d_{22}(\nu) = -Y_v - Y_{|v|v}|v|$, $d_{33}(\nu) = -N_r - N_{|v|r}|v| - N_{|r|r}|r|$, $Y_{\dot{r}} = N_{\dot{v}}$, where m is AMV mass, and I_z is the moment of inertia in yaw. X_*, Y_* and N_* are the hydrodynamic coefficients.

As shown in Figure 2, the desired path $(x_d(s), y_d(s))$ is a continuous parameterized curve with a time-independent variable s. For any moving point on the curve, a path-tangential angle in the earth-fixed frame is defined as

$$\alpha = atan2(y_d'(s), x_d'(s)) \tag{10}$$

where $y_d'(s) = \partial y_d/\partial s$, $x_d'(s) = \partial x_d/\partial s$. The errors between (x, y) and (x_d, y_d) can be formulated as

$$\begin{bmatrix} x_e \\ y_e \end{bmatrix} = \begin{bmatrix} \cos\alpha & -\sin\alpha \\ \sin\alpha & \cos\alpha \end{bmatrix}^T \begin{bmatrix} x - x_d(s) \\ y - y_d(s) \end{bmatrix} \tag{11}$$

where x_e is the along-track error, and y_e is the cross-track error.

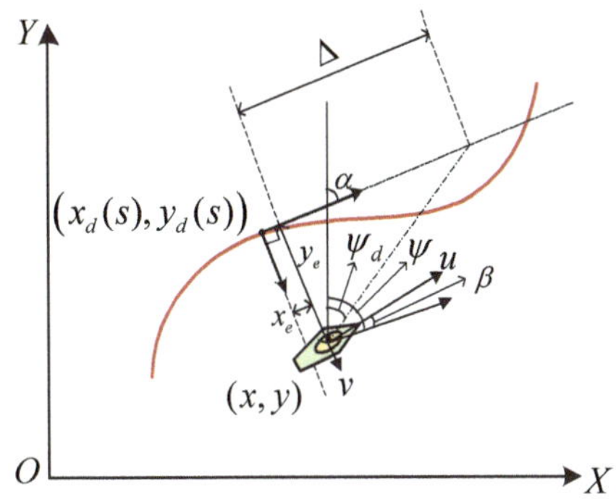

Figure 2. Diagram of horizontal path-following control.

In this article, our objective is to design the DRL-based path-following control scheme for an uncertain under-actuated AMV, such that the vehicle can follow the desired path with the desired velocities regardless of model uncertainties and unknown marine environment disturbances. To be specific, the objective can be formalized as

$$\begin{cases} \lim_{t\to\infty} x_e \leq \delta_x \\ \lim_{t\to\infty} y_e \leq \delta_y \end{cases} \tag{12}$$

where δ_x and δ_y are any small positive constants.

3. DRL-Based Path-Following Control Scheme

In this section, a DRL-based path-following control scheme is established for an under-actuated AMV in the presence of model uncertainties and unknown marine environment disturbances. The diagram of the proposed control scheme is shown in Figure 3, where kinematic guidance and dynamic control are designed, respectively.

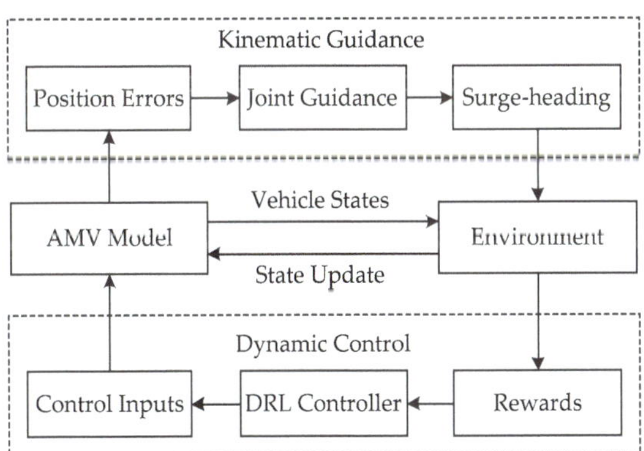

Figure 3. Diagram of the proposed path-following control scheme.

Within the kinematic level, according to the position errors and related motion states obtained by the AMV model, the desired surge velocity and heading angle are generated by the designed surge-heading joint guidance law based on the light-of-sight guidance. Within the dynamic level, DRL-based surge-heading controllers are presented for following the desired guidance signals. The reward function is designed to generate rewards by comparing the desired signals with the actual vehicle states in the environment. The controllers generate the control inputs based on the rewards, and the vehicle precisely tracks the desired signals based on the control inputs and the novel environment states to realize path-following control. By combining with the kinematic guidance and dynamic control, the objective (12) can be successfully completed.

3.1. Kinematic Guidance Design

Firstly, kinematic guidance is designed in this subsection, where desired surge velocities and heading angles are produced. Differentiating (11) along (5) yields

$$\begin{cases} \dot{x}_e = u\cos(\psi-\alpha) - v\sin(\psi-\alpha) + \dot{\alpha}y_e - u_s \\ \dot{y}_e = u\sin(\psi-\alpha) + v\cos(\psi-\alpha) - \dot{\alpha}x_e \end{cases} \tag{13}$$

where u_s is velocity of the virtual point along the desired path, which is defined by

$$u_s = \dot{s}\sqrt{x_d'^2(s) + y_d'^2(s)} \tag{14}$$

Define the sideslip angle of under-actuated vehicle as

$$\beta = \arctan\left(\frac{v}{u}\right) \tag{15}$$

In this context, path-following error dynamics (13) is rewritten as

$$\begin{cases} \dot{x}_e = u\cos(\psi-\alpha) - u\sin(\psi-\alpha)tan\beta + \dot{\alpha}y_e - u_s \\ \dot{y}_e = u\sin(\psi-\alpha) + u\cos(\psi-\alpha)tan\beta - \dot{\alpha}x_e \end{cases} \tag{16}$$

Then, select the Lyapunov function related to path following errors as

$$V = \frac{1}{2}\left(x_e^2 + y_e^2\right) \tag{17}$$

The time derivative of (17) along the solution (16) is

$$\begin{aligned} \dot{V} &= x_e\dot{x}_e + y_e\dot{y}_e \\ &= x_e\left(u\cos(\psi-\alpha) - u\sin(\psi-\alpha)tan\beta + \dot{\alpha}y_e - u_s\right) \\ &+ y_e\left(u\sin(\psi-\alpha) - \dot{\alpha}x_e + u\cos(\psi-\alpha)tan\beta\right) \end{aligned} \tag{18}$$

Thus, the surge-heading joint guidance law is designed as follows

$$\begin{cases} u_d = k_1\sqrt{y_e^2 + \Delta^2} \\ \psi_d = \alpha - \beta_d - \arctan\left(\frac{y_e}{\Delta}\right) \end{cases} \tag{19}$$

where $k_1 > 0$; $\Delta > 0$ is the look-ahead distance; $\beta_d = \arctan(v/u_d)$ and virtual velocity u_s is determined by

$$u_s = U_d\cos(\beta_d + \psi - \alpha) + k_2 x_e \tag{20}$$

with $k_2 > 0$ and $U_d = \sqrt{u_d^2 + v^2}$.

Using the fact

$$\begin{cases} \sin\left(\tan^{-1}\left(-\frac{y_e}{\Delta}\right)\right) = -\dfrac{y_e}{\left(y_e^2 + \Delta^2\right)^{1/2}} \\ \cos\left(\tan^{-1}\left(-\frac{y_e}{\Delta}\right)\right) = -\dfrac{\Delta}{\left(y_e^2 + \Delta^2\right)^{1/2}} \end{cases} \tag{21}$$

and substituting the guidance law (19) into (18) yield

$$\begin{aligned} \dot{V} &= x_e\left(\dot{\alpha}y_e - k_2x_e\right) + y_e\left(-\frac{k_1}{\cos\beta_d}y_e - \dot{\alpha}x_e\right) \\ &= -k_2x_e^2 - \frac{k_1}{\cos\beta_d}y_e^2 \end{aligned} \tag{22}$$

Since $0 < \cos\beta_d \leq 1$ renders $\dot{V} \leq -k_2x_e^2 - k_1y_e^2$, it indicates that path-following error x_e and y_e can globally asymptotically converge to the origin using the proposed guidance method.

3.2. Dynamic Control Design

Because of inaccurate measurements and environment disturbances, AMV model parameters cannot be obtained completely, thereby resulting in uncertainties of dynamics. To enhance the engineering practicality and reduce the complexity of controllers, a TD3-based reinforcement learning control method is presented within the dynamic level, where control inputs of the AMV are generated successfully.

The main purpose of the DRL algorithm is to make the vehicle take actions in the case of different path information. The proposed TD3 algorithm is based on the actor-critic structure, where policy functions are produced using actor networks and critic networks used to judge the performance of the actor [42]. Additionally, LSTM network layers are introduced into the TD3 and thereby enhance the utilization rate of historical states.

Firstly, the network structure of TD3 algorithm is shown in Figure 4. By virtue of initial environment states, actions of the AMV are generated using actor networks, and rewards are accordingly calculated using reward functions; thus, the states can be updated with the generated actions. The empirical value is defined as $e(t) = \{s, a, r, s_{t+1}\}$ and saved into buffer *MemoryD*. Through repeated training, the empirical replay sequence $D = \{e_1, e_2, \ldots, e_n\}$ can be formed. Considering that the adjacent actions of path-following have strong relevance, a batch of empirical sequences are selected randomly for training. The actor network of target generates the action a_{t+1} according to the state s_{t+1} in the empirical replay sequence, and the critic network of target calculates the $Q_{\min}$ value, where $Q_{\min}$ is the smaller of the two Q_{target} values generated by target networks. Two critic networks are updated based on Q, $Q_{\min}$ and loss functions. Actor networks generate actions using states. Critic networks generate the Q value using states and actions, and thus calculate the policy gradient and update actor networks using the Q value.

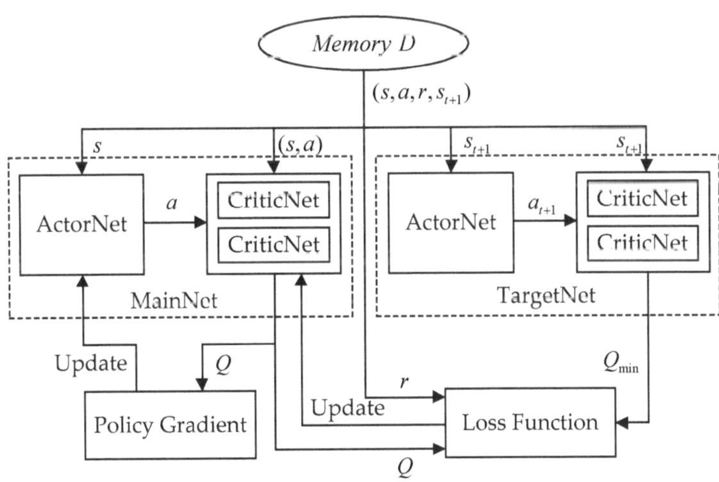

Figure 4. Network structure of the TD3.

The specific renewal process is as follows. Considering that TD3 algorithm is a deterministic policy and has the characteristic of target policy smoothing regularization, random noise ε is added into target actions. Therefore, there is $a_{t+1} = \mu'(s_{t+1}|w') + \varepsilon$, where μ' is the policy of target actor networks, and w' is the parameter of target actor networks.

The target value is calculated as

$$y = r + \gamma Q_{\min}(s_{t+1}, a_{t+1}|\theta_i') \tag{23}$$

where $i = 1,\ 2,\ \theta_i'$ represents parameters of target critic networks.

The loss function is defined as

$$L(\theta_i) = \frac{1}{N}(y - Q(s,a|\theta_i))^2 \tag{24}$$

where N represents the number of mini-batch, and θ_i represents parameters of critic networks. The gradient is updated by

$$\frac{\partial L(\theta_i)}{\partial \theta_i} = E\left[(y - Q(s,a|\theta_i))\frac{\partial Q(s,a|\theta_i)}{\partial \theta_i}\right] \tag{25}$$

Subsequently, the policy gradient of actor networks is updated by

$$\begin{cases} \dfrac{\partial J(w)}{\partial w} = E\left[\dfrac{\partial Q(s,a|\theta)}{\partial a}\dfrac{\partial \mu(s|w)}{\partial w}\right] \\ \nabla_w J \approx \dfrac{1}{N}\Sigma(\nabla_{\mathrm{a}} Q(s,a|\theta_1)|_{a=\mu(s)}\nabla_w \mu(s|w)) \end{cases} \tag{26}$$

After a couple of cycles, target parameters are soft updated by

$$\begin{cases} \theta_i' = \xi\theta_i + (1-\xi)\theta_i' \\ w' = \xi w + (1-\xi)w' \end{cases} \tag{27}$$

where $\xi \in (0,1)$ represents the learning rate.

Then, the LSTM network is introduced into actor-critic networks, thus contributing to the LSTM-TD3-based reinforcement learning controllers. The LSTM-TD3 network still retains the actor-critic structure, where LSTM inputs is a length of sequences. According to the real-time navigational information, the continuous states are saved into the sequences. The LSTM network layer is connected to generate the final hidden state h_t, where h_t is a one-dimensional array. Via the multi-layer perceptron (MLP) neural networks, the path-following control inputs of an AMV are generated, which include surge forces and yaw torques. The network structure of the actor is shown in Figure 5. Note that the critic has the similar network structure to the actor, and generates available actions.

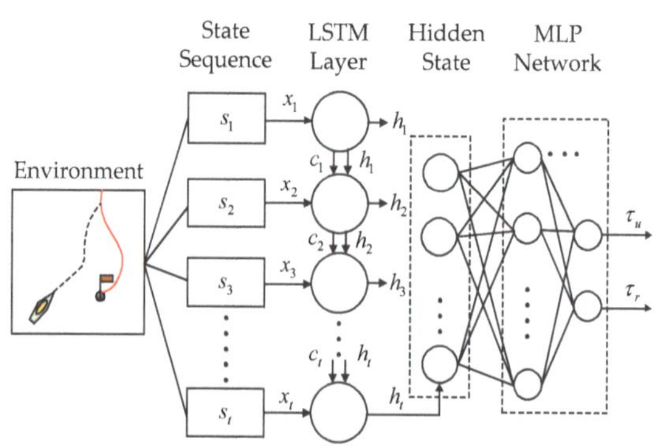

Figure 5. Network structure of Actor.

Finally, the state space, action space and reward function are designed as follows. To be specific, the state space represents perceived environment information of the AMV, which is the basis of decision-making and reward-evaluating. The action space represents control inputs of the AMV, including surge forces and yaw torques. The reward function is used to evaluate current state of the AMV.

In this context, the state space is defined as

$$s_t = \left[x_e,\ y_e,\ \psi,\ \psi_e,\ u,\ v,\ r,\ u_e,\ \tau_{u(t-1)},\ \tau_{r(t-1)}\right] \tag{28}$$

where $\tau_{u(t-1)}$ and $\tau_{r(t-1)}$ are control inputs at the previous moment. $u_e = u - u_d$ and $\psi_e = \psi - \psi_d$ with u_d, ψ_d are generated by the guidance law (19).

Taking path-following errors, surge velocity and heading angle errors into considerations, the reward function r_1 is designed as

$$r_1 = \lambda\left(2\exp^{-k_3|u_e|} - 1\right) + \left(2\exp^{-k_4|\psi_e|} - 1\right) + \left(2\exp^{-k_5|\sqrt{x_e^2+y_e^2}|} - 1\right) \tag{29}$$

where $k_*(* = 3,4,5) > 0$ and $\lambda > 0$. Note that the exponential function is used to calculate rewards, which limits the size of rewards and avoids high rewards.

Furthermore, the reward function r_2 is designed as

$$r_2 = \exp^{-k_6|\sigma_{\tau_u}|} + \exp^{-k_6|\sigma_{\tau_r}|} - 1 \tag{30}$$

where $k_6 > 0$; σ_{τ_u} and σ_{τ_r} are the standard deviation of two inputs, which are used to reduce the chattering of control inputs.

By combining with (29) and (30), the hybrid rewards of path-following control are established as

$$r = r_1 + k_7 r_2 \tag{31}$$

where $k_7 > 0$, and satisfies $k_7 \in (0,1)$.

The framework of the dynamic control algorithm is summarized in Algorithm 1.

Algorithm 1. Dynamic control algorithm of an AMV.

Inputs: Learning rate ξ, l_θ and l_w, regular factor ε, gradient threshold parameter g, discount factor γ, sequence length L, the maximum number of steps per training K, updating cycle of target network parameter d, training cycle F, mini-batch N.

Initialize: Critic network $Q(s,a|\theta_i)$ and actor network $\mu(s|w)$ with random parameters θ_1, θ_2 and w, target network $\theta_1' \leftarrow \theta_1$, $\theta_2' \leftarrow \theta_2$ and $w' \leftarrow w$, experience replay buffer *Memory D*, navigation environment of an AMV.

Procedure:

for $n_1 = 1,\ldots,F$ **do**
 for $n_2 = 1,\ldots,K$ **do**
  Select actions with exploration noise $a \sim \mu(s|w) + \varepsilon$ and obtain reward r and next moment state s_{t+1}
  Save transition tuple (s,a,r,s_{t+1}) into *Memory D*
  Sample N transitions (s,a,r,s_{t+1})
$$\begin{cases} a_{t+1} \leftarrow \mu'(s_{t+1}|w') + \varepsilon \\ y \leftarrow r + \gamma Q_{\min}(s_{t+1}, a_{t+1}|\theta_i') \end{cases}$$
  Update Critic networks parameters θ_i as
$\theta_i \leftarrow \mathrm{argmin}_{\theta_i} \frac{1}{N}(y - Q(s,a|\theta_i))^2$
  if t mod d **then**
   Update actor network parameters w as
$\nabla_w J \approx \frac{1}{N}\sum(\nabla_a Q(s,a|\theta_1)|_{a=\mu(s)} \nabla_w \mu(s|w))$
   Update target network as
$$\begin{cases} \theta_i' = \xi\theta_i + (1-\xi)\theta_i' \\ w' = \xi w + (1-\xi) w' \end{cases}$$
end if
 end for

Outputs: Actor network parameter w, critic network parameters θ_1 and θ_2.

4. Simulation Studies

In this section, simulation studies are shown to verify the effectiveness and superiority of the proposed DRL-based path-following control method. Consider an under-actuated AMV described by (5) with model uncertainties and unknown marine environment disturbances. Model parameters of the prototype AMV can be found in [43].

Within the kinematic guidance level, relative parameters are chosen as follows: $k_1 = 0.2$, $k_2 = 2$, $\Delta = 3$. Within the dynamic control level, relative parameters are chosen as follows: $k_3 = 1.5$, $k_4 = 6$, $k_5 = 1$, $k_6 = 1$, $\lambda = 0.8$, $k_7 = 0.3$. Training hyper parameters and network parameters of the LSTM-TD3 are shown in Tables 1 and 2, respectively.

Table 1. Training hyper parameters.

Parameters	Value
Discount factor γ	0.99
State sequence length L	20
Training cycle F	1000
Maximum number of steps K	1000
Capacity of buffer D	100,000
Learning rate l	0.001
Optimizer	Adam
Gradient threshold parameter g	1
Regular factor ε	0.00005
Mini-batch N	128

Table 2. LSTM-TD3 network parameters.

Parameters	Value
Input layer of actor network	11
Input layer of critic network	13
Fully connected layer	200
LSTM layer of actor network	100
LSTM layer of critic network	100
Output layer of actor	2
Output layer of critic	1

Episode rewards with different DRL algorithms of path-following control of an AMV are shown in Figure 6. It can be seen that the initial reward is extremely low since the vehicle explores the environment randomly during the initial training stage. After collecting enough data, the rewards converge to a stable value under the DRL control method. Compared to the asynchronous advantage of actor-critic (A3C) developed in [44], the TD3 and LSTM-TD3 can effectively increase the accumulated reward. Additionally, the proposed LSTM-TD3 shows a faster convergence rate and a more stable convergence process than the other two algorithms.

After training the algorithm for 1000 episodes, the optimal actor network parameters of the LSTM-TD3 and TD3 are saved and utilized. Simulation time is set as 200s. The initial positions and attitude are $\eta = [-10, 0, 0]^T$ and the initial velocities are $\nu = [0, 0, 0]^T$. Time-varying marine environment disturbances are defined as

$$\tau_d = \begin{bmatrix} 5\sin(0.1t + \pi/5) \\ 2.2\cos(0.1t + 6) \\ 1.2\cos(0.1t + 3) \end{bmatrix} \tag{32}$$

The desired path is defined as

$$\begin{cases} x_d = s \\ y_d = 10\sin(0.3s) + 2s \end{cases} \tag{33}$$

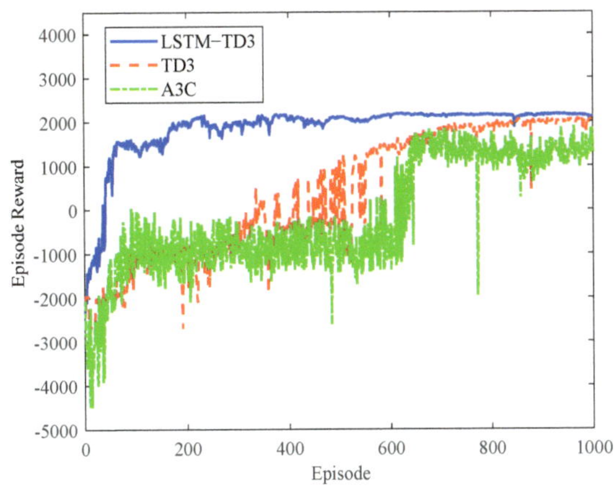

Figure 6. Episode rewards with different DRL algorithms.

Simulation results are shown in Figures 7–11. Figure 7 shows the path-following control performance of an under-actuated AMV where the desired path, actual path under the proposed LSTM-TD3 and the traditional TD3 are plotted. Obviously, the proposed control method has significant superiority in terms of transient responses and steady-state performance. Figure 8 shows the path-following errors of an AMV subject to model uncertainties and marine environment disturbances. It can be seen that the tangential error and the normal error can converge to the origin faster under the proposed control method. Figure 9 shows the surge velocity and heading angle, where desired signals are generated by guidance law (19). The actual velocities gradually converge to the desired value by the aid of the DRL controller. Note that the slight chattering of velocities is due to large path inflection point, and under-actuated AMV have to reduce their speed to follow the desired path. Figure 10 shows the velocity error and heading angle error of an AMV. It can be seen that the LSTM network considers historical states and thus enhances control performance. Figure 11 shows path-following control inputs of an AMV, including the surge force and the yaw torque. Because of the hybrid rewards with standard deviation, where 20 continuous inputs are set as a calculation group and dynamic sliding is introduced, the control chattering is effectively relieved.

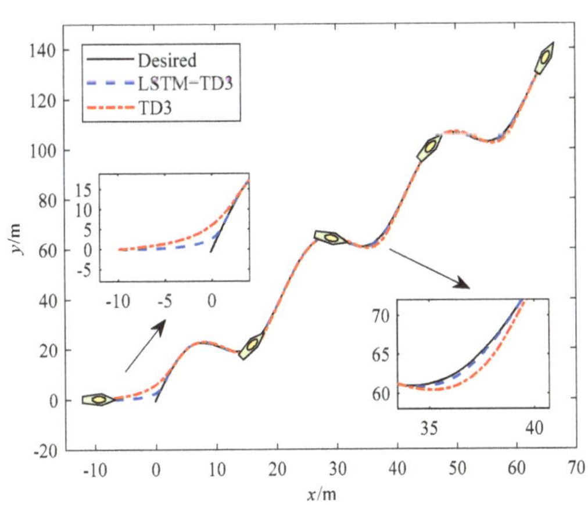

Figure 7. Path-following control performance of an AMV.

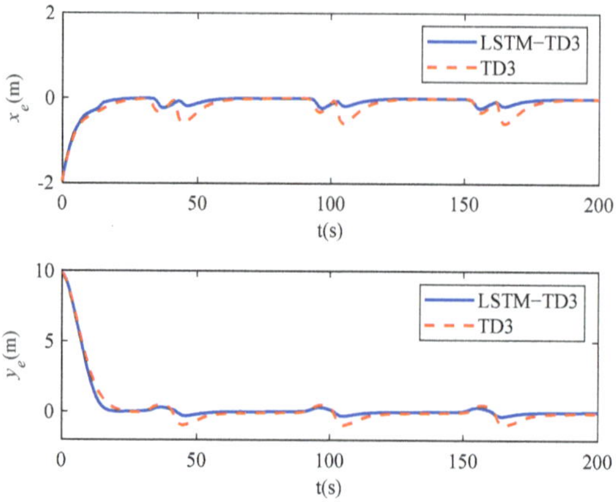

Figure 8. Path-following errors of an AMV under LSTM-TD3 and TD3.

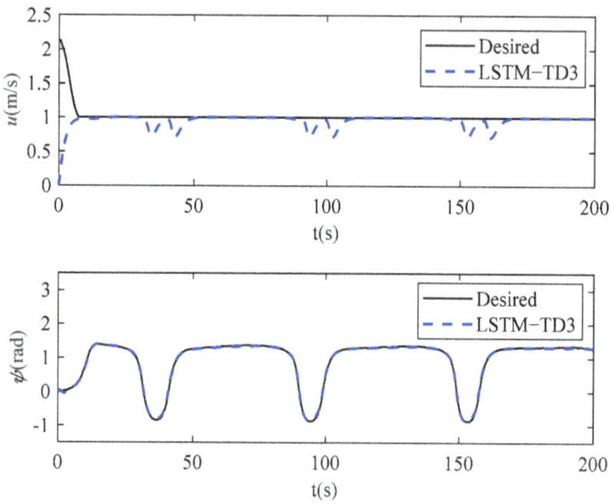

Figure 9. Surge velocity and heading angle of an AMV.

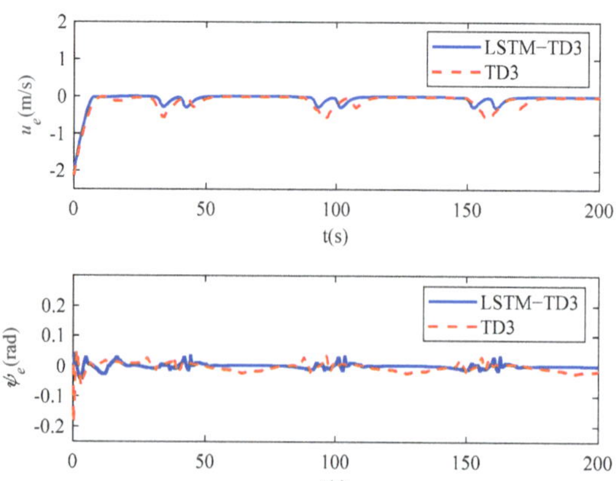

Figure 10. Velocity and heading errors of an AMV under LSTM-TD3 and TD3.

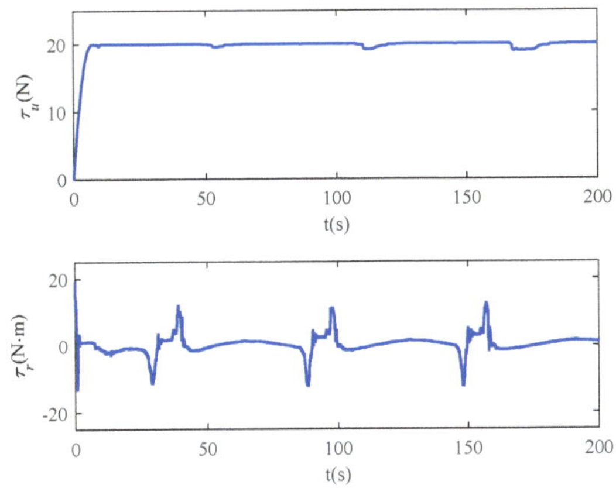

Figure 11. Surge force and yaw torque of an AMV.

5. Conclusions

This article studies the path-following control of an under-actuated AMV subject to unknown model parameters and marine environmental disturbances. Within the kinematic level, a surge-heading joint guidance law is presented, and makes the vehicle follow the desired path. Within the dynamic level, a LSTM-TD3-based reinforcement learning controller is presented, where vehicle actions are generated by the state space and hybrid reward. Additionally, actor-critic networks are developed using the LSTM network, and vehicles can make a decision by the aid of historical states, thus enhancing the convergence rate of controller networks. Simulation results and comprehensive comparisons demonstrate the remarkable effectiveness and superiority of the proposed path-following control method. By the aid of the proposed controller, the AMV can achieve path following regardless of marine environment disturbances.

Author Contributions: Conceptualization, X.Q. and Y.J.; methodology, X.Q.; software, Y.J.; validation, X.Q., Y.J. and R.Z.; formal analysis, F.L.; investigation, Y.J.; resources, X.Q.; data curation, X.Q.; writing—original draft preparation, X.Q.; writing—review and editing, X.Q.; visualization, F.L.; supervision, F.L.; project administration, R.Z.; funding acquisition, R.Z. All authors have read and agreed to the published version of the manuscript.

Funding: This research was funded by the National Natural Science Foundation of China (grant number 61673084).

Institutional Review Board Statement: Not applicable.

Informed Consent Statement: Not applicable.

Data Availability Statement: Not applicable.

Conflicts of Interest: The authors declare no conflict of interest.

References

1. Jorge, V.A.; Granada, R.; Maidana, R.G.; Jurak, D.A.; Heck, G.; Negreiros, A.P.; Dos Santos, D.H.; Gonçalves, L.M.; Amory, A.M. A Survey on Unmanned Surface Vehicles for Disaster Robotics: Main Challenges and Directions. *Sensors* **2019**, *19*, 702. [CrossRef]
2. Liu, T.; Dong, Z.; Du, H.; Song, L.; Mao, Y. Path Following Control of the Underactuated USV Based on the Improved Line-of-Sight Guidance Algorithm. *Pol. Marit. Res.* **2017**, *24*, 3–11. [CrossRef]
3. Mu, D.; Wang, G.; Fan, Y.; Bai, Y.; Zhao, Y. Fuzzy-Based Optimal Adaptive Line-of-Sight Path Following for underactuated unmanned surface vehicle with uncertainties and time-varying disturbances. *Math. Probl. Eng.* **2018**, *2018*, 7512606. [CrossRef]
4. Koh, S.; Zhou, B.; Fang, H.; Yang, P.; Yang, Z.; Yang, Q.; Guan, L.; Ji, Z. Real-time deep reinforcement learning based vehicle navigation. *Appl. Soft Comput.* **2020**, *96*, 106694. [CrossRef]
5. Mu, D.; Wang, G.; Fan, Y.; Bai, Y.; Zhao, Y. Path following for podded propulsion unmanned surface vehicle: Theory, simulation and experiment. *IEEJ Trans. Electr. Electron. Eng.* **2018**, *13*, 911–923. [CrossRef]

6. Lekkas, A.M.; Fossen, T.I. Integral LOS Path Following for Curved Paths Based on a Monotone Cubic Hermite Spline Parametrization. *IEEE Trans. Control Syst. Technol.* **2014**, *22*, 2287–2301. [CrossRef]
7. Fossen, T.I.; Lekkas, A.M. Direct and indirect adaptive integral line-of-sight path-following controllers for marine craft exposed to ocean currents. *Int. J. Adapt. Control Signal Process.* **2017**, *31*, 445–463. [CrossRef]
8. Fossen, T.I.; Pettersen, K.Y.; Galeazzi, R. Line-of-Sight Path Following for Dubins Paths with Adaptive Sideslip Compensation of Drift Forces. *IEEE Trans. Control Syst. Technol.* **2014**, *23*, 820–827. [CrossRef]
9. Liu, Z.; Song, S.; Yuan, S.; Ma, Y.; Yao, Z. ALOS-Based USV Path-Following Control with Obstacle Avoidance Strategy. *J. Mar. Sci. Eng.* **2022**, *10*, 1203. [CrossRef]
10. Rout, R.; Subudhi, B. Inverse optimal self-tuning PID control design for an autonomous underwater vehicle. *Int. J. Syst. Sci.* **2017**, *48*, 367–375. [CrossRef]
11. Yu, C.; Xiang, X.; Lapierre, L.; Zhang, Q. Nonlinear guidance and fuzzy control for three-dimensional path following of an underactuated autonomous underwater vehicle. *Ocean Eng.* **2017**, *146*, 457–467. [CrossRef]
12. Xiang, X.; Yu, C.; Zhang, Q. Robust fuzzy 3D path following for autonomous underwater vehicle subject to uncertainties. *Comput. Oper. Res.* **2017**, *84*, 165–177. [CrossRef]
13. Zhang, J.; Xiang, X.; Lapierre, L.; Zhang, Q.; Li, W. Approach-angle-based three-dimensional indirect adaptive fuzzy path following of under-actuated AUV with input saturation. *Appl. Ocean Res.* **2021**, *107*, 102486. [CrossRef]
14. Sahu, B.K.; Subudhi, B. Adaptive tracking control of an autonomous underwater vehicle. *Int. J. Autom. Comput.* **2014**, *11*, 299–307. [CrossRef]
15. Shin, J.; Kwak, D.J.; Lee, Y. Adaptive Path-Following Control for an Unmanned Surface Vessel Using an Identified Dynamic Model. *IEEE/ASME Trans. Mechatron.* **2017**, *22*, 1143–1153. [CrossRef]
16. Lamraoui, H.C.; Zhu, Q. Path following control of fully-actuated autonomous underwater vehicle in presence of fast-varying disturbances. *Appl. Ocean Res.* **2019**, *86*, 40–46. [CrossRef]
17. Zhang, H.; Zhang, X.; Cao, T.; Bu, R. Active disturbance rejection control for ship path following with Euler method. *Ocean Eng.* **2022**, *247*, 110516. [CrossRef]
18. Zhang, G.; Huang, H.; Wan, L.; Li, Y.; Cao, J.; Su, Y. A novel adaptive second order sliding mode path following control for a portable AUV. *Ocean Eng.* **2018**, *151*, 82–92. [CrossRef]
19. Zhang, H.; Zhang, X.; Bu, R. Radial Basis Function Neural Network Sliding Mode Control for Ship Path Following Based on Position Prediction. *J. Mar. Sci. Eng.* **2021**, *9*, 1055. [CrossRef]
20. Wang, J.; Wang, C.; Wei, Y.; Zhang, C. Three-Dimensional Path Following of an Underactuated AUV Based on Neuro-Adaptive Command Filtered Backstepping Control. *IEEE Access* **2018**, *6*, 74355–74365. [CrossRef]
21. Yan, Z.; Yang, Z.; Zhang, J.; Zhou, J.; Jiang, A.; Du, X. Trajectory tracking control of UUV based on backstepping sliding mode with fuzzy switching gain in diving plane. *IEEE Access* **2019**, *7*, 166788–166795. [CrossRef]
22. Zhou, J.; Zhao, X.; Chen, T.; Yan, Z.; Yang, Z. Trajectory tracking control of an underactuated AUV based on backstepping sliding mode with state prediction. *IEEE Access* **2019**, *7*, 181983–181993. [CrossRef]
23. Chen, X.; Liu, Z.; Zhang, J.; Zhou, D.; Dong, J. Adaptive sliding-mode path following control system of the underactuated USV under the influence of ocean currents. *J. Syst. Eng. Electron.* **2018**, *29*, 1271–1283. [CrossRef]
24. Liang, X.; Wan, L.; Blake, J.I.; Shenoi, R.A.; Townsend, N. Path Following of an Underactuated AUV Based on Fuzzy Backstepping Sliding Mode Control. *Int. J. Adv. Robot. Syst.* **2016**, *13*, 122. [CrossRef]
25. Qiu, B.; Wang, G.; Fan, Y.; Mu, D.; Sun, X. Path Following of Underactuated Unmanned Surface Vehicle Based on Trajectory Linearization Control with Input Saturation and External Disturbances. *Int. J. Control Autom. Syst.* **2020**, *18*, 2108–2119. [CrossRef]
26. Wang, N.; Sun, Z.; Yin, J.; Zou, Z.; Su, S. Fuzzy unknown observer-based robust adaptive path following control of underactuated surface vehicles subject to multiple unknowns. *Ocean Eng.* **2019**, *176*, 57–64. [CrossRef]
27. Havenstrøm, S.T.; Rasheed, A.; San, O. Deep reinforcement learning controller for 3D path following and collision avoidance by autonomous underwater vehicles. *Front. Robot. AI* **2021**, *7*, 211. [CrossRef]
28. Meyer, E.; Heiberg, A.; Rasheed, A.; San, O. COLREG-compliant collision avoidance for unmanned surface vehicle using deep reinforcement learning. *IEEE Access* **2020**, *8*, 165344–165364. [CrossRef]
29. Sola, Y.; Le Chenadec, G.; Clement, B. Simultaneous control and guidance of an auv based on soft actor–critic. *Sensors* **2022**, *22*, 6072. [CrossRef]
30. Fang, Y.; Huang, Z.; Pu, J.; Zhang, J. AUV position tracking and trajectory control based on fast-deployed deep reinforcement learning method. *Ocean Eng.* **2022**, *245*, 110452. [CrossRef]
31. Zhang, T.; Tian, R.; Wang, C.; Xie, G. Path-Following Control of Fish-like Robots: A Deep Reinforcement Learning Approach. *IFAC-PapersOnLine* **2020**, *53*, 8163–8168. [CrossRef]
32. Woo, J.; Yu, C.; Kim, N. Deep reinforcement learning-based controller for path following of an unmanned surface vehicle. *Ocean Eng.* **2019**, *183*, 155–166. [CrossRef]
33. Han, Z.; Wang, Y.; Sun, Q. Straight-Path Following and Formation Control of USVs Using Distributed Deep Reinforcement Learning and Adaptive Neural Network. *IEEE/CAA J. Autom. Sin.* **2023**, *10*, 572–574. [CrossRef]
34. Sun, Y.; Ran, X.; Zhang, G.; Wang, X.; Xu, H. AUV path following controlled by modified Deep Deterministic Policy Gradient. *Ocean Eng.* **2020**, *210*, 107360. [CrossRef]

35. Zheng, Y.; Tao, J.; Sun, Q.; Sun, H.; Chen, Z.; Sun, M.; Xie, G. Soft Actor–Critic based active disturbance rejection path following control for unmanned surface vessel under wind and wave disturbances. *Ocean Eng.* **2022**, *247*, 110631. [CrossRef]
36. Liang, Z.; Qu, X.; Zhang, Z.; Chen, C. Three-Dimensional Path-Following Control of an Autonomous Underwater Vehicle Based on Deep Reinforcement Learning. *Pol. Marit. Res.* **2022**, *29*, 36–44. [CrossRef]
37. Liu, Y.; Peng, Y.; Wang, M.; Xie, J.; Zhou, R. Multi-usv system cooperative underwater target search based on reinforcement learning and probability map. *Math. Probl. Eng.* **2020**, *2020*, 7842768. [CrossRef]
38. Havenstrøm, S.T.; Sterud, C.; Rasheed, A.; San, O. Proportional integral derivative controller assisted reinforcement learning for path following by autonomous underwater vehicles. *arXiv* **2020**, arXiv:2002.01022. [CrossRef]
39. Zhang, W.; Wu, P.; Peng, Y.; Liu, D. Roll motion prediction of unmanned surface vehicle based on coupled CNN and LSTM. *Future Internet* **2019**, *11*, 243. [CrossRef]
40. Li, J.; Tian, Z.; Zhang, G.; Li, W. Multi-AUV Formation Predictive Control Based on CNN-LSTM under Communication Constraints. *J. Mar. Sci. Eng.* **2023**, *11*, 873. [CrossRef]
41. Fossen, T.I. *Handbook of Marine Craft Hydrodynamics and Motion Control*; John Wiley & Sons: Hoboken, NJ, USA, 2011.
42. Chu, Z.; Sun, B.; Zhu, D.; Zhang, M.; Luo, C. Motion control of unmanned underwater vehicles via deep imitation reinforcement learning algorithm. *IET Intell. Transp. Syst.* **2020**, *14*, 764–774. [CrossRef]
43. Wang, N.; Gao, Y.; Yang, C.; Zhang, X. Reinforcement learning-based finite-time tracking control of an unknown unmanned surface vehicle with input constraints. *Neurocomputing* **2022**, *484*, 26–37. [CrossRef]
44. Xie, S.; Chu, X.; Zheng, M.; Liu, C. A composite learning method for multi-ship collision avoidance based on reinforcement learning and inverse control. *Neurocomputing* **2020**, *411*, 375–392. [CrossRef]

Journal of *Marine Science and Engineering*

MDPI

Article

Dynamic Data-Driven Application System for Flow Field Prediction with Autonomous Marine Vehicles

Qianlong Jin [1,2,3], Yu Tian [1,2,4,*], Weicong Zhan [1,2,3], Qiming Sang [1,2,3], Jiancheng Yu [1,2] and Xiaohui Wang [1,2]

1 State Key Laboratory of Robotics, Shenyang Institute of Automation, Chinese Academy of Sciences, Shenyang 110016, China; jinqianlong@sia.cn (Q.J.); zhanweicong@sia.cn (W.Z.)
2 Institutes for Robotics and Intelligent Manufacturing, Chinese Academy of Sciences, Shenyang 110169, China
3 University of Chinese Academy of Sciences, Beijing 100049, China
4 China-Portugal Belt and Road Joint Laboratory on Space & Sea Technology Advanced Research, Shanghai 201304, China
* Correspondence: tiany@sia.cn

Abstract: Efficiently predicting high-resolution and accurate flow fields through networked autonomous marine vehicles (AMVs) is crucial for diverse applications. Nonetheless, a research gap exists in the seamless integration of data-driven flow modeling, real-time data assimilation from flow sensing, and the optimization of AMVs' sensing strategies, culminating in a closed-loop dynamic data-driven application system (DDDAS). This article presents a novel DDDAS that systematically integrates flow modeling, data assimilation, and adaptive flow sensing using networked AMVs. It features a hybrid data-driven flow model, uniting a neural network for trend prediction and a Gaussian process model for residual fitting. The neural network architecture is designed using knowledge extracted from historic flow data through tidal harmonic analysis, enhancing its capability in flow prediction. The Kriged ensemble transform Kalman filter is introduced to assimilate spatially correlated flow-sensing data from AMVs, enabling effective model learning and accurate spatiotemporal flow prediction, while forming the basis for optimizing AMVs' flow-sensing paths. A receding horizon strategy is proposed to implement non-myopic optimal path planning, and a distributed strategy of implementing Monte Carlo tree search is proposed to solve the resulting large-scale tree searching-based optimization problem. Computer simulations, employing underwater gliders as sensing networks, demonstrate the effectiveness of the proposed DDDAS in predicting depth-averaged flow in nearshore ocean environments.

Keywords: dynamic data-driven application system; autonomous marine vehicles; flow field prediction; data assimilation; adaptive sampling

Citation: Jin, Q.; Tian, Y.; Zhan, W.; Sang, Q.; Yu, J.; Wang, X. Dynamic Data-Driven Application System for Flow Field Prediction with Autonomous Marine Vehicles. *J. Mar. Sci. Eng.* **2023**, *11*, 1617. https://doi.org/10.3390/jmse11081617

Academic Editor: Sergei Chernyi

Received: 28 July 2023
Revised: 16 August 2023
Accepted: 16 August 2023
Published: 18 August 2023

1. Introduction

The advancement of affordable, long-range autonomous marine vehicle (AMV) technologies, including unmanned surface vehicles (USVs), autonomous underwater vehicles (AUVs), underwater gliders, and wave gliders [1], has presented opportunities for the deployment of persistent, robotic, and autonomous Lagrangian networks in ocean environment sensing. In contrast to conventional ocean observation networks such as the ARGO array [2], the utilization of mobile and autonomous networks composed of a collection of AMVs offers a distinctive technical capability: controllable maneuverability in complex and dynamic ocean environments. Therefore, effectively leveraging their controllable maneuverability to enable the networks to acquire cost-effective and information-rich sensing data within nonuniform and dynamic ocean environments, aligning with the requirements of observation missions, is a key issue within the research community focused on autonomous ocean sensing with AMV networks.

The tracking of dynamic ocean features to gather in situ data for characterizing and predicting the evolution of these features, such as thermoclines [3], plumes [4], and mesoscale eddies [5], is an aspect of applying AMV networks for ocean sensing. Besides feature tracking, a foundational application objective in utilizing AMV networks is to autonomously and adaptively collect a variety of informative Lagrangian data streams, such as temperature, salinity, and flow velocity of seawater, that can be assimilated into predictive ocean models. The assimilation of these data streams can lead to significant improvement in the accuracy of temporal predictions for spatial fields in ocean environments [6]. Consequently, the real-time optimization of the data streams for assimilation using numerical ocean models is crucial in maximizing model prediction performance, and it plays an essential role in both research and practical applications of ocean sensing with AMV networks. One key approach to addressing this optimization problem is through the utilization of feedback control theory and the formulation of a sensor management technical framework [7]. This framework establishes a closed-loop system by integrating a numerical ocean model with an AMV sensing network [8] and uses iterative feedback and optimization processes to continuously adapt the network's sensing strategies and improve prediction performance. Different from conventional feedback control of dynamic systems, the feedback control in the closed-loop adaptive ocean sensing and prediction system involves cooperative control of the sensing actions of the networked AMVs, taking into account the constraints imposed by ocean environments. And the primary objective of this feedback control is to collect informative sensing data of ocean environments, which can be utilized in ocean models to improve prediction performance. The AOSN II and ASAP projects have made significant contributions to the research on autonomous and adaptive ocean sensing with AMV networks, particularly regarding the utilization of underwater gliders as part of the sensing network [9,10]. These projects have laid an important foundation for the further advancements in this field.

In the context of ocean sensing and prediction systems, predictive ocean models play an important role. Alongside traditional methods based on numerically solving partial differential equations of geophysical fluid dynamics, data-driven techniques have emerged as powerful and complementary tools for ocean modeling and prediction. Leveraging machine learning and statistical approaches, data-driven methods extract patterns and relationships from observed data, enabling them to make high-resolution and accurate predictions with reduced computational costs. Extensive research efforts have been dedicated to developing and applying data-driven techniques for ocean modeling and spatial and temporal predictions. Notable approaches include objective analysis (OA) [11,12], Gaussian process (GP) [13], Kriging geostatistical prediction [14], compressive sensing [15], radial basis functions (RBFs) [16], proper orthogonal decomposition [17,18], dynamic mode decomposition (DMD) [19], and neural networks [20,21]. Compared to conventional ocean models, data-driven techniques rely more heavily on the information content within the data for their implementation. Building upon the aforementioned data-driven techniques, researchers have devised methodologies to optimize sensing strategies for both Eulerian and Lagrangian platforms, as demonstrated in [11–13,16–18]. These optimized strategies enable the collection of informative data, leading to enhancements in data-driven prediction performance. Through the integration of data-driven techniques and adaptive observation strategies within the closed-loop system, the ocean sensing and prediction capabilities of AMVs can be significantly improved. This approach holds great potential in achieving high-resolution and accurate ocean prediction with AMV sensing networks.

High-resolution and accurate flow field prediction is a fundamental component in ocean prediction. The spatiotemporal flow information not only offers insights into the multi-scale dynamic behavior of ocean currents but also plays a crucial role in path planning [22,23], navigation [24,25], feature prediction and tracking [26], and motion control for both individual and swarms of AMVs [11,27] implementing ocean sensing missions. Many studies have been conducted on data-driven flow prediction [22,24,25,28–30], addressing flow prediction in diverse oceanic environments where tidal or non-tidal flow is dominant. However, a research gap exists in the specific area of integrating data-driven flow modeling, online data assimilation of flow sensing data into data-driven flow models, and optimizing AMVs' flow sensing strategy, based on the closed-loop feedback ocean prediction and adaptive sensing technical framework.

Based on the above review and discussions, as well as recognizing the significance of high-resolution and accurate flow field prediction in ocean prediction and AMVs' operations, this article is dedicated to exploring this data-driven closed-loop approach and systematically developing methods for data-driven flow modeling, data assimilation, and optimizing AMVs' observations. The systematic development establishes a dynamic data-driven application system (DDDAS) [31] for flow prediction and sensing with AMVs. The DDDAS seamlessly integrates data-driven modeling, data assimilation, and an optimal decision-making process, as depicted in Figure 1, thereby enhancing high-resolution and accurate flow field prediction capabilities and enabling adaptive and efficient flow sensing using AMVs.

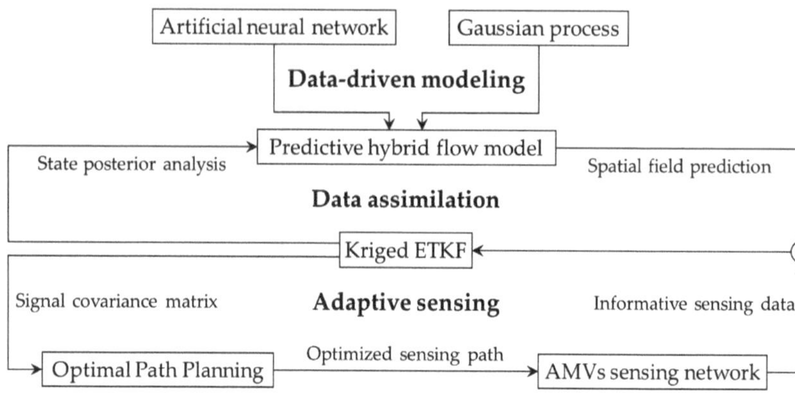

Figure 1. Illustration of the DDDAS for flow field prediction and sensing with AMVs.

The main contributions and novelties of this article are as follows: (1) Pioneering closed-loop DDDAS development: This article represents an advancement by introducing the first-ever closed-loop data-driven DDDAS designed for predicting flow fields. The system seamlessly integrates data-driven flow modeling, data assimilation, and adaptive sensing with AMVs. (2) A hybrid data-driven flow model: This article proposes an innovative hybrid data-driven flow model that combines the predictive power of a neural network with the statistical estimation of a Gaussian process model. Notably, the neural network architecture is designed to incorporate prior knowledge of tidal components, contributing to accurate flow predictions. This hybrid model not only captures flow trends accurately but also effectively handles residuals. (3) The Kriged ensemble transform Kalman filter (ETKF) data assimilation: This article introduces a pioneering application of the Kriged ETKF method for assimilating spatially correlated flow-sensing data from AMVs. This methodology demonstrates exceptional efficacy in model learning and spatiotemporal flow prediction. It also enhances the optimization of AMVs' sensing paths. (4) A receding horizon strategy for optimal path planning: Addressing the challenge of non-myopic optimal path planning for a network of AMVs, this article introduces a strategic receding horizon approach. To tackle the resulting computational complexity of large-scale tree-searching-based optimization, this article proposes a novel distributed Monte Carlo Tree Search (MCTS) method. This innovative approach harnesses the collective computational capabilities of networked AMVs to explore the search space and derive effective planning solutions.

The rest of this paper is structured as follows. Section 2 presents the data-driven hybrid flow model, including the neural network model and the Gaussian process model. Section 3 elaborates on the state and observation models for data assimilation and the implementation of the Kriged ETKF method for model learning and flow prediction.

Section 4 presents the optimization of the AMVs' paths for collecting informative flow-sensing data to enhance flow prediction accuracy. Section 5 presents simulation studies to demonstrate the performance of the DDDAS. Finally, Section 6 presents the conclusions.

2. Data-Driven Flow Prediction Model

The flow model serves as a fundamental component within the developed DDDAS. The section introduces the developed hybrid data-driven model, which leverages the machine learning model of neural networks and the statical model of the Gaussian process. Section 2.1 provides an overview of the hybrid model. Section 2.2 presents the utilization of the tidal harmonic analysis method to extract essential prior knowledge of the flow, supporting the design of the neural network model. Section 2.3 elaborates on the development of the neural network model and Section 2.4 presents the Gaussian process model.

2.1. Model Overview

To enable the prediction of the flow speed at specific spatiotemporal locations $\boldsymbol{x} = (\boldsymbol{s}, t)$, where $\boldsymbol{s} \in R^2$ represents the spatial location and $t \in R$ represents the time, a data-driven spatiotemporal flow field model is established. This model comprises two submodels:

$$u(\boldsymbol{x}) = \mu_u(\boldsymbol{x}) + r_u(\boldsymbol{x}) \tag{1}$$

$$v(\boldsymbol{x}) = \mu_v(\boldsymbol{x}) + r_v(\boldsymbol{x}) \tag{2}$$

Equation (1) represents the submodel for the latitudinal flow speed $u(\boldsymbol{x})$, and Equation (2) represents the submodel for the longitudinal flow speed $v(\boldsymbol{x})$, at the given spatiotemporal locations $\boldsymbol{x}$.

The flow model consists of two fundamental components: the mean component, and the residual component. The mean component, denoted as $\mu_u(\boldsymbol{x})$ and $\mu_v(\boldsymbol{x})$, captures the average or expected flow speeds, providing the trends in flow at the given spatiotemporal location. In this article, a feedforward neural network is employed to approximate the continuous and nonlinear function of $\mu_u(\boldsymbol{x})$ and $\mu_v(\boldsymbol{x})$. The choice is motivated by the neural network's universal approximation capabilities, which enable it to effectively learn and represent the complex relationships between input parameters including longitude, latitude, time, and the corresponding flow speed. To implement the flow model, a specific neural network is designed, and its detailed architecture is described in Section 2.3. On the other hand, the residual component, denoted as $r_u(\boldsymbol{x})$ and $r_v(\boldsymbol{x})$, accounts for variations from the mean component. These residuals may arise due to unmodelled influences or other factors affecting the flow speed. In this article, the Gaussian process is employed to model $r_u(\boldsymbol{x})$ and $r_v(\boldsymbol{x})$, which will be described in detail in Section 2.4. The composition of the flow prediction model, take $u(\boldsymbol{x})$ for example, is depicted in Figure 2. By decomposing the flow into mean and residual components, the spatiotemporal flow model could effectively capture both the average behavior and the deviations from the average, which enhance the accuracy and reliability of the predictions.

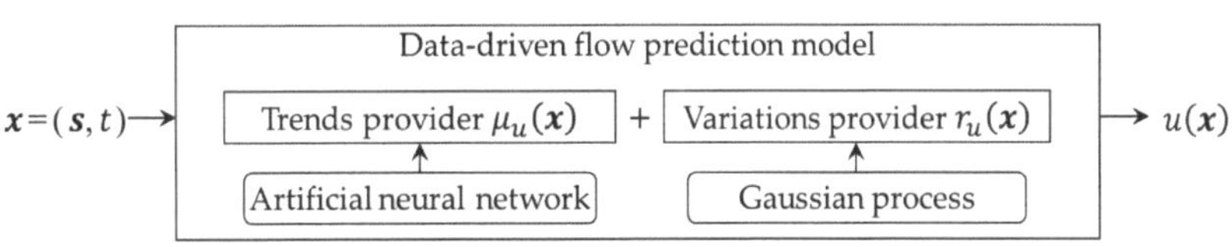

Figure 2. The composition of the data-driven flow prediction model.

Throughout the subsequent sections, the model for $u(\boldsymbol{x})$ will be focused on as an example for description and discussion. And to simplify expression and notation, the subscript u in $\mu_u(\boldsymbol{x})$ and $r_u(\boldsymbol{x})$ will be omitted. It is important to note that both $u(\boldsymbol{x})$ and

$v(\boldsymbol{x})$ share the same structural components and underlying principles. Therefore, the descriptions, methodologies, and discussions related to $u(\boldsymbol{x})$ can be directly applied to $v(\boldsymbol{x})$.

2.2. Decomposition of Flow Using Tidal Harmonic Analysis

Driven by tidal forces, flow fields in nearshore ocean environments often exhibit spatiotemporal periodic features, which can be characterized by the presence of multiple periodic components. These components possess distinct frequencies and amplitudes, playing significant roles in shaping the overall flow dynamics. Incorporating information about the tidal components in the design of the neural network architecture allows for capturing and learning about the relationships between the tidal components and the flow speed more effectively. Therefore, this article implements this integration of knowledge of tidal components into the neural network design, leveraging the periodic patterns present in the flow field to enhance the accuracy of flow prediction.

Tidal harmonic analysis is an effective approach for decomposing the flow into different constituents, represented by trigonometric functions with distinct frequencies. It processes the training data of a temporal sequence of flow fields to identify the dominant tidal frequencies and their corresponding amplitudes. The tidal frequencies can be calculated using methods such as least-squares harmonic estimation (LS-HE) [32] and multivariate LS-HE [33]. Once the frequencies are determined, the amplitudes and phases of the corresponding trigonometric function for each tidal constituent can be calculated using methods such as least-squares estimation. These constituents capture the fundamental periodic components of the flow.

In this article, the tidal harmonic analysis is implemented to obtain the tidal constituents. Each obtained tidal constituent corresponds to two orthogonal temporal basis functions, given by:

$$\begin{cases} \psi_i(t) = cos\left(\frac{2\pi t}{\mathscr{T}_i}\right) \\ \psi_{i+N_{tc}}(t) = sin\left(\frac{2\pi t}{\mathscr{T}_i}\right) \end{cases} \quad i = 1, \cdots, N_{tc} \tag{3}$$

where $\mathscr{T}_i$ represents the period of the i-th tidal constituent, and N_{tc} represents the total number of the chosen dominant constituents. These basis functions form a basis set that captures the periodic variations in the flow over time.

For a specific spatial location $\boldsymbol{s}$, the flow contributed by the i-th tidal constituent can be expressed as follows:

$$\mu_i(\boldsymbol{s}, t) = \beta_i(\boldsymbol{s})cos\left(\frac{2\pi t}{\mathscr{T}_i}\right) + \beta_{i+N_{tc}}(\boldsymbol{s})sin\left(\frac{2\pi t}{\mathscr{T}_i}\right) \quad i = 1, \cdots, N_{tc} \tag{4}$$

where the coefficients $\beta_i(\boldsymbol{s})$ and $\beta_{i+N_{tc}}(\boldsymbol{s})$ represent the amplitudes of the basis functions $\psi_i(\cdot)$ and $\psi_{i+N_{tc}}(\cdot)$ at location $\boldsymbol{s}$, respectively. Considering the contributions of all tidal constituents, the overall flow at the spatial location $\boldsymbol{s}$ can be expressed as:

$$\mu(\boldsymbol{s}, t) = \beta_0(\boldsymbol{s}) + \sum_{i=1}^{N_{tc}} \mu_i(\boldsymbol{s}, t) = \boldsymbol{\beta}^T(\boldsymbol{s})\boldsymbol{\psi}(t) \tag{5}$$

where $\beta_0(\boldsymbol{s})$ represents the non-tidal constant component of the flow at spatial location $\boldsymbol{s}$, $\boldsymbol{\beta}(\cdot) = [\beta_0(\cdot), \beta_1(\cdot), \cdots, \beta_{2N_{tc}}(\cdot)]^T$ varies with the locations, and $\boldsymbol{\psi}(\cdot) = [1, \psi_1(\cdot), \cdots, \psi_{2N_{tc}}(\cdot)]^T$.

2.3. Neural Network Model for the Mean Component

Based on the flow model in Equation (5), a feedforward neural network is designed to model $\mu(\boldsymbol{x})$ in this article, and its architecture is depicted in Figure 3. The main objective of the network is to learn about the relationship between $\mu(\boldsymbol{x})$ and $\boldsymbol{x} = (\boldsymbol{s}, t)$, enabling the prediction of the flow field.

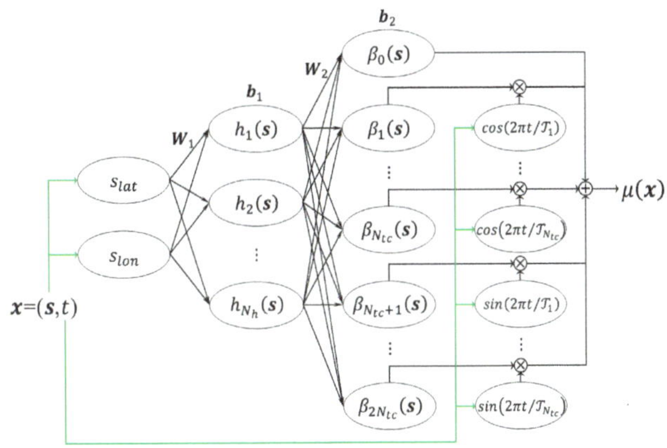

Figure 3. The architecture of the neural network.

In the model, the neural network takes the latitudinal coordinate s_{lat} and the longitudinal coordinate s_{lon} of the spatial location $\mathbf{s}$, along with the time t as input parameters. It then outputs the predicted flow speed $\mu(\mathbf{x})$ at $\mathbf{x} = (\mathbf{s}, t)$. The neural network consists of three hidden layers: the h-layer, the β-layer, and the ψ-layer. The h-layer consists of N_h nodes that are fully connected to the model inputs s_{lat} and s_{lon}. The β-layer has $2N_{tc} + 1$ nodes that are fully connected to the h-layer. And the ψ-layer is a hidden multiplicative layer with $2N_{tc}$ nodes, which is connected to the model input t. The model output is obtained by calculating the inner product of the outputs from the β-layer and the ψ-layer. The parameters of the neural network are denoted as follows: the connection weight matrix between the h-layer nodes and the input nodes is denoted as $\mathbf{W}_1$, and the bias vector of the h-layer nodes is denoted as $\boldsymbol{b}_1$. Similarly, the connection weight matrix between the β-layer nodes and the h-layer nodes is denoted as $\mathbf{W}_2$, and the bias vector of the β-layer nodes is denoted as $\boldsymbol{b}_2$. Additionally, the connection weight between the node $\beta_i(\cdot)$ and the node $h_j(\cdot)$ is denoted as $w_{2,i,j}$, and the bias of the node $\beta_i(\cdot)$ is denoted as $b_{2,i}$, where i ranges from 0 to $2N_{tc}$, and j ranges from 1 to N_h.

To introduce nonlinearity into the neural network, the S-type hyperbolic tangent function $g_h(\cdot)$ is selected as the activation function for the h-layer, i.e., $\boldsymbol{h}(\mathbf{s}) = [h_1(\mathbf{s}), \cdots, h_{N_h}(\mathbf{s})]^T = g_h(\mathbf{W}_1\mathbf{s} + \boldsymbol{b}_1) = tanh(\mathbf{W}_1\mathbf{s} + \boldsymbol{b}_1)$. For the β-layer, the linear activation function $g_\beta(\cdot)$ is chosen, i.e., $\boldsymbol{\beta}(\mathbf{s}) = [\beta_0(\mathbf{s}), \cdots, \beta_{2N_{tc}}(\mathbf{s})]^T = g_\beta(\mathbf{W}_2\boldsymbol{h}(\mathbf{s}) + \boldsymbol{b}_2) = \mathbf{W}_2\boldsymbol{h}(\mathbf{s}) + \boldsymbol{b}_2$. The temporal basis functions are chosen as the corresponding activation functions for the ψ-layer.

For the offline learning of the network parameters using the data of flow and its learned tidal constituents, the Levenberg–Marquardt backpropagation optimization method is employed. This optimization technique efficiently adjusts the connection weights and biases of the neural network to minimize the mean squared error function, which is chosen as the objective function. Once the network is trained, it can effectively predict the flow field at new spatiotemporal locations using the learned relationships from the training data.

2.4. Gaussian Process Model for the Residual Component

In this article, the residual component $r(\mathbf{x})$ of the flow model is modeled using a Gaussian process model. GPs, renowned in machine learning and statistics, prove invaluable for representing complex data relationships and providing predictions along with uncertainty estimations. In this article, the Gaussian process model adopts a zero mean and an assumption is made that in the model the residuals do not exhibit temporal dependencies, meaning that they are temporally uncorrelated, thus $r(\mathbf{x}) \sim GP(\mathbf{0}, k((\mathbf{s}_i, t_k), (\mathbf{s}_j, t_k)))$. This assumption allows the model to capture the spatial correlation between flow measurements at different spatial locations.

In the model, the kernel function $\mathcal{k}((s_i, t_k), (s_j, t_k)) = E[(r(s_i, t_k))(r(s_j, t_k))]$ quantifies the covariance between the spatiotemporal locations (s_i, t_k) and (s_j, t_k), thereby measuring the similarity or dissimilarity between the flow at different spatiotemporal locations. Here, $E[\cdot]$ represents the expectation operator. In this article, the anisotropic squared exponential kernel function is selected to describe the spatial correlation of the residual component of the model at time t_k:

$$\begin{aligned} & \mathcal{k}((s_i, t_k), (s_j, t_k)) \\ = \; & COV(r(s_i, t_k), r(s_j, t_k)) \\ = \; & \theta_{k,0}^2 exp\left\{ -\frac{1}{2}(s_i - s_j)^T \begin{bmatrix} \theta_{k,1}^2 & 0 \\ 0 & \theta_{k,2}^2 \end{bmatrix} (s_i - s_j) \right\} \end{aligned} \tag{6}$$

Here, $\theta_{k,0}$, $\theta_{k,1}$, and $\theta_{k,2}$ are the model hyper-parameters at time t_k, where $\theta_{k,0}^2$ is the overall variance, $\theta_{k,1}$ is the latitudinal length scale, and $\theta_{k,2}$ is the longitudinal length scale. And $\boldsymbol{\theta}_k = [\theta_{k,0}, \theta_{k,1}, \theta_{k,2}]^T$ represents the vector of the model hyper-parameters.

The accuracy of the hyper-parameters $\boldsymbol{\theta}_k$ significantly impacts the prediction accuracy of the model. To estimate the model hyper-parameters $\boldsymbol{\theta}_k$, the maximum likelihood estimation method is employed. This estimation method uses the available data at time t_k to find the hyper-parameters that maximize the likelihood of the observed residuals. The implementation details of this estimation process are provided in Section 3.5.

It should be noted that the modeling performance can be further enhanced by employing more advanced and sophisticated kernel functions. This article, however, focuses on the widely used anisotropic squared exponential kernel function. Comparisons involving different kernel functions fall beyond the scope of this article.

3. Data Assimilation with the Kriged ETKF

To achieve accurate predictions of flow speeds using the presented flow model, it is essential to accurately estimate the model parameters, including the weights and biases in the neural network model, as well as the hyper-parameters in the residual model. While the parameters of the flow model can be learned from historical data, there are limitations and uncertainties associated with the data used for training. These challenges include data sparsity, noise, and discrepancies in the actual flow conditions that the model aims to predict. To address these issues and improve the accuracy of flow predictions, this article implements online real-time learning of the parameters of the model using in situ sensing data of flow speeds provided by AMVs, employing the sequential Bayesian filtering method. This real-time learning enables the model to adapt to changing conditions to closely align with the actual flow conditions and to refine its predictions based on the most recent observations, thereby leading to more accurate and reliable predictions of flow in nearshore ocean environments.

This section delves into the implementation of data assimilation, a crucial aspect of the developed DDDAS. Sections 3.1 and 3.2 provide the state model and observation model, respectively, which serve as the foundation for the subsequent implementation of the sequential data assimilation process. Section 3.3 presents the learning of the neural network model using the ETKF method, and Section 3.4 presents the flow field prediction using contributions from both the learned neural network model and the Gaussian process model. Finally, Section 3.5 addresses the estimation of the hyper-parameters within the Gaussian process model, thereby completing the comprehensive data assimilation process with the Kriged ETKF method.

3.1. State Model

For the neural network model, all the connection weights $\{w_{2,i,j} | i = 0, \cdots, 2N_{tc}; j = 1, \cdots, N_h\}$, as well as all the biases $\{b_{2,i} | i = 0, \cdots, 2N_{tc}\}$, play a crucial role in establishing the nonlinear regression relationship between $\boldsymbol{\beta}(s)$ and s for the spatiotemporal flow field model. These model parameters are selected as model state variables. And the model state vector, denoted as $\boldsymbol{\alpha}$, consists of all model state variables, represented as follows:

$$\boldsymbol{\alpha} = \left[w_{2,0,1}, \cdots, w_{2,2N_{tc},N_h}, b_{2,0}, \cdots, b_{2,2N_{tc}}\right]^T \tag{7}$$

And the temporal state evolution is assumed to be driven by white noise, and it conforms to the following equation:

$$\boldsymbol{\alpha}_k^f = \boldsymbol{\alpha}_{k-1}^a + \boldsymbol{\eta}_k \tag{8}$$

In this equation, $\boldsymbol{\alpha}_{k-1}^a$ represents the posterior analysis state at the previous time step t_{k-1}, which is the estimated state based on the available observations up to time step t_{k-1}, $\boldsymbol{\alpha}_k^f$ represents the forecast state at time step t_k based on $\boldsymbol{\alpha}_{k-1}^a$, and $\boldsymbol{\eta}_k$ represents the Gaussian white process noise at time step t_k.

3.2. Observation Model

This section describes the observation model for the implementation of the online learning with the sequential Bayesian filtering method, which establishes the relationship between the observed flow speeds from AMVs and the underlying state variables.

Assuming that there are N_s flow speed observations $\left\{y_1, \cdots, y_{N_s}\right\}$ at N_s spatial locations $\{s_1, \cdots, s_{N_s}\}$ by N_s AMVs, and the N_s-dimensional observation vector $\boldsymbol{y}_k = \left[y_1, \cdots, y_{N_s}\right]^T$ at time t_k for all the N_s spatial locations is given by:

$$\boldsymbol{y}_k = \boldsymbol{u}_k + \boldsymbol{\varepsilon}_k \tag{9}$$

where $\boldsymbol{u}_k = [u(s_1, t_k), \cdots, u(s_{N_s}, t_k)]^T$ is the true flow speed at time t_k for all N_s spatial locations, and $\boldsymbol{\varepsilon}_k$ is the observation noise at t_k. $\boldsymbol{\varepsilon}_k$ follows a Gaussian distribution with zero-mean vector and covariance matrix $\sigma_\varepsilon^2 \boldsymbol{I}$, where σ_ε^2 is the observation noise variance, and $\boldsymbol{I}$ is the N_s-dimensional identity matrix.

In Equation (9), the mean component $\boldsymbol{\mu}_k$ of $\boldsymbol{u}_k$ can be expressed as follows:

$$\boldsymbol{\mu}_k = \boldsymbol{H}_k \boldsymbol{\alpha}_k \tag{10}$$

where $\boldsymbol{H}_k$ is the observation matrix. The observation matrix maps the flow from the model state space to the observation space at t_k for all N_s spatial observation locations and is defined based on the architecture of the neural network as follows:

$$\boldsymbol{H}_k = \begin{bmatrix} \boldsymbol{\hbar}(s_1) & \mathscr{H}_1(s_1, t_k) & \cdots & \mathscr{H}_{2N_{tc}}(s_1, t_k) & \boldsymbol{\psi}^T(t_k) \\ \vdots & \vdots & \ddots & \vdots & \vdots \\ \boldsymbol{\hbar}(s_{N_s}) & \mathscr{H}_1(s_{N_s}, t_k) & \cdots & \mathscr{H}_{2N_{tc}}(s_{N_s}, t_k) & \boldsymbol{\psi}^T(t_k) \end{bmatrix} \tag{11}$$

where $\boldsymbol{\hbar}(s_i) = \left[h_1(s_i), \cdots, h_{N_h}(s_i)\right]$, and $\mathscr{H}_j(s_i, t_k) = \boldsymbol{\hbar}(s_i)\psi_j(t_k)$, in which $i = 1, \cdots, N_s$ and $j = 1, \cdots, 2N_{tc}$.

Based on the above equations, the measurement equation can be expressed as follows:

$$\boldsymbol{y}_k = \boldsymbol{H}_k \boldsymbol{\alpha}_k + \boldsymbol{r}_k + \boldsymbol{\varepsilon}_k \tag{12}$$

3.3. Filtering with ETKF

Based on the structure of the presented flow model and the need for model learning and flow prediction using flow speed data from AMVs, this article employs the Kriged Kalman filter method [34] to implement the data assimilation and flow prediction. The Kriged Kalman Filter integrates Kriging's spatial interpolation capabilities and the Kalman filter's optimal state estimation to effectively combine spatial correlations and temporal dynamics. In the DDDAS system, the data assimilation process not only improves flow predictions but also supports the optimization of AMVs' observation strategies. To achieve this, the Kriged ETKF is introduced. The implementation of the Kriged ETKF is detailed as follows, and the optimization strategy which further enhances the model's predictive capabilities will be elaborated on in Section 4.

In the ensemble-based filtering approach, a total of N_{em} ensemble members are considered. The initial model state vector is perturbed to generate the initial ensemble of the model state vectors $\left\{\boldsymbol{\alpha}_0^{(i)}\middle| i=1,\cdots,N_{em}\right\}$. These perturbations introduce variability and diversity in the initial state estimates, which is crucial for robust data assimilation. The forecast equation then drives the evolution of each ensemble member over time as follows:

$$\boldsymbol{\alpha}_k^{f,(i)} = \boldsymbol{\alpha}_{k-1}^{a,(i)} + \boldsymbol{\eta}_k^{(i)} \quad i=1,\cdots,N_{em} \tag{13}$$

where $\left\{\boldsymbol{\alpha}_k^{f,(i)}\middle| i=1,\cdots,N_{em}\right\}$ represents the forecast ensemble at t_k, $\left\{\boldsymbol{\alpha}_{k-1}^{a,(i)}\middle| i=1,\cdots,N_{em}\right\}$ is the perturbed posterior analysis ensemble at t_{k-1}, and $\left\{\boldsymbol{\eta}_k^{(i)} \sim \mathcal{N}_{N_\alpha}(\mathbf{0},\boldsymbol{\tau})\middle| i=1,\cdots,N_{em}\right\}$ is the ensemble of forecast noise at t_k, where $N_\alpha = (2N_{tc}+1)\times(N_h+1)$ is the total number of model state variables. This noise follows a Gaussian distribution with a zero-mean vector and a covariance matrix $\boldsymbol{\tau}$.

By subtracting the forecast ensemble mean from each ensemble member, the forecast ensemble perturbations matrix at t_k can be obtained as follows [35]:

$$\boldsymbol{Z}_k^f = \frac{1}{\sqrt{N_{em}-1}}\left[\boldsymbol{\alpha}_k^{f,(1)} - \overline{\boldsymbol{\alpha}}_k^f,\cdots,\boldsymbol{\alpha}_k^{f,(N_{em})} - \overline{\boldsymbol{\alpha}}_k^f\right] \tag{14}$$

where $\overline{\boldsymbol{\alpha}}_k^f$ is the mean forecast state at t_k, which is defined as the ensemble mean of the forecast state as follows:

$$\overline{\boldsymbol{\alpha}}_k^f = \frac{1}{N_{em}}\sum_{i=1}^{N_{em}} \boldsymbol{\alpha}_k^{f,(i)} \tag{15}$$

The forecast error covariance matrix at t_k can be obtained as follows:

$$\begin{aligned}\boldsymbol{P}_k^f &= \tfrac{1}{N_{em}-1}\sum_{i=1}^{N_{em}}\left(\boldsymbol{\alpha}_k^{f,(i)} - \overline{\boldsymbol{\alpha}}_k^f\right)\left(\boldsymbol{\alpha}_k^{f,(i)} - \overline{\boldsymbol{\alpha}}_k^f\right)^T \\ &= \boldsymbol{Z}_k^f \boldsymbol{Z}_k^{fT}\end{aligned} \tag{16}$$

where $\boldsymbol{P}_k^f$ represents the forecast error covariance matrix at time t_k, and $\boldsymbol{Z}_k^f$ is the forecast ensemble perturbations matrix as defined in Equation (14). The forecast error covariance matrix provides a measure of uncertainty in the forecast state, which is essential for accurately estimating the flow field and making reliable predictions.

The mean posterior analysis state at time t_k can be obtained as follows:

$$\overline{\boldsymbol{\alpha}}_k^a = \overline{\boldsymbol{\alpha}}_k^f + \boldsymbol{K}_k\left(\boldsymbol{y}_k - \boldsymbol{H}_k\overline{\boldsymbol{\alpha}}_k^f\right) \tag{17}$$

where $\boldsymbol{K}_k$ represents the Kalman gain matrix at time t_k as follows:

$$\boldsymbol{K}_k = \boldsymbol{P}_k^f \boldsymbol{H}_k^T\left(\boldsymbol{H}_k \boldsymbol{P}_k^f \boldsymbol{H}_k^T + \boldsymbol{\Sigma}_k\right)^{-1} \tag{18}$$

and $\boldsymbol{\Sigma}_k$ is the observation error covariance matrix at time t_k given by:

$$\boldsymbol{\Sigma}_k = COV(\boldsymbol{r}_k,\boldsymbol{r}_k) + \sigma_\varepsilon^2 \boldsymbol{I} \tag{19}$$

where $\boldsymbol{r}_k = [r(s_1,t_k),\cdots,r(s_{N_s},t_k)]^T$ is the residual vector at time t_k, and $\boldsymbol{I}$ is the N_s-dimensional identity matrix.

The ETKF [35] introduces a transformation matrix that efficiently transforms the forecast ensemble perturbations into the posterior analysis ensemble perturbations, thereby reducing the complexity of the posterior analysis state calculation. At time t_k, the transformation matrix is given by:

$$T_k = C_k(\Gamma_k + I)^{-1/2} \tag{20}$$

where C_k is a matrix containing all the orthogonal eigenvectors of the matrix $Z_k^{f^T} H_k^T \Sigma_k^{-1} H_k Z_k^f$, Γ_k is a diagonal matrix with all corresponding eigenvalues as its main diagonal elements, and I is the N_{em}-dimensional identity matrix. This transformation matrix plays a crucial role in the data assimilation process by efficiently adjusting the forecast ensemble perturbations to be consistent with the observations, leading to more accurate and reliable posterior analysis state estimates.

The posterior analysis ensemble perturbations matrix at time t_k can be obtained by applying the transformation matrix T_k as follows [35]:

$$Z_k^a = Z_k^f T_k \tag{21}$$

By adding the mean posterior analysis state to each column of the posterior analysis ensemble perturbation matrix Z_k^a, the perturbed posterior analysis ensemble at time t_k can be obtained as follows:

$$\left\{ \alpha_k^{a,(i)} = \overline{\alpha}_k^a + \sqrt{N_{em} - 1} Z_k^{a,(i)} \middle| i = 1, \cdots, N_{em} \right\} \tag{22}$$

The posterior analysis error covariance matrix at t_k can be obtained as follows:

$$\begin{aligned} P_k^a &= \frac{1}{N_{em}-1} \sum_{i=1}^{N_{em}} \left(\alpha_k^{a,(i)} - \overline{\alpha}_k^a \right) \left(\alpha_k^{a,(i)} - \overline{\alpha}_k^a \right)^T \\ &= Z_k^a Z_k^{aT} \end{aligned} \tag{23}$$

Based on the above calculations, the posterior analysis of the flow and the corresponding uncertainty for observed spatial locations under given observations at time t_k can be obtained, respectively, as follows:

$$u_k^a = H_k \overline{\alpha}_k^a \tag{24}$$

$$COV(u_k^a | y_k) = H_k P_k^a H_k^T + \Sigma_k \tag{25}$$

3.4. Flow Prediction

After obtaining the posterior analysis state by assimilating the sensing data of flow speeds, the Kriging method is employed to further estimate the flow speeds at unobserved locations.

Let $\{s_i^* | i - 1, \cdots, N_{s^*}\}$ represent the collection of spatial locations without sensing data, and $r_k^* = \left[r(s_1^*, t_k), \cdots, r\left(s_{N_{s^*}}^*, t_k\right) \right]^T$ be the corresponding residual components of the model at time t_k, where N_{s^*} is the total number of unobserved locations considered for flow prediction. Define $\Sigma_k^* = COV(r_k, r_k^*)$ as the $N_s \times N_{s^*}$-dimensional matrix of covariances of residual components of the model evaluated at all pairs of spatial observed and unobserved locations at t_k, and $\Sigma_k^{**} = COV(r_k^*, r_k^*)$ as the $N_{s^*} \times N_{s^*}$-dimensional matrix of covariances of residual components of the model evaluated at all pairs of spatial unobserved locations at time t_k. The observation matrix at time t_k for all spatial unobserved locations is defined as follows:

$$H_k^* = \begin{bmatrix} \hbar(s_1^*) & \mathcal{H}_1(s_1^*, t_k) & \cdots & \mathcal{H}_{2N_{tc}}(s_1^*, t_k) & \psi^T(t_k) \\ \vdots & \vdots & \ddots & \vdots & \vdots \\ \hbar\left(s_{N_{s^*}}^*\right) & \mathcal{H}_1\left(s_{N_{s^*}}^*, t_k\right) & \cdots & \mathcal{H}_{2N_{tc}}\left(s_{N_{s^*}}^*, t_k\right) & \psi^T(t_k) \end{bmatrix} \tag{26}$$

Based on the abovementioned definition, the posterior analysis of the flow and the corresponding covariance for all unobserved spatial locations under given observations at time t_k can be obtained, respectively, as follows [34]:

$$u_k^{*,a} = H_k^* \overline{\alpha}_k^a + \Sigma_k^{*T} \Sigma_k^{-1} (y_k - H_k \overline{\alpha}_k^a) \tag{27}$$

$$COV(u_k^{*,a} | y_k) = \Sigma_k^{**} - \Sigma_k^{*T} \Sigma_k^{-1} \Sigma_k^* + \Phi_k P_k^a \Phi_k^T \tag{28}$$

where $u_k^* = \left[u\left(s_1^*, t_k\right), \cdots, u\left(s_{N_s^*}^*, t_k\right)\right]^T$, and $\Phi_k = H_k^* - \Sigma_k^{*T} \Sigma_k^{-1} H_k$.

3.5. Estimation of Hyper-Parameters

To achieve flow prediction, it is essential to estimate the hyper-parameters in the residual model at each data assimilation time. In this article, the maximum likelihood estimation method is employed to estimate the model hyper-parameters using the flow speed data collected by the AMVs at time t_k.

According to Equation (12), the observed flow speed vector y_k follows a multivariate normal distribution, i.e., $y_k \sim \mathcal{N}_{N_s}\left(H_k \alpha_k, H_k P_k H_k^T + \Sigma_k\right)$. The likelihood function, denoted as $\mathscr{L}(\cdot)$, with respect to the model hyper-parameters θ_k at t_k is expressed as follows:

$$\begin{aligned} & \mathscr{L}\left(\theta_k \middle| y_k, \overline{\alpha}_k^f, P_k^f\right) \\ = & \, p\left(y_k \middle| \theta_k, \overline{\alpha}_k^f, P_k^f\right) \\ = & \, \frac{1}{(2\pi)^{N_s/2} |D_k|^{1/2}} exp\left\{-\frac{1}{2} \xi_k^T D_k^{-1} \xi_k\right\} \end{aligned} \tag{29}$$

where $D_k = H_k P_k^f H_k^T + \Sigma_k$, $\xi_k = y_k - H_k \overline{\alpha}_k^f$, and $p(\cdot)$ is the probability density function of the observations. The likelihood function $\mathscr{L}\left(\theta_k \middle| y_k, \overline{\alpha}_k^f, P_k^f\right)$ quantifies the likelihood of observing the data y_k given the model hyper-parameters θ_k, the mean forecast state $\overline{\alpha}_k^f$, and the forecast error covariance matrix P_k^f. By maximizing this likelihood function, the optimal values for the model hyper-parameters θ_k could be obtained. The logarithmic likelihood function, denoted as $ln\mathscr{L}$, is commonly used for numerical optimization to solve the optimal values of θ_k, as follows:

$$ln\mathscr{L} = -\frac{1}{2} ln|D_k| - \frac{1}{2} \xi_k^T D_k^{-1} \xi_k - \frac{N_s}{2} ln(2\pi) \tag{30}$$

To find the numerical solution for the model hyper-parameters θ_k, the genetic algorithm is employed to solve the following objective function:

$$\theta_k = \underset{\theta_k \in R^3}{arg\ min} \left(-ln\mathscr{L}\left(\theta_k \middle| y_k, \overline{\alpha}_k^f, P_k^f\right)\right) \tag{31}$$

By using the genetic algorithm, the numerical solution for the model hyper-parameters θ_k can be efficiently obtained, enabling accurate estimation of the model parameters and improving the flow prediction performance.

4. Optimization of AMVs' Sensing Paths

In the context of the DDDAS for flow prediction, the effective optimization of the AMVs' sensing paths is of utmost importance. The assimilation of the informative flow-sensing data streams from AMVs enables improved estimation of model parameters, leading to enhanced accuracy in flow prediction. The neural network model serves as the backbone of the flow model, and this section focuses on the optimization of AMVs sensing paths to improve the estimation of the neural network parameters. Through this optimization process, the developed DDDAS is completed, enabling adaptive flow sensing using AMVs.

In the Kriged ETKF data assimilation, the ETKF method provides an effective means to quantify the reduction in error variance of the forecast state variables and to access the impact of different observation schemes. In the article, the optimization is built upon the ETKF method, and two optimization scenarios are investigated. In the first scenario, the optimization focuses on finding optimal AMVs' sensing locations for improving the estimation performance only in the subsequent data assimilation time step. This strategy is relatively straightforward to implement. However, it has a myopic nature, as it only considers the immediate impact on the subsequent time step. On the other hand, the second scenario investigated extends the optimization to consider multiple data-assimilation time steps over future time horizon. Compared to the greedy optimization in the first scenario, the non-myopic optimization problem in the second scenario presents more complexities. This article places a greater emphasis on non-myopic optimization.

This section elaborates on the optimization. Section 4.1 states the optimization problems for the two scenarios. Section 4.2 presents the strategies for solving the optimization problems, with a particular emphasis on the proposed receding horizon strategy for the non-myopic optimization scenario. And Section 4.3 presents a proposed distributed strategy for implementing the MCTS method to solve the receding horizon optimization.

4.1. Problem Statement

In the Kriged ETKF data assimilation described in Section 3, the signal covariance matrix at time t_k, denoted as S_k and defined in Equation (32), as follows [35]

$$S_k = P_k^f - P_k^a = Z_k^f C_k \Gamma_k (\Gamma_k + I)^{-1} C_k^T Z_k^{fT} \tag{32}$$

provides a means to assess the impact of assimilating observational information from various observation schemes $H_k(\mathcal{S}_k)$ at time t_k on the reduction in error variance of the forecast state variables and posterior analysis, where $\mathcal{S}_k = \{s_1, \cdots, s_{N_s}\}$ represents the spatial locations of N_s AMVs. This difference represents the information gained from the observations, which leads to an improvement in state estimation.

The trace of S_k, denoted as $tr\{S_k\}$, serves as a measure of the improvement in the model's state estimation when assimilating the flow-sensing data $y_k = \left[y_1, \cdots, y_{N_s}\right]^T$ obtained at $\mathcal{S}_k$. A larger value of S_k indicates a more significant reduction in error variance, suggesting that the assimilated flow-sensing data have a substantial positive impact on enhancing the accuracy of the model estimation. Therefore, in this article, the optimization of the AMVs' flow-sensing paths is based on maximizing the trace of the signal covariance matrix.

For the first scenario of greedy optimization performed at time t_k, the optimization of AMVs' sensing locations to be reached at time t_{k+1} is expressed as follows:

$$\mathcal{S}_{LO}^* = \underset{\mathcal{S}_{LO} \in RS_{k+1}(\mathcal{S}_k)}{arg\,max} \quad tr(S_{k+1}) \tag{33}$$

In this expression, $\mathcal{S}_{LO}^*$ represents the optimized locations, and $RS_{k+1}(\mathcal{S}_k)$ represents the set of possible sensing locations for the AMVs at time t_{k+1}, which are constrained to be within the reachable region determined by the previous sensing locations $\mathcal{S}_k$ at time t_k. The objective of the optimization is to find $\mathcal{S}_{LO}^*$ to maximize the data assimilation performance at time t_{k+1}. By implementing the optimization continuously at each data assimilation time step, the DDDAS forms the optimal paths of AMVs that consistently improve the data assimilation performance.

For the second scenario of non-myopic optimization performed at time t_k, the objective is to find the optimal paths of AMVs, covering multiple time steps from time t_{k+1} to time

t_{k+N_T}, with the aim to enhance the data assimilation performance during the time step t_{k+1} to a user-defined future time step t_{k+N_T}. The optimization problem is expressed as follows:

$$\mathcal{S}^*_{PA} = \underset{\mathcal{S}_{PA} \in RS_{k+N_T}(\mathcal{S}_k)}{arg\ max} \sum_{i=1}^{N_T} tr(w_{k+i} S_{k+i}) \tag{34}$$

In this expression, $\mathcal{S}^*_{PA}$ represents the optimized paths from time t_{k+1} to time t_{k+N_T}, where N_T is the number of time steps in the optimization horizon. The weight w_{k+i} reflects the user's preference for each time step from t_{k+1} to t_{k+N_T}. $RS_{k+N_T}(\mathcal{S}_k)$ represents the set of possible sensing paths for the AMVs from time t_{k+1} to time t_{k+N_T}. These paths are constrained to be within the reachable region determined by the sensing locations $\mathcal{S}_k$ at time t_k.

In the optimization process of AMVs' sensing paths, S_{k+1} may consist of two components, one for the latitudinal direction and the other for the longitudinal direction, calculated using the two submodels. This can be expressed as $S_{k+1} = w_{lat} S^{lat}_{k+1} + w_{lon} S^{lon}_{k+1}$, where w_{lat} and w_{lon} are user-defined weights with the constraint that $w_{lat} + w_{lon} = 1$. The choice of these weights depends on the specific objectives and requirements of the mission.

4.2. Strategies for Solving the Optimization Problems

In the implementation of the optimization in this article, the AMVs are assumed to navigate at a constant speed relative to the sea bottom, and the optimization focuses on finding the optimal commanded heading angles for the AMVs, i.e., the commanded heading angles are selected as the decision variables. With the optimized commanded heading angles, the AMVs could navigate to desired flow-sensing locations using guidance and control algorithms [36]. To solve the optimization problems, the commanded heading angle of an AMV is discretized by dividing the range of heading angles into N_H equal parts. The heading angle $i \times 360° / N_H$ is selected as one of the possible decision choices, where i ranges from 1 to N_H. The user defined constant value of N_H determines the level of discretization. Selecting a higher value will achieve fine-grained exploration of heading angle options, with an increase in computational costs.

For the optimization problem in Equation (33), the optimal solution can be obtained by evaluating the objective function for each combination of the discretized heading angles of AMVs and selecting the one that maximizes the $tr(S_{k+1})$. For a small number of possible combinations, an exhaustive search could be performed. For a large number of possible combinations, an exhaustive search becomes computationally expensive and time-consuming. In such cases, this article employs the genetic algorithm to efficiently implement the optimization.

For the optimization problem in Equation (34), the objective is to find the optimal sequences of commanded heading angles for the AMVs, which will determine the paths of the AMVs over multiple time steps. The temporal sequence of the possible combinations of commanded heading angles of AMVs forms a tree structure, where each node in a layer represents a possible combination of commanded heading angles of the AMVs at a specific time step. The tree expands from one time step to the next, representing the different possibilities for AMVs' paths over the optimization horizon. To efficiently search for the optimal solution in the tree structure, this article employs the MCTS method [37]. MCTS is a heuristic search method commonly used in decision-making problems, particularly in scenarios with large search spaces. It is suitable for the optimization problem addressed in this article.

Considering the presence of uncertainties in the flow model, real flow conditions, and the AMVs' navigation and control, this article employs the receding horizon optimization to implement the path planning considering multiple time steps. In this strategy, the optimization is performed at time t_k using the MCTS method, and the computed commanded heading angles for time t_{k+1} are provided to the AMVs for implementation. This approach implements the optimization process repeatedly at each time step t_k. By this approach, the

AMVs continuously update their paths based on the most recent information, making the flow sensing and prediction capabilities more adaptive and robust.

It should be noted that the receding horizon strategy could be implemented in two different ways to achieve continuous AMV paths and optimize the data assimilation performance over a future time window, based on different mission requirements. In the first approach, the process involves repeating the optimization with the same time horizon at each data assimilation–optimization time step, i.e., at each time t_k, the optimization of $\mathcal{S}^*_{PA}$ in Equation (34) is implemented. This results in continuous AMV paths, where the AMVs' locations at each time step are chosen to be optimal for the future N_T time steps. In contrast, in the second approach, the paths of the AMVs are optimized for a fixed user-defined time window, and at each time step the optimization only considers the remaining time steps within the window. Specifically, at one time t_k, the optimization of maximization of $\sum_{i=1}^{N_T} tr(w_{k+i}S_{k+i})$ is implemented, whereas at the subsequent time t_{k+1}, the optimization of $\sum_{i=2}^{N_T} tr(w_{k+i}S_{k+i})$ is implemented, which considers the optimization of the remaining time steps from t_{k+2} to t_{k+N_T}. This optimization process terminates at t_{k+N_T-1}, resulting in continuous AMVs paths terminated at t_{k+N_T}. And the AMVs' locations at each time step are chosen to be optimal for the future remaining time steps within the time window.

4.3. A Distributed Strategy for Implementing MCTS

When implementing MCTS on a single processor with limited computational capability to solve large search problems within a constrained decision-making time frame, obtaining an optimal solution may be difficult or impractical. To tackle this issue, and considering the operation scenario of the DDDAS where the computation is implemented on all N_s AMVs without the support of onshore stations, this article proposes a distributed strategy to implement the optimization process. This strategy involves partitioning the search tree into subtrees and assigning each subtree to a different computing device on AMVs for simultaneous exploration. By conducting parallel searches on individual subtrees simultaneously and combining their results, the strategy can leverage the collective computational power of multiple devices of the networked AMVs, facilitating an efficient and optimal decision-making process.

The proposed distributed strategy partitions the original search tree into N_s subtrees. Each subtree is then assigned to an AMV for independent execution. The final result is obtained by combining the N_s outcomes of all subtrees. The initial state of the root nodes $node^i_{sub}$ of the subtrees, $i = 1, 2, \cdots, N_s$, is the same as that of the root node $node_r$ in the original tree. The number of child nodes for each $node^i_{sub}$ is equal to N_a/N_s, where N_a denotes the number of child nodes of each node in the original search tree.

After splitting the tree, all subtrees are executed simultaneously on the corresponding computing devices. Once all the subtree searches are completed, the final optimal decision is determined by selecting the decision with the highest quality among the best child nodes of each subtree, as follows:

$$a^* = \underset{a^{i^*}}{arg\,max} q^{i^*} \quad i = 1, 2, \cdots, N_s \tag{35}$$

where a^{i^*} represents the decision of the best child of $node^i_{sub}$, and q^{i^*} represents the quality of that child.

Existing distributed MCTS methods often necessitate frequent information interactions between multiple threads [38,39], and thus may not be suitable for scenarios like distributed AMV networks with wireless communication, where frequent communication among network nodes is unfavorable or impractical. The proposed distributed strategy in this article requires minimal communication between devices, limited to allocating subtrees and collecting results for determining the final optimal decision. With low data transmission needs, it is well-suited for scenarios with limited wireless communication capabilities, such as the AMV networks studied in this article.

In the implementation of the MCTS, to balance exploitation and exploration in the search process, nodes are selected based on the UCB1 (upper confidence bound 1) method [40] commonly used in MCTS. By using the UCB1, the MCTS method ensures a balance between exploring promising nodes and exploiting nodes with high estimated quality.

5. Simulation Results

In this article, a comprehensive simulation study is conducted to validate and evaluate the performance of the developed DDDAS, which is presented in this section.

Section 5.1 describes the simulation scenario, where four autonomous underwater gliders are utilized to predict the depth-averaged flow in nearshore ocean environments with tidal features. These gliders serve as the flow-sensing network, and their paths are optimized to collect informative flow-sensing data. Section 5.2 presents a comparison of three flow prediction approaches: (1) using the neural network alone; (2) incorporating the data assimilation into the neural network; and (3) further incorporating the Gaussian process. This comparison enables an assessment of the improvement in flow prediction accuracy with each successive approach. Section 5.3 focuses on comparing three glider sensing strategies: virtually moored at fixed locations, random motion, and optimized motion. The accuracy of flow prediction is evaluated under each strategy, revealing the impact of gliders' flow-sensing locations on the performance of flow prediction. Lastly, in Section 5.4, the results of implementing the receding horizon optimization with the distributed MCTS are presented, comparing the results of greedy optimization and non-myopic optimization and demonstrating the utilization of multiple AMVs' computational power for decision-making, when the computation of the DDDAS is implemented on the AMVs without support from onshore stations.

5.1. Simulation Scenario

Underwater gliders are highly valuable autonomous mobile platforms for large-scale and long-term ocean observation. Driven by buoyancy engines, gliders commonly navigate at relatively slow speeds, thus their motion can be significantly influenced by ocean flow. Therefore, accurate depth-averaged flow prediction information is crucial for navigation and control of underwater gliders. In the simulation study of the developed DDDAS in this article, the underwater gliders are employed as the sensing network, and their paths are optimized to collect flow-sensing data for predicting the depth-averaged flow in the nearshore ocean environment.

In the simulation, a designated region of $1^\circ \times 1^\circ$ in the South China Sea is chosen as the operational area. Within this operational area, four underwater gliders are deployed to navigate and perform the sensing mission. For the simulation, the depth-averaged flow data predicted by a numerical ocean model of POM (Princeton Ocean Model) during March and June 2019 are employed as the true flow. This flow dataset serves as the basis for supporting the glider sensing and acts as the ground truth for comparison and evaluation. The flow data have a temporal resolution of 1 h and a horizontal spatial resolution of $1/15^\circ$ in both the latitudinal and longitudinal directions. In Figure 4a, the flow field at 31 March 2019, 18:00 (UTC + 08:00) is depicted, where the flow is predicted on the 16 (latitudinal direction) $\times 16$ (longitudinal direction) $= 256$ uniformly distributed grid points. For the initialization of the neural network model, the flow data covering the time range from 1 March 2019, 00:00 to 31 March 2019, 23:00 (UTC + 08:00) is utilized. The tidal analysis method is then applied to this dataset to extract the tidal constituents and a total of primary $N_{tc} = 29$ tidal constituents are selected, resulting in $2N_{tc} = 58$ temporal basis functions and 59 corresponding coefficients taking into account the non-tidal component for the mean flow. Based on these tidal constituents, the neural network model is constructed, with $N_h = 7$ nodes in the hidden layer, leading to a total number of $N_\alpha = 472$ model state variables. It should be noted that the flow data used for the tidal initialization of the neural network with the gradient descent method is subsampled, with a spatial resolution of 0.2°, as shown in Figure 4b. This approach aims to demonstrate the capability of predicting high-resolution flow fields using a model taught with relatively sparse training data.

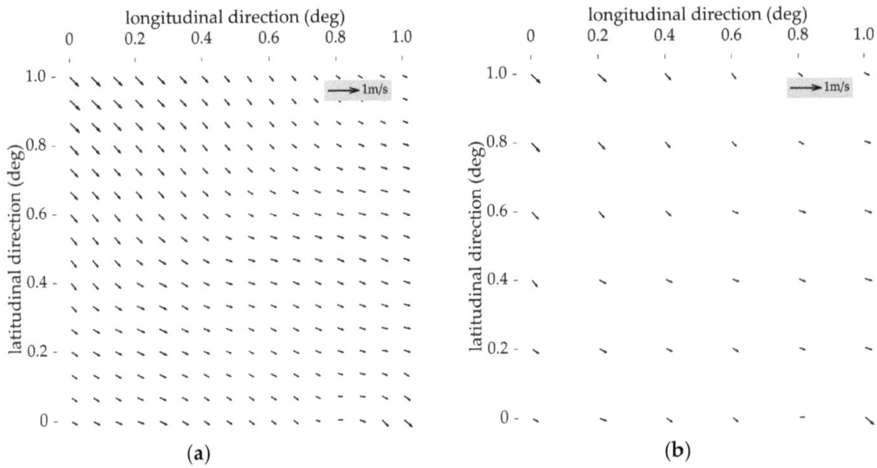

Figure 4. (**a**) The flow field predicted by the POM. (**b**) The subsampled flow field for model initialization.

In the simulation, four underwater gliders are deployed in the operational area, i.e., $N_s = 4$. The gliders are set to navigate at a constant speed of 0.7 kn relative to the sea bottom, and the influence of the flow on the motion of the glider is not considered. Each glider completes one saw-tooth dive cycle in one hour and reports its sensed depth-averaged flow speed upon surfacing. The sensed depth-averaged flow is calculated by linearly interpolating the four predicted flow values around the glider's surfacing location, with the addition of sensing noise having a variance of $\sigma_\varepsilon^2 = 10^{-4}$. The model assimilates the sensing data from the four gliders every hour. During the assimilation, the perturbation covariance matrix of the initial ensemble of the model state vector is set as $10^{-6}\boldsymbol{I}$, and the total number of ensemble members is set as $N_{em} = 100$.

For the optimization of the gliders' paths, both $\boldsymbol{S}_{k+1}^{lat}$ and $\boldsymbol{S}_{k+1}^{lon}$ are considered, and the weights w_{lat} and w_{lon} are both set as 0.5, implying an equal contribution from both components in determining the optimized sensing paths. And two sets of heading angles of $\{0, 45, 90, 135, 180, 225, 270, 325\}$ and $\{0, 90, 180, 270\}$ are used as the decision choices for the greedy and the non-myopic optimization, respectively. In the simulation, the gliders' motion is constrained within the operational area. If a heading angle leads a glider out of the operational area, that specific heading angle is not considered during the optimization process, and the optimal heading angle is selected from the remaining feasible choices to ensure that the glider's navigation remains within the operational area.

During the simulation, the flow prediction, data assimilation, and adaptive sampling processes are conducted for a total of 129 consecutive hours, specifically from 1 June 2019, 00:00 to 6 June 2019, 08:00 (UTC + 08:00). To simplify the description, the time steps are labeled starting from 0 h, which corresponds to the 1 June 2019, 00:00, and each subsequent time step represents one hour of the simulation. The predicted flow values from the DDDAS are then compared with the true flow values at the 256 grid points in the designated operational area, as shown in Figure 4a. To evaluate the prediction performance, the root mean squared error (RMSE) is calculated for the 256 grid points at each hour, as follows:

$$RMSE(t) = \sqrt{\frac{1}{256}\sum_{i=1}^{256}\left(u^{pre}(\boldsymbol{s}_i, t) - u^{true}(\boldsymbol{s}_i, t)\right)^2} \tag{36}$$

where $u^{pre}(\cdot)$ represents the model's prediction of the flow speed, and $u^{true}(\cdot)$ represents the corresponding true flow speed from the POM dataset at the same spatiotemporal location. The RMSE serves as a metric to measure the accuracy of the flow field prediction.

5.2. Comparision of Three Flow Prediction Approaches

In the comparison study described in this section, the paths of the gliders are optimized using the greedy optimization strategy. Over the 129 consecutive hours, continuous paths of the gliders are formed, and these trajectories will be depicted in Section 5.3. By assimilating the optimized sensing data at each time step, the flow field is predicted and updated at each time step. Figure 5 presents examples of the predicted flow field at the 1st hour and the 9th hour, along with the corresponding variance of the predicted flow speeds in both the latitudinal and longitudinal directions. The variance information is important for many tasks such as the robust path planning or motion control of AMVs. Thanks to the capabilities of the data-driven flow model, flow field predictions could be obtained with any desired level of spatial resolution.

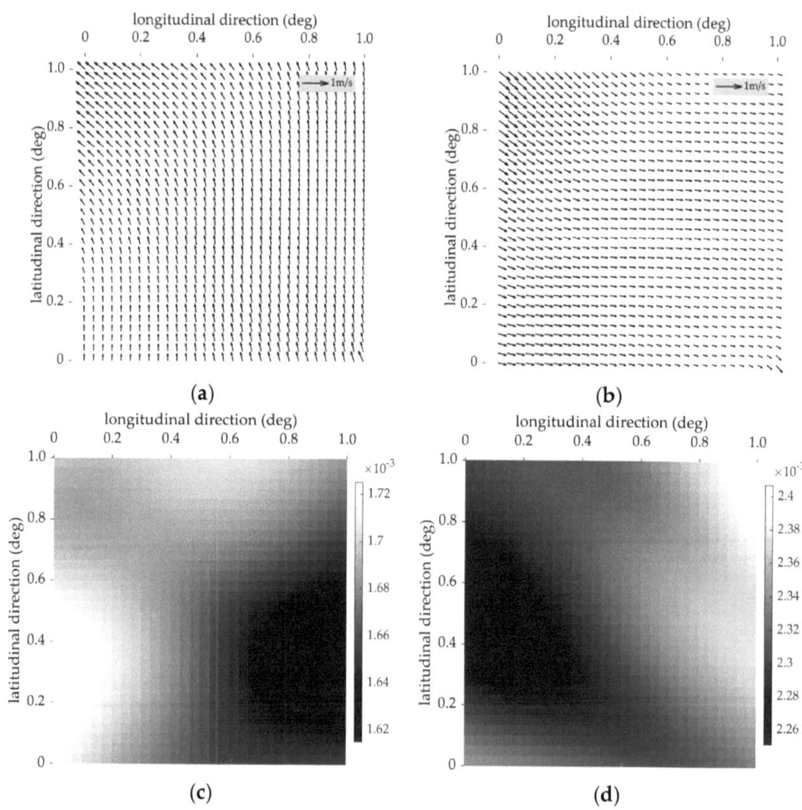

Figure 5. (**a**) Predicted flow field at the 1st hour. (**b**) Predicted flow field at the 9th hour. (**c**) Variance of predicted flow speed along the latitudinal direction at the 9th hour. (**d**) Variance of predicted flow speed along the longitudinal direction at the 9th hour.

To demonstrate the performance of the hybrid data-driven flow model and the data assimilation, the RMSE is compared for the results predicted by the three approaches: (1) using the neural network alone; (2) incorporating the data assimilation into the neural network; and (3) further incorporating the Gaussian process. Figure 6 demonstrates the variations in RMSE over the 129 time steps, where NN, NN + DA, and NN + GP + DA represent the three flow prediction approaches, respectively (NN: neural network alone, NN + DA: neural network with data assimilation, NN + GP + DA: neural network with Gaussian process and data assimilation). Figure 6a presents the RMSE results of the flow speed component in the latitudinal direction, and Figure 6b presents the corresponding RMSE results for the flow speed component in the longitudinal direction. The unit of measurement is meters per second (m/s).

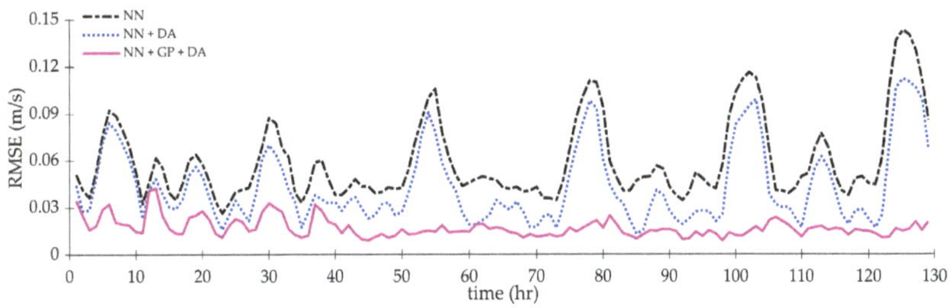

(a)

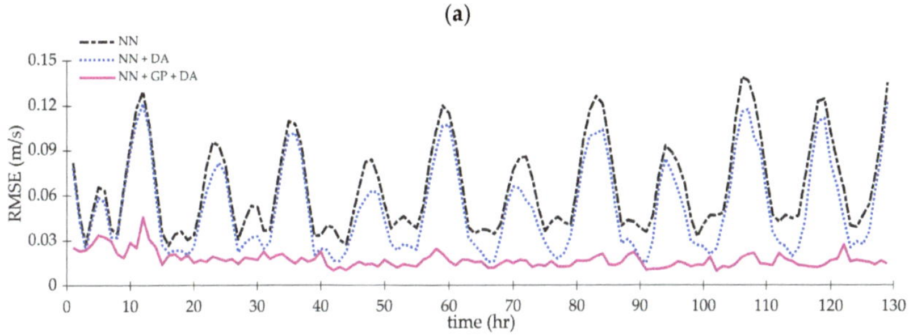

(b)

Figure 6. RMSE of the flow field prediction using three different flow prediction approaches. (**a**) RMSE results of the flow speed component along the latitude direction. (**b**) RMSE Results of the flow speed component along the longitude direction.

Table 1 shows the maximum, the minimum, and the maximum and the mean reduction (compared with the NN approach) values of the RMSE shown in Figure 6 for each respective flow prediction approach.

Table 1. Performance of RMSE for three flow prediction approaches, corresponding to Figure 6.

Approach	Direction	Maximum Value	Minimum Value	Max Reduction Value	Mean Reduction Value
NN	Latitudinal	0.143	0.026	NA	NA
	Longitudinal	0.139	0.027	NA	NA
NN + DA	Latitudinal	0.112	0.013	0.037	0.016
	Longitudinal	0.124	0.015	0.028	0.013
NN + GP + DA	Latitudinal	0.043	0.009	0.128	0.042
	Longitudinal	0.046	0.010	0.121	0.048

The results presented in Figure 6 and Table 1 highlight the significant impact of data assimilation and the Gaussian process on improving flow field predictions. When compared to the predictions without data assimilation, the inclusion of data assimilation leads to a notable reduction in RMSE for both latitudinal and longitudinal flow field predictions. This reduction indicates that data assimilation effectively incorporates observed data from the gliders into the model. Furthermore, the integration of the residual component Gaussian process model with data assimilation yields more accurate predictions. The combination of these techniques contributes to a substantial reduction in RMSE and enhances the model's ability to achieve accurate flow predictions. This study suggests that the hybrid data-driven modeling approach, together with data assimilation, provides an effective framework for obtaining high-resolution and accurate flow field predictions in nearshore ocean environments.

5.3. Comparision of Three Flow Sensing Strategies

In this section, three glider sensing strategies are compared to evaluate their performance in flow prediction. The first strategy involves virtually mooring the gliders at fixed locations to sense the temporal variations in the flow. In the second strategy, the commanded heading angle of each glider is randomly selected from the set of $\{0, 45, 90, 135, 180, 225, 270, 325\}$ at each time step, representing a more exploratory approach. In contrast, the third strategy optimizes the paths of gliders using the greedy optimization method, allowing for adaptive and optimized sampling based on the current situations. In Figure 7, the paths of the gliders generated by the second (random heading angles) and the third (optimized paths) sensing strategies are depicted. The starting locations of the optimized paths are set as the same locations where the gliders are virtually moored for the first sensing strategy.

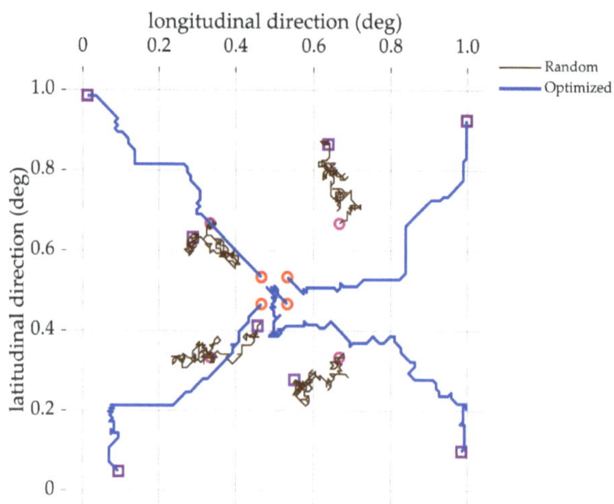

Figure 7. Paths of underwater gliders for the three sensing strategies. The circles represent the start locations, and the rectangles represent the end locations of the glider paths.

The data collected along these paths from the three sensing strategies are then assimilated into the hybrid data-driven flow model, enabling flow field predictions for each strategy at every time step. Figure 8 illustrates the variations in RMSE over the course of 129 consecutive hours, with "static", "random", and "optimized" representing the three sensing strategies, respectively. Specifically, Figure 8a plots the RMSE results of the flow component in the latitudinal direction, while Figure 8b plots the RMSE results of the flow component in the longitudinal direction. The unit of measurement is meters per second (m/s).

Table 2 shows the maximum, the minimum, and the maximum and the mean reduction (compared with the "static" strategy) values of the RMSE shown in Figure 8 for each respective flow sensing strategy.

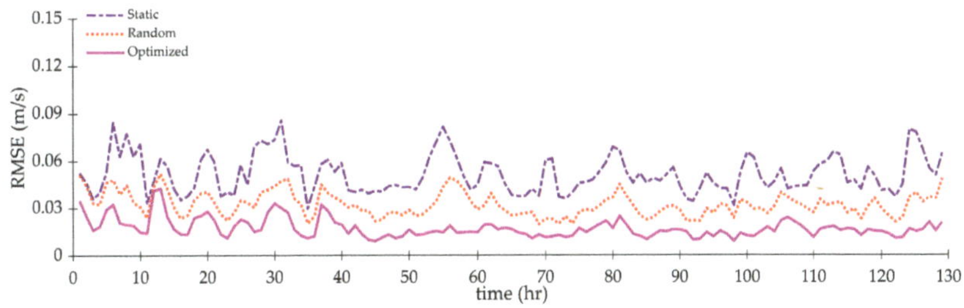

(**a**)

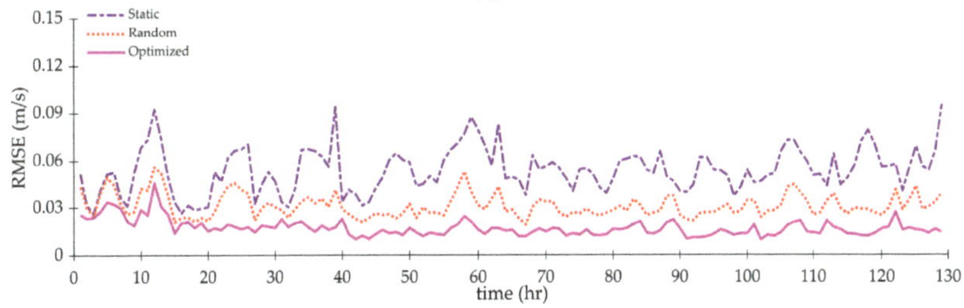

(**b**)

Figure 8. RMSE of the flow field prediction using three different flow sensing strategies. (**a**) RMSE results of the flow speed component along the latitudinal direction. (**b**) RMSE results of the flow speed component along the longitudinal direction.

Table 2. Performance of RMSE for three flow prediction approaches, corresponding to Figure 8.

Strategy	Direction	Maximum Value	Minimum Value	Max Reduction Value	Mean Reduction Value
Static	Latitudinal	0.086	0.030	NA	NA
	Longitudinal	0.101	0.027	NA	NA
Random	Latitudinal	0.052	0.020	0.043	0.019
	Longitudinal	0.056	0.019	0.066	0.023
Optimized	Latitudinal	0.043	0.009	0.067	0.034
	Longitudinal	0.046	0.010	0.081	0.037

The results above highlight the significant impact of using mobile and optimized observations in the flow prediction process. When compared to the fixed observation locations, the RMSE of the prediction results using observations from different spatiotemporal locations is notably reduced. More notably, the accuracy of the prediction results corresponding to the optimal observation locations selected by the optimization method is the most promising.

By optimizing data collection locations, more relevant and informative flow-speed data are provided for the data assimilation process, which, in turn, accelerates the optimization of model state variables and enhances the accuracy of continuous flow field predictions. The simulation study validates the utility of the adaptive sensing approach, demonstrating its potential to improve the overall performance of the hybrid data-driven flow model in predicting flow fields.

5.4. Non-Myopic Optimization with Distributed MCTS

In certain applications of the DDDAS, the main objective may be to provide flow information solely for the AMVs themselves to facilitate tasks such as path planning and cooperative control in dynamic flow fields. These scenarios may not involve onshore stations or external infrastructures for support, and as a result, the DDDAS is directly integrated into the networked AMVs. When implementing the DDDAS on AMVs, the optimization considering multi-step information gain using the MCTS approach on a single AMV can become time-consuming. To achieve timely decision-making, this article proposes a distributed strategy to leverage the collective computational power of AMVs to reduce tree-searching time. This section conducts a simulation study on non-myopic optimization using the distributed MCTS approach.

In the simulation, three underwater gliders ($N_s = 3$) are deployed, and the data assimilation–optimization process occurs at a time step of 6 h. The optimization process involves planning gliders' paths within a fixed time window of 18 h, i.e., 3 time steps, with each step aiming to maximize the weighted sum of predicted information gain for future remaining time steps within the fixed time window. This setup corresponds to the second approach discussed in Section 4.2. Specifically, at time t_k, the optimization aims to maximize the weighted sum of $1/6 \times tr(S_{k+6}) + 2/6 \times tr(S_{k+12}) + 3/6 \times tr(S_{k+18})$, prioritizing flow prediction accuracy at time t_{k+18}. Subsequently, at the time t_{k+6}, the optimization shifts to $1/3 \times tr(S_{k+12}) + 2/3 \times tr(S_{k+18})$, and at time t_{k+12}, the optimization considers $tr(S_{k+18})$. The weights are set accordingly to prioritize improving flow prediction accuracy at the targeted time t_{k+18}. At each time step, the distributed MCTS is utilized to solve for the optimal solution, where three subtrees are allocated to the three gliders. Then the receding horizon strategy is implemented, and this optimization process concludes at time t_{k+18}. This process results in continuous paths over 18 h, facilitating the accurate flow field prediction at time t_{k+18}. Figure 9 shows the simulation result of the paths of three gliders from a simulation run.

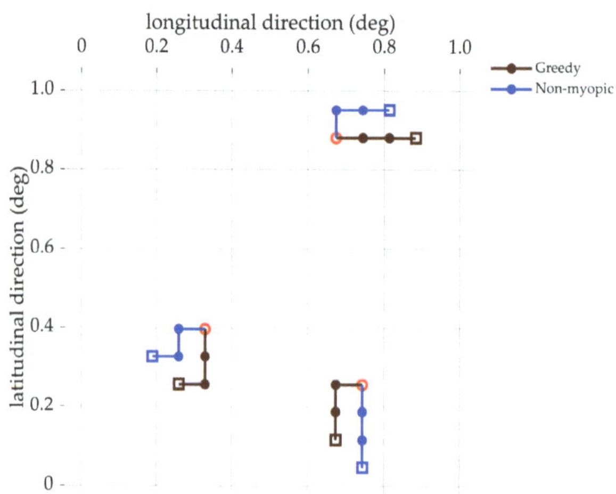

Figure 9. The paths of three underwater gliders correspond to the ID #1 simulation run in Table 3. The circles represent the start locations, the solid dots represent the locations of gliders at the second and the third data assimilation time steps, and the rectangles represent the end locations of paths.

In this simulation, ten simulation runs are performed, each starting from the time t_0. The initial starting locations of the three gliders are randomly set within the operational area. For each run, both the non-myopic optimization and the greedy optimization are performed with the same simulation setup. Figure 9 also shows the simulation result of the paths of three gliders from a greedy optimization. After each simulation run, the $tr(P_0^a - P_{18}^a)$ is calculated for both the cases of optimization considering greedy and non-myopic information gain. The results are presented in Table 3. Additionally, in the last column of Table 3, the percentage improvement in $tr(P_0^a - P_{18}^a)$ achieved through the non-myopic optimization compared to the greedy optimization is also provided.

Table 3. The $tr(P_0^a - P_{18}^a)$ values calculated in ten simulation runs.

ID of Runs	Greedy ($\times10^{-5}$)	Non-Myopic ($\times10^{-5}$)	Improvement
1	5.714	6.519	14.1%
2	5.939	6.751	13.7%
3	4.904	5.378	9.7%
4	5.772	6.346	9.9%
5	4.396	4.835	10.0%
6	4.864	5.391	10.8%
7	4.551	5.074	11.5%
8	4.661	5.059	8.5%
9	4.654	5.198	11.7%
10	5.835	6.759	15.8%

Upon analysis of the results, it can be observed that the non-myopic optimization consistently yields higher $tr(P_0^a - P_{18}^a)$ values compared to the greedy optimization for all ten simulation runs. The improvement percentages range from approximately 9.7% to 15.8%, with an average improvement of around 11.6% across the ten simulation runs. By considering information gains from future time steps, the gliders can better plan their paths and collect more informative data, leading to more accurate flow field predictions at a future targeted time.

The results also validate the effectiveness of the distributed strategy for implementing MCTS, enabling effective decision-making with about 1/3 of computation time compared to the implementation of the MCTS on a single vehicle. This reduction in computation time is crucial for the real-time decision-making of AMVs. It is worth noting that while the distributed MCTS method has demonstrated promising results in this simulation study, there is great potential for further performance improvement. One area of future research work for the authors is to explore adaptively allocating computational budgets to the subtrees, to further enhance the capability of distributed MCTS in obtaining optimal and reliable decisions.

Overall, the consideration of multiple future time steps, the receding horizon optimization, and the distributed MCTS demonstrate a promising approach to optimize the non-myopic paths of AMVs and achieve more accurate flow field predictions for a targeted time.

6. Conclusions

This research article presents a novel data-driven closed-loop dynamic data-driven application system that effectively addresses the challenge of high-resolution and accurate flow field prediction in ocean environments. The DDDAS leverages a hybrid data-driven flow model for trend prediction, along with a Gaussian process model for residual fitting. The assimilation of spatially correlated flow sensing data from AMVs using the Kriged ETKF further enhances model online learning and achieves accurate spatiotemporal flow prediction. The proposed receding horizon strategy and distributed strategy of implementing Monte Carlo tree search is demonstrated to be effective in optimizing AMVs' coordinated flow-sensing paths, enabling non-myopic path planning, and solving large-scale tree searching-based optimization problems in a timely manner. The proposed DDDAS offers a comprehensive and effective solution for achieving high-resolution and accurate flow field prediction in ocean environments using AMV networks.

As future research directions, several potential avenues can be explored. Firstly, investigating the integration of other machine learning models, such as deep neural networks, and data-driven techniques into the DDDAS may lead to further improvements in computational efficiency and prediction accuracy of complex oceanic processes. Additionally, the application of the DDDAS can be extended to incorporate a wider range of ocean data, such as ocean temperature and salinity, to achieve a more comprehensive prediction of ocean environments. Furthermore, optimizing the number of AMVs for environmental sensing is crucial, as it can lead to ocean prediction with the required level of accuracy with a minimum number of AMVs needed. This optimization has the potential to reduce costs and resource requirements for the system, considering the high costs associated with deploying and operating AMVs in ocean environments.

Author Contributions: Conceptualization, Q.J. and Y.T.; methodology, Q.J. and Y.T.; software, Q.J., W.Z. and Q.S.; validation, Q.J. and Y.T.; formal analysis, Q.J. and Y.T.; writing—original draft preparation, Y.T.; writing—review and editing, Y.T.; visualization, Q.J. and Y.T.; supervision, Y.T., J.Y. and X.W.; project administration, Y.T. and J.Y.; funding acquisition, Y.T. and J.Y. All authors have read and agreed to the published version of the manuscript.

Funding: This research was funded in part by the National Natural Science Foundation of China (Grant No. 52271353), the Liaoning Revitalization Talents Program, China (Grant No. XLYC2007035), the Natural Science Foundation of Shenyang, China (Grant No. 22-315-6-08), the Program of the State Key Laboratory of Robotics at Shenyang Institute of Automation, Chinese Academy of Sciences (Grant No. 2022-Z11), the International Partnership Program of the Chinese Academy of Sciences (Grant No. 173321KYSB20180011), the Fundamental Research Program of Shenyang Institute of Automation, Chinese Academy of Sciences (Grant No. 2022JC3K05), and the National Key R&D Program of China (Grant No. 2022YFE0204600).

Institutional Review Board Statement: Not applicable.

Informed Consent Statement: Not applicable.

Data Availability Statement: Not applicable.

Conflicts of Interest: The authors declare no conflict of interest.

References

1. Chen, Z.; Yu, J.; Zhang, A. Overview on observation-oriented unmanned marine vehicles with high cruising ability: Development status and prospect. *J. Ocean Technol.* **2016**, *35*, 122–130.
2. Riser, S.C.; Freeland, H.J.; Roemmich, D.; Wijffels, S.; Troisi, A.; Belbéoch, M.; Gilbert, D.; Xu, J.; Pouliquen, S.; Thresher, A.; et al. Fifteen years of ocean observations with the global Argo array. *Nat. Clim. Chang.* **2016**, *6*, 145–153. [CrossRef]
3. Han, G.; Zhou, Z.; Zhang, T.; Wang, H.; Liu, L.; Peng, Y.; Guizani, M. Ant-Colony-based complete-coverage path-planning algorithm for underwater gliders in ocean areas with thermoclines. *IEEE Trans. Veh. Technol.* **2020**, *69*, 8959–8971. [CrossRef]
4. Petillo, S.; Schmidt, H.; Balasuriya, A. Constructing a distributed AUV network for underwater plume-tracking operations. *Int. J. Distrib. Sens. Netw.* **2012**, *8*, 191235. [CrossRef]
5. Zamuda, A.; Hernández Sosa, J.D. Differential evolution and underwater glider path planning applied to the short-term opportunistic sampling of dynamic mesoscale ocean structures. *Appl. Soft. Comput.* **2014**, *24*, 95–108. [CrossRef]
6. Peng, S.; Zhu, Y.; Li, Z.; Li, Y.; Xie, Q.; Liu, S.; Luo, Y.; Tian, Y.; Yu, J. Improving the real-time marine forecasting of the northern South China Sea by assimilation of glider-observed T/S profiles. *Sci. Rep.* **2019**, *9*, 17845. [CrossRef] [PubMed]
7. Hero, A.O.; Cochran, D. Sensor management: Past, present, and future. *IEEE Sens. J.* **2011**, *11*, 3064–3075. [CrossRef]
8. Zhang, F. Cyber-maritime cycle: Autonomy of marine robots for ocean sensing. *Found. Trends Robot.* **2016**, *5*, 1–115. [CrossRef]
9. Leonard, N.E.; Paley, D.A.; Davis, R.E.; Fratantoni, D.M.; Lekien, F.; Zhang, F. Coordinated control of an underwater glider fleet in an adaptive ocean sampling field experiment in Monterey Bay. *J. Field Robot.* **2010**, *27*, 718–740. [CrossRef]
10. Curtin, T.B.; Bellingham, J.G. Progress toward autonomous ocean sampling networks. *Deep Sea Res. Part II Top. Stud. Oceanogr.* **2009**, *56*, 62–67. [CrossRef]
11. Leonard, N.E.; Paley, D.A.; Lekien, F.; Sepulchre, R.; Fratantoni, D.M.; Davis, R.E. Collective motion, sensor networks, and ocean sampling. *Proc. IEEE* **2007**, *95*, 48–74. [CrossRef]
12. L'Hévéder, B.; Mortier, L.; Testor, P.; Lekien, F. A glider network design study for a synoptic view of the oceanic mesoscale variability. *J. Atmos. Ocean. Technol.* **2013**, *30*, 1472–1493. [CrossRef]
13. Fossum, T.O.; Eidsvik, J.; Ellingsen, I.; Alver, M.O.; Fragoso, G.M.; Johnsen, G.; Mendes, R.; Ludvigsen, M.; Rajan, K. Information-driven robotic sampling in the coastal ocean. *J. Field Robot.* **2018**, *35*, 1101–1121. [CrossRef]
14. Gregório Ramos, P.A. Geostatistical prediction of ocean outfall plume characteristics based on an autonomous underwater vehicle. *Int. J. Adv. Robot. Syst.* **2013**, *10*, 289. [CrossRef]
15. Sun, J.; Liu, S.; Zhang, F.; Song, A.; Yu, J.; Zhang, A. A Kriged compressive sensing approach to reconstruct acoustic fields from measurements collected by underwater vehicles. *IEEE J. Ocean. Eng.* **2021**, *46*, 294–306. [CrossRef]
16. Munafò, A.; Simetti, E.; Turetta, A.; Caiti, A.; Casalino, G. Autonomous underwater vehicle teams for adaptive ocean sampling: A data-driven approach. *Ocean Dyn.* **2011**, *61*, 1981–1994. [CrossRef]
17. Grasso, R.; Braca, P.; Fortunati, S.; Gini, F.; Greco, M.S. Dynamic underwater glider network for environmental field estimation. *IEEE Trans. Aerosp. Electron. Syst.* **2016**, *52*, 379–395. [CrossRef]
18. Yildirim, B.; Chryssostomidis, C.; Karniadakis, G.E. Efficient sensor placement for ocean measurements using low-dimensional concepts. *Ocean Model.* **2009**, *27*, 160–173. [CrossRef]

19. Heuss, J.P.; Haley, P.J.; Mirabito, C.; Coelho, E.; Schönau, M.C.; Heaney, K.; Lermusia, P.F.J. Reduced order modeling for stochastic prediction onboard autonomous platforms at sea. In Proceedings of the Global Oceans 2020: Singapore—U.S. Gulf Coast, Biloxi, MS, USA, 5–30 October 2020.
20. Arcucci, R.; Zhu, J.; Hu, S.; Guo, Y.-K. Deep data assimilation: Integrating deep learning with data assimilation. *Appl. Sci.* **2021**, *11*, 1114. [CrossRef]
21. Yang, Y.; Dong, J.; Sun, X.; Lima, E.; Mu, Q.; Wang, X. A CFCC-LSTM model for sea surface temperature prediction. *IEEE Geosci. Remote Sens. Lett.* **2018**, *15*, 207–211. [CrossRef]
22. Ma, Y.; Mao, Z.; Wang, T.; Qin, J.; Ding, W.; Meng, X. Obstacle avoidance path planning of unmanned submarine vehicle in ocean current environment based on improved firework-ant colony algorithm. *Comput. Electr. Eng.* **2020**, *87*, 106773. [CrossRef]
23. Lolla, T.; Lermusiaux, P.F.J.; Ueckermann, M.P.; Haley, P.J. Time-optimal path planning in dynamic flows using level set equations: Theory and schemes. *Ocean Dyn.* **2014**, *64*, 1373–1397. [CrossRef]
24. Liang, X.; Wu, W.; Chang, D.; Zhang, F. Real-time modelling of tidal current for navigating underwater glider sensing networks. *Procedia Comput. Sci.* **2012**, *10*, 1121–1126. [CrossRef]
25. Chang, D.; Zhang, F.; Edwards, C.R. Real-time guidance of underwater gliders assisted by predictive ocean models. *J. Atmos. Ocean. Technol.* **2015**, *32*, 562–578. [CrossRef]
26. Smith, R.N.; Chao, Y.; Jones, B.H.; Caron, D.A.; Li, P.P.; Sukhatme, G.S. Trajectory design for autonomous underwater vehicles based on ocean model predictions for feature tracking. In *Field and Service Robotics: Results of the 7th International Conference*; Howard, A., Iagnemma, K., Kelly, A., Eds.; Springer: Berlin/Heidelberg, Germany, 2010; Volume 62, pp. 263–273.
27. Lekien, F.; Mortier, L.; Testor, P. Glider coordinated control and Lagrangian coherent structures. *IFAC Proc. Vol.* **2008**, *41*, 125–130. [CrossRef]
28. Chang, D.; Wu, W.; Edwards, C.R.; Zhang, F. Motion tomography: Mapping flow fields using autonomous underwater vehicles. *Int. J. Robot. Res.* **2017**, *36*, 320–336. [CrossRef]
29. Saha, D.; Deo, M.C.; Joseph, S.; Bhargava, K. A combined numerical and neural technique for short term prediction of ocean currents in the Indian Ocean. *Environ. Syst. Res.* **2016**, *5*, 4. [CrossRef]
30. Thongniran, N.; Vateekul, P.; Jitkajornwanich, K.; Lawawirojwong, S.; Srestasathiern, P. Spatio-temporal deep learning for ocean current prediction based on HF radar data. In Proceedings of the 2019 16th International Joint Conference on Computer Science and Software Engineering (JCSSE), Chonburi, Thailand, 10–12 July 2019.
31. Darema, F. Dynamic data driven applications systems: A new paradigm for application simulations and measurements. In *Computational Science—ICCS 2004*; Bubak, M., Albada, G.D., Sloot, P.M.A., Dongarra, J., Eds.; Springer: Berlin/Heidelberg, Germany, 2004; Volume 3038, pp. 662–669.
32. Mousavian, R.; Hossainali, M.M. Detection of main tidal frequencies using least squares harmonic estimation method. *J. Geod. Sci.* **2012**, *2*, 224–233. [CrossRef]
33. Amiri-Simkooei, A.R.; Zaminpardaz, S.; Sharifi, M.A. Extracting tidal frequencies using multivariate harmonic analysis of sea level height time series. *J. Geod.* **2014**, *88*, 975–988. [CrossRef]
34. Mardia, K.V.; Goodall, C.; Redfern, E.J.; Alonso, F.J. The Kriged Kalman filter. *Test* **1998**, *7*, 217–282. [CrossRef]
35. Bishop, C.H.; Etherton, B.J.; Majumdar, S.J. Adaptive sampling with the ensemble transform Kalman filter. Part I: Theoretical aspects. *Mon. Weather Rev.* **2001**, *129*, 420–436. [CrossRef]
36. Tian, Y.; Zhang, A. Development of a guidance and control system for an underwater plume exploring AUV. In Proceedings of the 2010 8th World Congress on Intelligent Control and Automation, Jinan, China, 7–9 July 2010.
37. Browne, C.B.; Powley, E.; Whitehouse, D.; Lucas, S.M.; Cowling, P.I.; Rohlfshagen, P.; Tavener, S.; Perez, D.; Samothrakis, S., Colton, S. A survey of Monte Carlo tree search methods. *IEEE Trans. Comput. Intell. AI Games* **2012**, *4*, 1–43. [CrossRef]
38. Graf, T.; Lorenz, U.; Platzner, M.; Schaefers, L. Parallel Monte Carlo tree search for HPC systems. In *Euro-Par 2011 Parallel Processing*; Jeannot, E., Namyst, R., Roman, J., Eds.; Springer: Berlin/Heidelberg, Germany, 2011; Volume 6853, pp. 365–376.
39. Schaefers, L.; Platzner, M. Distributed Monte Carlo tree search: A novel technique and its application to Computer Go. *IEEE Trans. Comput. Intell. AI Games* **2015**, *7*, 361–374. [CrossRef]
40. Auer, P.; Cesa-Bianchi, N.; Fischer, P. Finite-time analysis of the multiarmed bandit problem. *Mach. Learn.* **2002**, *47*, 235–256. [CrossRef]

Journal of
Marine Science and Engineering

MDPI

Article

Critical Node Identification of Multi-UUV Formation Based on Network Structure Entropy

Yi Chen [1], Lu Liu [1,2,*], Xiaomeng Zhang [1], Wei Qiao [1], Ranzhen Ren [1], Boyu Zhu [1], Lichuan Zhang [1,2], Guang Pan [1,2] and Yang Yu [1,2]

[1] School of Marine Science and Technology, Northwestern Polytechnical University, Xi'an 710072, China; yich@mail.nwpu.edu.cn (Y.C.)
[2] Research & Development Institute of Northwestern Polytechnical University, Shenzhen 518057, China
* Correspondence: liulu12201220@nwpu.edu.cn

Abstract: In order to identify and attack the multi-UUV (unmanned underwater vehicle) groups, this paper proposes a method for identifying the critical nodes of multi-UUV formations. This method helps in combating multi-UUV formations by identifying the key nodes to attack them. Moreover, these multi-UUV formations are considered to have an unknown structure as the research object. Therefore, the network structure of the formation is reconstructed according to its space–time trajectory, and the importance of nodes is determined based on network structure entropy. As for the methodology, firstly, based on the swarm intelligence behavior method, the motion similarity of multi-UUV nodes in the formation is analyzed in pairs; furthermore, the leader–follower relationship and the network structure of the formation are calculated successively. Then, based on this network structure, the importance of the network nodes is further determined by the network structure entropy method. Finally, through simulation and experiments, it is verified that the algorithm can accurately construct the network structure of the unknown multi-UUV formation, and the accuracy of the calculated time delay data reaches 84.6%, and compared with the traditional information entropy algorithm, the ordering of the important nodes obtained by this algorithm is more in line with the underwater formation network.

Keywords: critical node; multi-UUV formation; network structural entropy; formation identification; network reconstruction

Citation: Chen, Y.; Liu, L.; Zhang, X.; Qiao, W.; Ren, R.; Zhu, B.; Zhang, L.; Pan, G.; Yu, Y. Critical Node Identification of Multi-UUV Formation Based on Network Structure Entropy. *J. Mar. Sci. Eng.* **2023**, *11*, 1538. https://doi.org/10.3390/jmse11081538

Academic Editor: Tieshan Li

Received: 7 July 2023
Revised: 29 July 2023
Accepted: 30 July 2023
Published: 1 August 2023

1. Introduction

Compared to a single UUV, multi-UUV formations have the advantages of modularity, high fault tolerance, high efficiency, etc., and they can also complete more challenging work through cooperation between them [1]. While the technology of using multiple UUV formations for coordinated search and exploration operations is becoming more sophisticated, this technology poses a threat to national maritime security. Facing the increasingly complex UUV formation structure, it is of great significance to effectively configure the UUV nodes in different positions according to the position relationship and the importance of UUV nodes in the cluster, save the equipment costs, improve the formation operation efficiency, or strike the important nodes of the enemy's UUV formation to reduce the efficiency of the formation operation at a minimum cost [2].

The coordinated behavior of the underwater formations can be viewed as a form of grouping of collaborative intelligence, which represents the traits of a group intelligence behavior displayed by individuals with simple intelligence through mutual cooperation and organization while maintaining the naturally distributed and self-organizing characteristics [3,4]. In nature, there are several groups of cooperative intelligent behaviors, such as flocks of birds and fish migrating in groups to adapt to air or seawater [5,6]. For instance, using high-precision GPS tracking of pairs of pigeons, Biro et al. found that if the conflict between two birds' directional preferences was small, individuals averaged their routes [7]. This study shows that there is a leadership relationship exists between pairs of pigeons when the directional relationship between them exceeds a certain value. Based on the behavioral characteristics of crowd intelligence, several scholars have carried out a lot of research on this topic. Based on the leadership stability of the flock, Huo et al. designed a control model suitable for heterogeneous formation flight of UAVs, aiming to effectively avoid obstacles in unknown environments [8]. Moreover, Park and Kahng proposed a synchronous leader–follower switching method by observing the migration pattern of birds [9]. Furthermore, for hierarchical leader–follower networks with time-varying layer-to-layer delay, Xu et al. propose a new Hierarchical Event-based Control (HEC) algorithm [10].

With the increase in UUV formation members, a complex network, along with complex interaction relationships and a large number of nodes, has been formed; therefore, it is necessary to further explore the impact of nodes on the entire network and improve the management and the control efficiency of the actual UUV network [11]. Currently, the identification methods of the critical nodes are mainly divided into adjacent node methods and path propagation methods, such as the degree centrality method [12], the local centrality method [13], and the mesocentric method [14]. For instance, Kitsak et al. believed that the location of a node in the center of the network indicates that the node is more critical, and they proposed a K-core decomposition method based on this theory [15]. Moreover, Yu et al. identified the key nodes from the perspective of entropy by using the impact of node clustering coefficient and the number of first and second-order neighbors on the node importance [16]. In addition, Wang et al. proposed a novel community-based method to identify a set of vital nodes for influence maximization in the attributed networks [17]. Finally, Jiang et al. developed an attenuation-based supra-adjacency matrix (ASAM) modeling method to further evaluate the importance of the nodes by calculating the similarity between adjacent layers and the cross-layer networks [18].

For underwater confrontation scenarios, Liu et al. proposed a multi-UUV maneuvering decision-making algorithm for a counter-game with a dynamic target scenario [19]. The algorithm uses interval intuitionistic fuzzy rules to model the game and uses fractional order recurrent neural networks (RNN) to achieve optimal maneuvering strategies for the confrontation. From another point of view, considering the characteristics of large delay of underwater communication, the algorithm proposed in this paper is to reconstruct the network structure of enemy formation based on the time delay from the perspective of identification-strike, and then rank the importance of nodes based on the network, in order to strike the critical nodes of enemy formation to maximize the destruction of enemy combat effectiveness for confrontation. The innovations of this paper mainly include: 1. proposing a network reconstruction algorithm for unknown structure underwater formation, and reconstructing the formation network structure through its spatiotemporal trajectory; 2. based on the formation network, proposing a critical node identification algorithm with comprehensive importance network structure entropy to analyze the importance of each node in the multi-UUV formation.

To sum up, this paper is divided as follows: in Section 2, the formation network reconstruction is provided whereas the key node identification algorithm is presented in Section 3. As for Section 4, it represents the results and the discussion and finally, the conclusion and some future works are proposed in Section 5.

2. Formation Network Reconstruction

2.1. Leader–Follower Relationship

According to the analysis method of group intelligence behavior, the formation with group intelligence has some similarity with the leader in behavioral actions, and there is a hierarchical relationship characterized by the time delay between individuals [20]. For example, in a pigeon flock, the follower pigeons will observe the movements of the leader pigeons visually and make corresponding movements to maintain the consistency of the formation. In the underwater formation, other UUVs follow the trajectory of the leader UUV and keep the relative angle and distance to stabilize the formation, i.e., there are also motion similarities between individuals and leaders in the group. In reverse analysis, the leader–follower relation between UUVs in underwater formations can be derived based on their motion similarity analysis. The spatiotemporal trajectory data of UUV nodes is analyzed in arbitrary pairs by using the correlation function, and the obtained correlation coefficient can be used to characterize the motion similarity between the paired UUVs. Moreover, when the corresponding motion correlation coefficient remains at a fairly high value at any given moment, it can be regarded as the behavior of one UUV being "inherited" by another one, that is, it is considered that there is a leader–follower relation between the UUVs. Among them, the "inherited" behavior is to follow the UUV whereas the other is to pilot the UUV. To quantify this link, the spatiotemporal trajectory function of any UUV in the formation and other UUVs is analyzed to determine the motion correlation, and the motion similarity function is established as follows:

$$R_{ij}^{t}(\tau) = dot\left(\vec{u}_i(t), \vec{u}_j(t+\tau)\right) \tag{1}$$

where $\vec{u}_i(t)$ represents the UUV$_i$'s normalized speed at moment t, $\vec{u}_j(t+\tau)$ represents the UUV$_j$'s normalized speed at moment $t+\tau$, τ is time delay which is a variable, *dot* is the inner product operator sign, and $R_{ij}^{t}(\tau)$ is the function of the motion correlation coefficient between UUV$_i$ and UUV$_j$ at time t but for different delay times.

In addition, we set a threshold R_T: when the motion correlation coefficient at a certain point in time is bigger than the threshold (e.g., $R_{ij}^{t}(\tau) > R_T$), it is believed that there is a leader–follower relation between the two UUVs at this moment. Therefore, in order to better determine the appropriate delay time, in the actual calculation process, individual motion correlation coefficients, having smaller values than the threshold, and mainly caused by the instability of the data or the error of acquiring the data, are used. Thus, the average motion correlation coefficient is determined as follows:

$$R_{ij}(\tau) = \frac{1}{m_t - 1}\left(\sum_{t=1}^{m_t-1} R_{ij}^{t}(\tau)\right) \tag{2}$$

where m_t is the number of spatiotemporal trajectories, $t \in [1, m_t - 1]$, that is, a total of $m_t - 1$ motion correlation coefficients are generated in m_t trajectory points at time delay τ. $R_{ij}(\tau)$ is the average motion correlation coefficient between the UUV$_i$ and UUV$_j$ at different delay times τ. Moreover, the similarity of motion between the pairs of UUV is determined.

By establishing the above similar motion model, the correlation coefficient $R_{ij}(\tau)$ relative to the action between the paired UUVs under different time delays τ is obtained, and the threshold to R_T is set. Moreover, if $R_{ij}(\tau) > R_T$ is obtained for any value of τ, it is considered that there is a leader–follower relationship between the paired vehicle; therefore, the delay time τ resulting in the maximum value of $R_{ij}(\tau)$ is defined to be the relevant time delay between the paired UUV$_i$ and UUV$_j$, and it will be denoted by τ_{ij}^{*}. Thus, the next step consists of setting the time delay matrix $T_{n\times n} = \left(\tau_{ij}^{*}\right)$ to represent the delay relationship between the formation UUVs: when τ_{ij}^{*} is positive, it means that the navigation direction of UUV$_i$ is ahead of the UUV$_j$, that is, UUV$_i$ is the leader, and UUV$_j$ is the following UUV. However, if τ_{ij}^{*} is negative, the roles of the UUVs are the opposite.

2.2. Formation Hierarchy

Limited by the narrow bandwidth of the underwater acoustic channel and the UUV formation method, the UUV formation movement mostly adopts a hierarchical interaction structure [21]. Moreover, it is faster and more efficient in navigation and command execution than in the equal interaction structure. While using hierarchical interaction in the underwater unmanned cluster formation, there is a hierarchical structure relationship between UUVs, and the higher the level of UUVs, the better its control and the greater the importance in the formation network will be.

Based on the similar motion model, the leader–follower relationship between all UUVs and other UUVs in the formation is obtained and the hierarchical relationship of the entire network is analyzed. In the leader–following model, the movement command is issued according to the hierarchy, and the UUV at the top of the hierarchy has a certain leadership relationship with the other UUVs. However, when both UUV_i and UUV_j of the previous level have a leadership relationship with the UUV_k pair of the next level, returning to the judgment of the time delay τ^* is performed: if $\tau^*_{ji} > \tau^*_{ki} > 0$, it Is considered that it has a leadership following relationship of the UUV with a smaller delay, that is, UUV_k has a leadership relationship with UUV_j.

2.3. Network Weight Matrix

In the network topology diagram, there is a certain weight coefficient between the two nodes to characterize the location proximity relationship between them. For example, when vertices represent some physical locations, the weight of the edge between two vertices can be set to the actual distance. During network formation, the distance between the different UUVs reflects the proximity relationship between the nodes, and here the normalized distance is used to represent the weight of the connected edges of nodes, usually the closer the distance the more reliable the interaction between two nodes is, and the weight of the connected edges is considered to be higher. Moreover, the distance d_{ij} between UUV_i and UUV_j is normalized as one of the factors affecting the weights on the edges; thus, one can get the following equation:

$$D_{ij} = \frac{d_{ij} - \min\limits_{i,j}(d_{ij})}{\max\limits_{i,j}(d_{ij}) - \min\limits_{i,j}(d_{ij})} \quad (3)$$

where D_{ij} is the result of normalized distance d_{ij}. In addition, considering that the correlation between the different UUVs is related to the motion similarity, higher motion similarity implies higher inheritance, and the more important the connected edge is considered to be. Thus, the average motion correlation coefficient $R_{ij}(\tau^*)$ of UUV_i and UUV_j at the time delay τ^*_{ij} is introduced into the weights on the edges of both nodes i and j in the network topology. It is then combined with the above-normalized distance in order to obtain the weight on the edge formed by both nodes:

$$w_{ij} = \frac{1}{D_{ij}} + R_{ij}(\tau^*) \quad (4)$$

2.4. Mobile Formation Network Structure

According to the above steps, the leader–follower relationship, the hierarchical structure, and the weights on the edges of two nodes in the network were obtained. Therefore, the adjacency matrix $A_{n\times n} = (a_{ij})$, pointed to the network nodes, and the weight matrix $W_{n\times n} = (\omega_{ij})$ representing the weights of the nodes, was established. In this matrix, if there is a leader–following relationship between nodes, the value of the cell will be equal to one (e.g., $a_{ij} = 1$), and vice versa, the absence of a relation yields a null value (e.g., $a_{ij} = 0$). Based on these definitions, we attained a map of the network topology of the mobile UUV formation. The following guidelines are made when creating the network topology diagram, though, in order to be more in line with the characteristics of the underwater cluster formation. This is because underwater communication has a distance limit, each UUV in the unmanned cluster formation has a different task, and the sensors carried on the boat are also different. Information is only dealt with layer by layer across neighboring levels; each UUV only receives instructions provided by one UUV, but can issue instructions to several UUVs. Thus, through regular filtering, the network structure reconstruction of the mobile UUV formation is completed.

3. Key node Identification Algorithm

3.1. Network Structure Model

In order to represent the connection between the different individuals in the mobile UUV formation in a more intuitive and clear way, the use of diagrams, where the nodes represent UUVs and the edges represent the interconnections between UUVs, was the applied solution. This network structure model is usually expressed as $G = (V, E, W)$, where $V = \{v_1, v_2, \cdots, v_n\}$ is the set of network nodes and $n = |V|$ is the total number of nodes in the network. Moreover, $E = \{e_1, e_2, \cdots, e_m\}$ is the set of edges between nodes, and $m = |E|$ is the total number of edges in the network. Finally, $W = (w_{ij})_{N\times N}$ represents the weight matrix where w_{ij} represents the weight value on the edge of nodes i and j, and generally has $w_{ij} \neq w_{ji}$ in directed networks. As for a weighted network, it can be thought of as a weighted network with all the weight values of 1. Finally, it is important to mention that there are four basic types of networks: undirected networks, weighted undirected networks, unweighted directed networks, and weighted directed networks [22]. All types are shown in Figure 1:

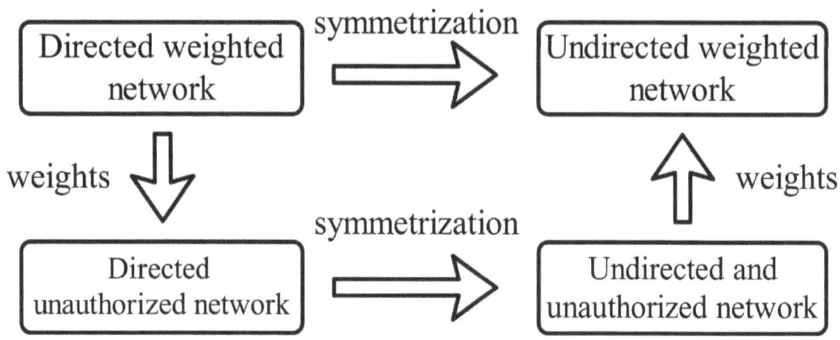

Figure 1. Four network types.

3.2. Network Structure Entropy

In the network topology, the performance of the network scalelessness is considered a kind of network "heterogeneity", and this "heterogeneity" of complex networks can be described using the concept of "entropy", that is, the entropy of the network structure [23].

In order to better establish the entropy model based on the network structure, the following keywords have been defined:

(1) Degree value. The degree value of a node is called the node strength, and the degree k_i of node v_i is defined as the number of nodes directly connected to the node v_i. Moreover, k_i is expressed using the following relation:

$$k_i = \sum_{j\in \Gamma_i}^{n} v_{ij} \quad (5)$$

where Γ_i is the collection of neighbor nodes of the particle node v_i. As for the weighted network,

$$k_i = \sum_{j\in \Gamma_i} w_{ij} \quad (6)$$

where w_{ij} is the weight on the edge connecting node v_i to node v_j. In a directed network, the degree value of a node is divided into an outdegree value and an in-degree value, and it is generally believed that both values have different effects on the node, that is

$$k_i = \lambda k_i^{in} + (1-\lambda)k_i^{out} \quad (7)$$

where k_i^{in} and k_i^{out} are the indegree and outdegree values of node v_i, respectively, and λ is the influence coefficient. When $\lambda > 0.5$, the relevance of the node is thought to be more influenced by its input strength.

(2) Adjacency. To more accurately reflect the impact of a node on its connected neighbor nodes, the adjacency of a node is defined as follows:

$$Q_i = \sum_{w\in \Gamma_i} k_{iw} + \sum_{w\in \Gamma_i} k_{wi} \tag{8}$$

where k_{iw} and k_{wi} are the degree value of the node pointing to v_i and the node pointed by v_i in the neighboring nodes of node v_i, Γ_i is the set of neighbor nodes of node v_i, and Q_i is the degree of adjacency of node v_i (the greater the value of Q_i is, the higher its impact on neighboring nodes will be).

(3) Importance. The nodes in the network affect each other, and considering only the degree value will lose the influence of indirect neighbors on the nodes, and considering the global nodes will increase the complexity of the algorithm, and the effect may not be very good. The influence of a node is limited, and it only has a large influence on its nearby neighbors. The probability function is used to describe the chance to select a given node among its neighbors, which is defined as:

$$p_i = \frac{k_i}{Q_j}, j \in \Gamma_i \tag{9}$$

In the entropy of network structures based on the node degree, the probability functions can be used to express the importance of nodes. However, in underwater mobile formations, often the higher the level of UUV is, the greater its importance will be. Therefore, considering the importance of the degree and level of nodes, the comprehensive importance is introduced to express the importance of the network nodes. Considering that the control commands and the information transmission in the formation are carried out layer by layer, the high-level UUV will have an impact on the low-level UUV; therefore, for an N-level network, the importance of the hierarchical nodes should be continuously reduced, and the weight factor δ_i of the nodes on the level n is expressed as follows:

$$\delta_i = \frac{1}{N}(N - n + 1) \tag{10}$$

The comprehensive importance of the node is calculated as follows:

$$H_i = p_i \cdot \delta_i = \frac{k_i}{Q_j} \cdot \frac{1}{N}(N - n + 1) \tag{11}$$

(4) Network structure entropy. Information entropy uses probabilistic and statistical methods to measure the complexity of a system, which represents the expectation of the amount of information brought by all possible events, and it can be used as well to measure the importance of network nodes. Consider the unrelated events x and y to be equal to the sum of the information obtained when the observed events occur at the same time, that is $h(x,y) = h(x) + h(y)$, and $p(x,y) = p(x) \times p(y)$. Therefore, it can be obtained that $h(x)$ must be related to the logarithm of $p(x)$. Thus, the relation between both variables can be written as follows:

$$h(x) = -\log_2 p(x) \tag{12}$$

Moreover, the expected amount of information is defined as follows:

$$E = E(h(x)) = -\sum_{i=1}^{n} p(x) \log_2 p(x) \tag{13}$$

Make sure, at this level, that the information entropy value is always positive, take the absolute value of the node information entropy when calculating it, and replace the probability function with the comprehensive importance H_i to obtain the entropy E_i of the collar network structure. Therefore, the resulting equation is as follows:

$$E_i = \sum_{j \in \Gamma_i} |(-H_i \log_2 H_i)| \tag{14}$$

3.3. Critical Node Identification Algorithm

By analyzing the interaction between the nodes and their indirect nodes, the entropy of the network structure is used to measure the importance of several nodes in the network. Considering that in the formation network structure, the instruction is transmitted from a high level to a low-level UUV, the node importance of the directed network is only analyzed; therefore, it is considered that the entry value of the node is smaller than the influence of the degree value on the node, and the impact factor is $\lambda = 0.45$. In more detail, the milestone algorithm steps are presented as follows:

Step 1: Analyze the formation network according to the mobile UUV formation spatiotemporal trajectory meter and get the adjacency matrix A and the weight matrix $\mathbf{W}$.

Step 2: Calculate the node degree value according to the difference between the node outdegree and indegree:

$$k_i^{\text{in}} = \sum_{j \in \Gamma_i} w_{ji} \tag{15}$$

$$k_i^{out} = \sum_{j \in \Gamma_i} w_{ij} \tag{16}$$

$$k_i = \lambda k_i^{in} + (1-\lambda) k_i^{out} \tag{17}$$

Step 3: Calculate the degree of adjacency:

$$Q_j = \lambda \sum_{w \in \Gamma_j} k_{wj} + (1-\lambda) \sum_{w \in \Gamma_j} k_{wj} \tag{18}$$

Step 4: Calculate the overall importance:

$$p_i = \frac{k_i}{Q_j}, j \in \Gamma_i \tag{19}$$

$$H_i = p_i \cdot \delta_i \tag{20}$$

Step 5: Calculate the entropy of the network structure:

$$E_i = \sum_{j \in \Gamma_i} |(-H_i \log_2 H_i)| \tag{21}$$

The calculation process is shown in Figure 2. Based on the above steps, the network structure entropy of each node in the network can be calculated. According to the size of entropy, each node is ordered, and the node entropy value is classified from large to small corresponding to the importance of this node.

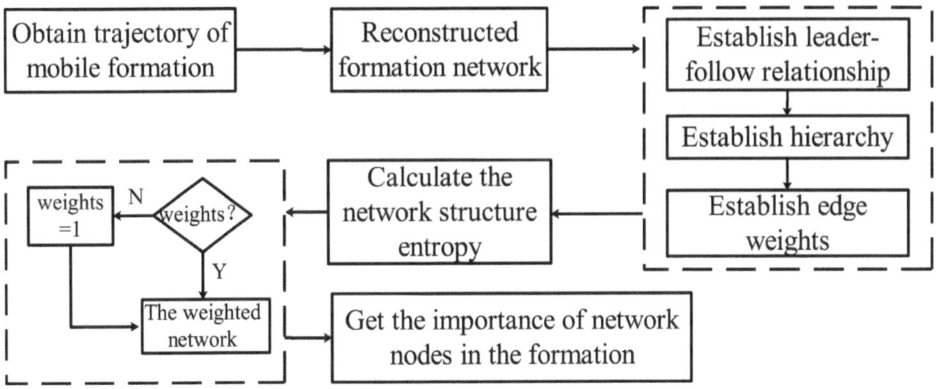

Figure 2. Algorithm flow of critical node identification.

4. Validation and Analysis

The identification of the key nodes of the mobile UUV formation is to establish the network topology of the mobile formation by analyzing spatiotemporal trajectories in order to further rank the importance of the nodes by the key node identification algorithm. Therefore, this section sets up the simulation experiments and the lake experiments to verify the efficiency of the proposed algorithm.

4.1. Simulation and Experiments Analysis

In order to verify the effectiveness of the key node identification algorithm based on the entropy of the network structure, this section uses Matlab© to perform the simulation experiments. Based on the leader–follower formation control model, we considered one leader and seven followers to navigate a "U" trajectory in a triangular formation to verify the discrimination effect of the algorithm. In this simulation, the distance matrix d and delay matrix T are set as follows. (In the matrix d A W, the element 0 indicates that the two nodes are not directly related and have no real physical significance):

$$d = \begin{bmatrix} 0 & 10 & 10 & 0 & 0 & 0 & 0 & 0 \\ 10 & 0 & 0 & 0 & 6 & 0 & 6 & 0 \\ 10 & 0 & 0 & 6 & 0 & 6 & 0 & 0 \\ 0 & 0 & 6 & 0 & 0 & 0 & 0 & 5 \\ 0 & 6 & 0 & 0 & 0 & 0 & 0 & 0 \\ 0 & 0 & 6 & 0 & 0 & 0 & 0 & 0 \\ 0 & 6 & 0 & 0 & 0 & 0 & 0 & 0 \\ 0 & 0 & 0 & 5 & 0 & 0 & 0 & 0 \end{bmatrix}, T = \begin{bmatrix} 0 & 1 & 2 & 3 & 3 & 3 & 3 & 4 \\ -1 & 0 & 0 & 1 & 2 & 1 & 2 & 3 \\ -2 & 0 & 0 & 1 & 1 & 1 & 1 & 2 \\ -3 & -1 & -1 & 0 & 0 & 0 & 0 & 1 \\ -3 & -2 & -1 & 0 & 0 & 0 & 0 & 1 \\ -3 & 1 & -1 & 0 & 0 & 0 & 0 & 1 \\ -3 & -2 & -1 & 0 & 0 & 0 & 0 & 1 \\ -4 & -3 & -1 & -1 & -1 & -1 & 1 & 0 \end{bmatrix} \quad (22)$$

In addition, we added the trajectory of an unrelated UUV to the formation in the simulation to compare and judge the effects of this additional feature. Therefore, the simulation results are displayed in Figure 3.

Figure 3 shows that the follower trajectories the leader well at a predetermined angle and distance based on the influence of the controller. Then, a pairwise analysis was performed on the spatiotemporal trajectories of all UUVs using the aforementioned motion similarity model. The delay matrix and the motion correlation coefficient (refer to Table 1) of the UUV formation are calculated as follows:

$$T_U = \begin{bmatrix} 0 & 1 & 2 & 3 & 3 & 3 & 3 & 4 & 6 \\ -1 & 0 & 0 & 2 & 2 & 1 & 2 & 3 & 6 \\ -2 & 0 & 0 & 1 & 2 & 1 & 2 & 3 & 6 \\ -3 & -2 & -1 & 0 & 0 & 0 & 0 & 1 & 6 \\ -3 & -2 & -2 & 0 & 0 & 0 & 0 & 1 & 6 \\ -3 & -1 & -1 & 0 & 0 & 0 & 0 & 2 & 6 \\ -3 & -2 & -2 & 0 & 0 & 0 & 0 & 1 & 6 \\ -4 & -3 & -3 & -1 & -1 & -2 & -1 & 0 & 6 \\ -6 & -6 & -6 & -6 & -6 & -6 & -6 & -6 & 0 \end{bmatrix} \quad (23)$$

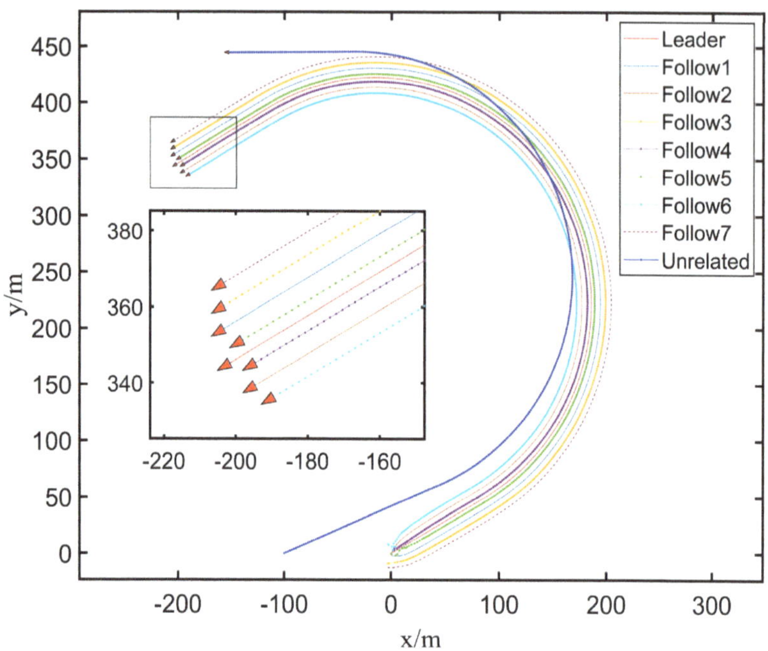

Figure 3. "U"−shaped formation trajectory. It shows the trajectories of each node navigating in formation in the simulation, and the enlarged portion of the figure shows the structure of the formation.

By comparing the time delay matrix T and T_U, the accuracy of the time delay data obtained by the algorithm was 84.6%, and in addition, the erroneous time delay data did not appear between nodes with the direct leader–follower relationships, and the erroneous data did not affect the accuracy of the subsequent reconstruction of the formation network structure. Based on the above delay matrix, the time delay between UUVs with a leader–following relationship represents an antisymmetric transfer, and the positive and negative delays indicate whether the UUV is following or being followed. Moreover, these values also specified the delay time for the follower to receive the leader's movement information and take action, which is in line with the law of following a relationship. When the time delay was null, it means that there was no leader–following relationship between the paired UUVs.

Referring to Table 1, the correlation coefficient corresponds to the time delay, and the motion correlation coefficient between the UUVs is varying at different time delays, as shown in Figure 4. Since the leader following the model does not introduce errors, such as propulsion, hydroacoustic delay, and complex environmental interference, the motion correlation coefficient between each UUV at the corresponding delay time was very large; however, when being compared to the motion correlation coefficient of the unrelated-UUV, there was a significant gap, and the unrelated UUV's motion correlation coefficient was much smaller than others. Setting the threshold value $R_{\min} = 0.9$, and as $\forall R_{unretaed}(\tau^*) < R_{\min}$, the correlation coefficient of the motion of the unrelated UUV and any other UUV was less than the threshold; therefore, it is considered that the unrelated UUV does not have a leader–follower relationship with any other UUV.

Table 1. "U" shape trajectory motion correlation coefficient.

Leader \ Follower	Leader	Follow1	Follow2	Follow3	Follow4	Follow5	Follow6	Follow7	Unrelated
Leader	1	1.0000	1.0000	1.0000	1.0000	1.0000	1.0000	1.0000	0.8271
Follow1	1.0000	1	1.0000	1.0000	1.0000	1.0000	1.0000	1.0000	0.8347
Follow2	1.0000	1.0000	1	1.0000	1.0000	1.0000	1.0000	1.0000	0.8363
Follow3	1.0000	1.0000	1.0000	1	1.0000	1.0000	1.0000	1.0000	0.8438
Follow4	1.0000	1.0000	1.0000	1.0000	1	1.0000	1.0000	1.0000	0.8444
Follow5	1.0000	1.0000	1.0000	1.0000	1.0000	1	1.0000	1.0000	0.8424
Follow6	1.0000	1.0000	1.0000	1.0000	1.0000	1.0000	1	1.0000	0.8448
Follow7	1.0000	1.0000	1.0000	1.0000	1.0000	1.0000	1.0000	1	0.8507
Unrelated	0.8250	0.8326	0.8341	0.8416	0.8423	0.8402	0.8427	0.8486	1

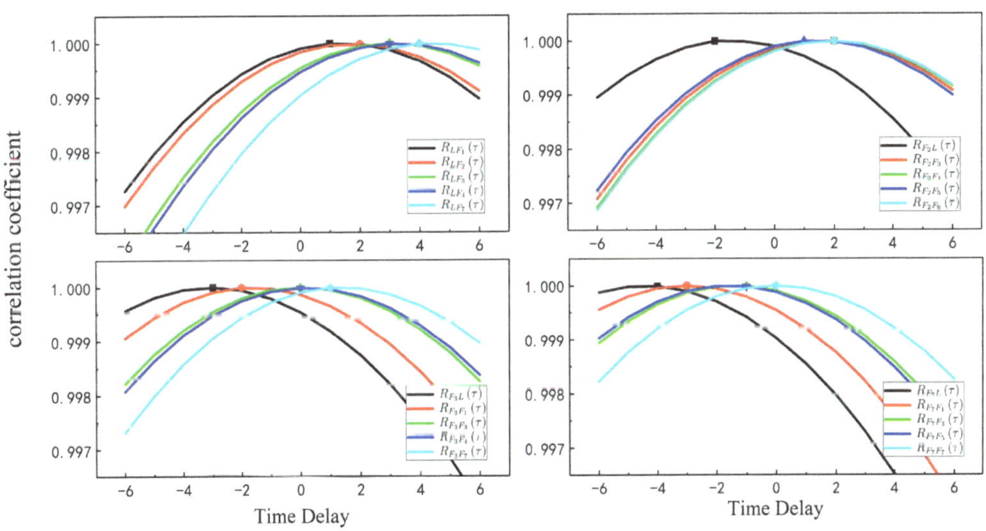

Figure 4. Correlation coefficient and time delay function.

Referring to the T_U matrix Equation (23), and according to the size of the delay, the leader–following relationship of UUVs between pairs was judged, and it was sorted according to the leader–follower level, and the following points were calculated: all followers had a following relationship for the leader, whereas Follow1 and Follow2 had a leadership relationship for the remaining UUVs, and Follow3, Follow4, Follow5, and Follow6 had a leadership relationship for Follow7.

Therefore, the formation network hierarchy was obtained: the UUV leader belonged to the first level, Follow1, Follow2 were part of the second level, Follow3, Follow4, Follow5, and Follow6 belonged to the third level, and, finally, Follow7 belonged to the fourth level. The network relationship, obtained through the above analysis, was still relatively complex where one UUV had a following relationship with multiple UUVs at the same time. Considering the communication restrictions of the underwater formation, etc., it was considered to have a following relationship with the nearest vehicle. Through the analysis of the network again using this rule, the network structure of the formation can be obtained, as shown in Figure 5.

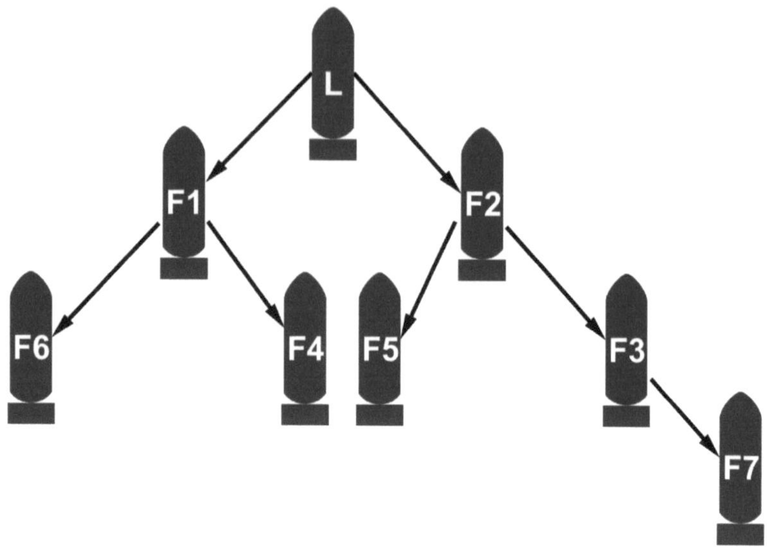

Figure 5. Mobile UUV formation network structure. The letters L and F stand for leader UUV and follower UUV; F1 denotes Follow1 UUV and F2 denotes Follow2 UUV and so on. The arrows indicate the direction of information transmission in the formation network.

Therefore, the collar matrix A of the formation network was obtained as follows, and, when being combined with the behavioral correlation coefficient matrix, it was brought into the established node edge weight coefficient model; hence, the weight coefficient matrix W of the formation network can be expressed as follows:

$$A = \begin{bmatrix} 0 & 1 & 1 & 0 & 0 & 0 & 0 & 0 \\ 1 & 0 & 0 & 0 & 1 & 0 & 1 & 0 \\ 1 & 0 & 0 & 1 & 0 & 1 & 0 & 0 \\ 0 & 0 & 1 & 0 & 0 & 0 & 0 & 1 \\ 0 & 1 & 0 & 0 & 0 & 0 & 0 & 0 \\ 0 & 0 & 1 & 0 & 0 & 0 & 0 & 0 \\ 0 & 1 & 0 & 0 & 0 & 0 & 0 & 0 \\ 0 & 0 & 0 & 1 & 0 & 0 & 0 & 0 \end{bmatrix} \tag{24}$$

$$W = \begin{bmatrix} 0 & 2 & 2 & 0 & 0 & 0 & 0 & 0 \\ 2 & 0 & 0 & 0 & 1.2 & 0 & 1.2 & 0 \\ 2 & 0 & 0 & 1.2 & 0 & 1.2 & 0 & 0 \\ 0 & 0 & 1.2 & 0 & 0 & 0 & 0 & 1 \\ 0 & 1.2 & 0 & 0 & 0 & 0 & 0 & 0 \\ 0 & 0 & 1.2 & 0 & 0 & 0 & 0 & 0 \\ 0 & 1.2 & 0 & 0 & 0 & 0 & 0 & 0 \\ 0 & 0 & 0 & 1 & 0 & 0 & 0 & 0 \end{bmatrix} \tag{25}$$

According to the formation network structure and its corresponding collar relationship and weight coefficient matrices, the node importance is calculated by using the network structure entropy model. Firstly, according to the adjacency and the weight matrices of the network, the input intensity value k_i^{in} and the output intensity value k_i^{out} of each node are calculated. Moreover, according to Equation (17), the comprehensive strength value of each node is calculated to get Table 2.

Table 2. Node comprehensive strength value.

Node	Leader	Follow1	Follow2	Follow3	Follow4	Follow5	Follow6	Follow7
k_i	0.6	2.06	2.06	1.17	1.02	1.02	1.02	0.85

Then, also taking $\lambda = 0.45$, according to Equation (18), the comprehensive adjacency strength value q of the node is calculated as shown in Table 3.

Table 3. Node adjacency strength value.

Node	Leader	Follow1	Follow2	Follow3	Follow4	Follow5	Follow6	Follow7
Q_i	0.9	2	2.15	2.7	2.55	2.55	2.55	1.7

Finally, according to Equation (21), the entropy of each network structure is calculated, and the following results are presented in Table 4.

Table 4. Node network structure entropy.

Node	Leader	Follow1	Follow2	Follow3	Follow4	Follow5	Follow6	Follow7
E_i	2.622	5.226	4.152	1.365	0.486	0.456	0.486	0.367

Based on the entropy of the network structure calculated above, the followers can be arranged using the following order: Follow2 > Follow1 > Leader > Follow3 > Follow5 > Follow4 = Follow6 > Follow7. According to the network structure, the relevance of the Follow1 and Follow2 nodes was higher, since Follow1 and Follow2 regulated information relative to the input and output flows. Moreover, Follow3 controlled Follow7; therefore, it was more critical than other followers of the same level. Finally, Follow4, Follow5, and Follow6 were all considered as edge nodes of the formation network; thus, their information entropy was basically the same, and this was conforming to the network structure law.

By using the traditional information entropy algorithm [24], the information entropy of each node under the network result was calculated as shown in Table 5.

Table 5. Node information entropy.

Node	Leader	Follow1	Follow2	Follow3	Follow4	Follow5	Follow6	Follow7
value	5.674	20.372	17.374	9.970	0.303	0.415	0.303	0.255

In the above table, the order of the information entropy of the nodes is: Follower1 > Follower2 > Follower3 > Leader > Follower5 > Follower4 = Follower6 > Follower7, but we believe that the node leader was more important than the node Follower3, and the node Follower2 was more important than the node Follower1. Compared with the network structure entropy results in Table 4, it can be obtained that the improved algorithm in this paper was more in line with the actual situation in ordering the important nodes of the underwater network structure than the traditional information entropy algorithm.

4.2. Lake Experiments and Analysis

In order to verify the effectiveness of critical node identification of the multi-UUV formation algorithm proposed in this paper, the lake formation experiment was carried out, and the real dead reckoning data were obtained to place the detection of UUV formations.

Therefore, three vehicles set up the trajectory of vehicles, not linked to the formation navigation, while sailing in linear and triangular formations on the Qiandao Lake in Hangzhou City as the test location in order to confirm that the algorithm can successfully recognize different network structures. Figure 6a shows the experiment platform and Figure 6b shows UUV formation sailing on the water. Two sets of trajectory points recorded by the UUV itself are shown in Figure 7:

(a)

(b)

Figure 6. Photos of lake experiments. (**a**) Experiment platform, (**b**) UUV formation navigation.

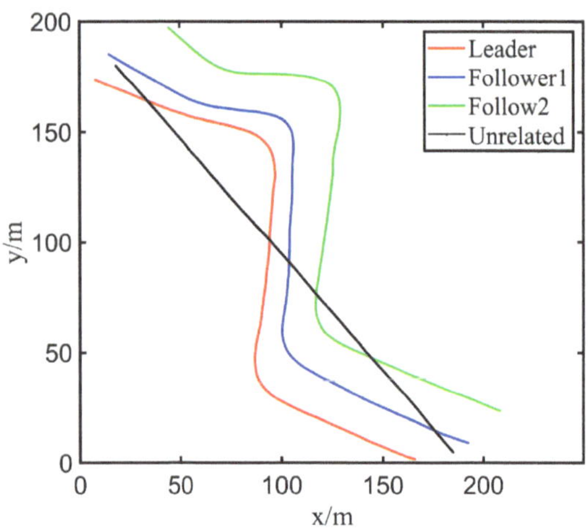

(a)

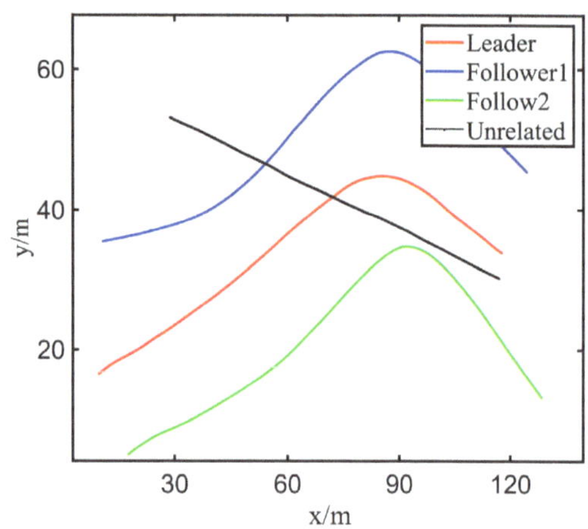

(b)

Figure 7. UUV formation waypoints in two groups of different formations. (**a**) Liner shape navigation trajectory, (**b**) triangular navigation trajectory.

Referring to the paths of Figure 7, the follower has several degrees of error with respect to the leader's trajectory, but the trajectory it follows has generally the same shape. The space–time UUV trajectories were substituted every two pairs with the motion similarity model to obtain the following motion correlation coefficient (refer to Equation (25)) and time delay tables (refer to Tables 6 and 7):

Table 6. Liner shape motion correlation coefficient.

Leader \ Follower	Leader	Follow1	Follow2	Unrelated
Leader	1	0.995	0.987	0.769
Follow1	0.995	1	0.990	0.748
Follow2	0.987	0.990	1	0.788
Unrelated	0.767	0.742	0.788	1

Table 7. Triangular motion correlation coefficient.

Leader \ Follower	Leader	Follow1	Follow2	Unrelated
Leader	1	0.996	0.997	0.883
Follow1	0.997	1	0.999	0.861
Follow2	0.997	0.999	1	0.858
Unrelated	0.856	0.839	0.828	1

In the above motion correlation coefficients table, the average motion similarity between the leaders and the followers was high; however, the motion similarity coefficients of spatiotemporal trajectories of unrelated UUV and other UUVs were quite different compared to others. Setting the threshold $R_{\min} = 0.9$, the similarity coefficient of the motion between the unrelated UUV and any other UUV was below the threshold; therefore, it is considered that there is no leader–follower relationship with any other UUV.

$$T_{line} = \begin{bmatrix} 0 & 3 & 5 & 6 \\ -3 & 0 & 3 & 6 \\ -5 & -3 & 0 & -6 \\ -6 & -6 & -6 & 0 \end{bmatrix}, T_{tri} = \begin{bmatrix} 0 & 2 & 1 & 5 \\ -2 & 0 & 0 & 5 \\ -1 & 0 & 0 & 5 \\ 5 & 5 & 5 & 0 \end{bmatrix} \quad (26)$$

In the above time delay matrix, when the formation was carried out in a liner shape, if the UUV leader is the leader, the time delay between it and Follow1 and Follow2 is greater than zero. However, when Follow1 was the leader, the time delay between it and the UUV leader was less than zero, and the time delay with Follow2 was greater than zero. Therefore, the UUV leader had the leadership relationships for Follow1 and Follow2, and Follow1 also had leadership relationships for Follow2, which resulted in the structural relationship shown in Figure 8a. When moving in a triangle, the UUV leader had a leadership relationship with Follow1 and Follow2. Regardless of whether Follow1 or Follow2 were leaders or followers, the time delay between them was equal to zero, that is, there was no leader following relationship, belonging to the same level, and the structural relationship, shown in Figure 8b, can be obtained, which was in line with the experimental setting.

Thus, the adjacency matrix A_1 and A_2 of the formation network was obtained:

$$A_1 = \begin{bmatrix} 0 & 1 & 0 \\ 1 & 0 & 1 \\ 0 & 1 & 0 \end{bmatrix}, A_2 = \begin{bmatrix} 0 & 1 & 1 \\ 1 & 0 & 0 \\ 1 & 0 & 0 \end{bmatrix} \quad (27)$$

In this experiment, the distances between the vehicles were all the same, and, combined with the motion similarity coefficient, the weight matrices under the two formations can be obtained:

$$W_1 = \begin{bmatrix} 0 & 1.005 & 0 \\ 1.005 & 0 & 1.01 \\ 0 & 1.01 & 0 \end{bmatrix}, W_2 = \begin{bmatrix} 0 & 1.004 & 1.003 \\ 1.004 & 0 & 0 \\ 1.003 & 0 & 0 \end{bmatrix} \quad (28)$$

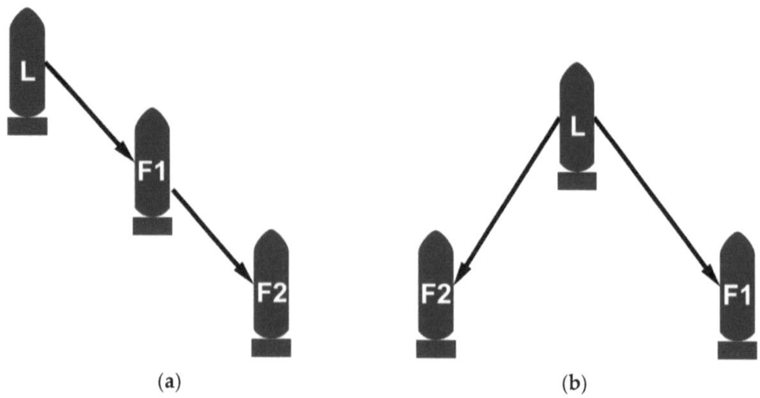

Figure 8. Multi-UUV formation. (**a**) Linear shape formation structure, (**b**) Triangular shape formation structure. The letters L and F stand for leader UUV and follower UUV; F1 denotes Follow1 UUV and F2 denotes Follow2 UUV. The arrows indicate the direction of information transmission in the formation network.

According to the weight coefficient matrix, we found out that the difference between the weight coefficients was very small, so the directed weightless network structure entropy algorithm was applied to calculate the importance between the nodes.

According to the Equation (17), the comprehensive strength value of each node in both networks is calculated to get Tables 8 and 9.

Table 8. Liner shape node comprehensive strength value.

Node	Leader	Follow1	Follow2
k_i	1.103	0.451	0.451

Table 9. Triangular node comprehensive strength value.

Node	Leader	Follow1	Follow2
k_i	0.553	1.009	0.456

Then, also taking $\lambda = 0.45$, according to the Equation (18), the comprehensive adjacency strength value of the node in both networks was calculated as shown in Tables 10 and 11.

Table 10. Liner shape node adjacency strength value.

Node	Leader	Follow1	Follow2
Q_i	0.496	0.496	0.496

Table 11. Triangular node adjacency strength value.

Node	Leader	Follow1	Follow2
Q_i	0.555	0.499	0.454

Finally, according to Equation (21), the entropy of each network structure was calculated, and the following results of both networks are presented in Tables 12 and 13.

Table 12. Liner shape node network structure entropy.

Node	Leader	Follow1	Follow2
E_i	0.162	4.128	0.120

Table 13. Triangular node network structure entropy.

Node	Leader	Follow1	Follow2
E_i	5.120	0.125	0.125

According to the node network structure entropy obtained in the above tables, the size of the entropy of each node was ordered, in the liner shape formation, as Follow1 was responsible for connecting the UUV leader and Follow2 in the middle position of the line shape; therefore, this position was more critical, and the UUV leader was responsible for piloting and sending data, so its importance was greater than that of Follow2.

As for the triangular formation, the UUV leader was responsible for connecting Follow1 and Follow2, and it was also responsible for calculating and sending the route data, which was more critical than the other two; moreover, the other two followers had the same position, the same role, and the same importance.

In this paper, the trajectory used was recorded by the aircraft itself. The experiment in this paper was mainly to prove that under a series of continuous spatiotemporal trajectories, the algorithm of this paper can be used to reconstruct the network structure of unknown formations and effectively rank the importance of nodes. However, in real situations, the results obtained when observing the formation's trajectory through sonar equipment or other methods will not be so dense, and the results obtained due to sensor interference will be biased. It is possible to consider adding an error model and using a filtering algorithm to process the tracks detected by the sonar.

5. Conclusions

Aiming at the critical node identification problem of UUV formation, this paper proposed a formation key node identification method, based on network structure entropy, which establishes the network structure of mobile UUV formation by presenting the motion similarity model, and then calculating the information entropy of network nodes by using the weighted network structure entropy algorithm to determine the importance of each node. The simulation experiments and lake experiments in this paper fully verify the effectiveness of the identification algorithm, which can be calculated from the spatiotemporal trajectory of the formation to calculate the importance ranking of the formation nodes, and also verify that it is feasible to use this method for underwater cluster countermeasures. As for future works, it should be considered to use sonar equipment to acquire formation trajectory data with disturbances to further validate the effectiveness of the algorithm. In the next step, it is planned to further enhance the structure reconstruction of unknown multi-UUV formations in other complex situations for discontinuous multi-UUV spatiotemporal trajectories, considering the impact of the hydroacoustic communication packet's loss and other effects on the control formation for more accurate critical node identification.

Author Contributions: Conceptualization, L.L.; Data curation, Y.C. and X.Z.; Funding acquisition, L.L.; Methodology, Y.C.; Project administration, L.Z.; Resources, Y.Y.; Software, B.Z.; Supervision, G.P.; Validation, L.L. and R.R.; Writing—original draft, Y.C.; Writing—review & editing, W.Q. and R.R. All authors have read and agreed to the published version of the manuscript.

Funding: This work was supported in part by the Local Science and Technology Special Foundation under the Guidance of the Central Government of Shenzhen under Grant 2021Szvup111; in part by the Shenzhen Science and Technology Program under Grant JCYJ20210324122010027 and JCYJ20210324122406019; in part by the National Natural Science Foundation of China under Grant 52001259, Grant 11902252, and Grant 51979229; in part by the National Research and Development

Project under Grant 2021YFC2803001; in part by the Maritime Defense Technology Innovation Center Innovation Fund under Grant JJ-2021-702-09; and in part by the China Postdoctoral Science Foundation under Grant 2020M673484.

Institutional Review Board Statement: Not applicable.

Informed Consent Statement: Not applicable.

Data Availability Statement: The data that support the findings of this study are available within the article.

Conflicts of Interest: The authors declare no conflict of interest.

References

1. Das, B.; Subudhi, B.; Pati Bibhuti, B. Cooperative Formation Control of Autonomous Underwater Vehicles: An Overview. *Int. J. Autom. Comput.* **2016**, *13*, 199–225. [CrossRef]
2. Khawaja, W.; Semkin, V.; Ratyal, N.I.; Yaqoob, Q.; Gul, J.; Guvenc, I. Threats from and Countermeasures for Unmanned Aerial and Underwater Vehicles. *Sensors* **2022**, *22*, 3896. [CrossRef] [PubMed]
3. Wang, G.-Y.; Cheng, D.-D.; Xia, D.-Y.; Jiang, H.-H. Swarm Intelligence Research: From Bio-inspired Single-population Swarm Intelligence to Human-machine Hybrid Swarm Intelligence. *Mach. Intell. Res.* **2023**, *20*, 121–144. [CrossRef]
4. Kumoye, A.O.; Prasad, R.; Fonkam, M. Swarm Intelligence Algorithm and its Application: A Critical Review. In Proceedings of the 2020 International Conference in Mathematics, Computer Engineering and Computer Science (ICMCECS), Ayobo, Nigeria, 18–21 March 2020.
5. Jung, N.; Weon, B.M.; Kim, P. Effects of adaptive acceleration response of birds on collective behaviors. *J. Phys. Complex.* **2022**, *3*, 015014. [CrossRef]
6. Mugica, J.; Torrents, J.; Cristin, J.; Puy, A.; Miguel, M.C.; Pastor-Satorras, R. Scale-free behavioral cascades and effective leadership in schooling fish. *Sci. Rep.* **2022**, *12*, 10783. [CrossRef]
7. Biro, D.; Sumpter, D.J.T.; Meade, J.; Guilford, T. From Compromise to Leadership in Pigeon Homing. *Curr. Biol.* **2006**, *16*, 2123–2128. [CrossRef]
8. Huo, M.Z.; Duan, H.B.; Ding, X.L. Manned Aircraft and Unmanned Aerial Vehicle Heterogeneous Formation Flight Control via Heterogeneous Pigeon Flock Consistency. *Unmanned Syst.* **2021**, *9*, 227–236. [CrossRef]
9. Park, J.; Kahng, B. Synchronization in leader-follower switching dynamics. *Phys. Rev. Res.* **2020**, *2*, 032061. [CrossRef]
10. Xu, G.H.; Xu, M.; Ge, M.F.; Ding, T.F.; Qi, F.; Li, M. Distributed Event-Based Control of Hierarchical Leader-Follower Networks with Time-Varying Layer-To-Layer Delays. *Energies* **2020**, *13*, 1808. [CrossRef]
11. Ugurlu, O. Comparative analysis of centrality measures for identifying critical nodes in complex networks. *J. Comput. Sci.* **2022**, 62. [CrossRef]
12. Cai, B.; Zeng, L.N.; Wang, Y.P.; Li, H.J.; Hu, Y.M. Community Detection Method Based on Node Density, Degree Centrality, and K-Means Clustering in Complex Network. *Entropy* **2019**, *21*, 1145. [CrossRef]
13. Yang, X.H.; Xiong, Z.; Ma, F.N.; Chen, X.Z.; Ruan, Z.Y.; Jiang, P.; Xu, X.L. Identifying influential spreaders in complex networks based on network embedding and node local centrality. *Phys. A-Stat. Mech. Its Appl.* **2021**, *573*, 125971. [CrossRef]
14. Wandelt, S.; Shi, X.; Sun, X.Q. Approximation of Interactive Betweenness Centrality in Large Complex Networks. *Complexity* **2020**, *2020*, 4046027. [CrossRef]
15. Kitsak, M.; Gallos, L.K.; Havlin, S.; Liljeros, F.; Muchnik, L.; Stanley, H.E.; Makse, H.A. Identification of influential spreaders in complex networks. *Nat. Phys.* **2010**, *6*, 888–893. [CrossRef]
16. Yu, Y.; Zhou, B.; Chen, L.J.; Gao, T.; Liu, J.Z. Identifying Important Nodes in Complex Networks Based on Node Propagation Entropy. *Entropy* **2022**, *24*, 275. [CrossRef]
17. Wang, Y.; Zheng, Y.; Liu, Y. Identifying vital nodes for influence maximization in attributed networks. *Sci. Rep.* **2022**, *12*, 22630. [CrossRef]
18. Jiang, J.-L.; Fang, H.; Li, S.-Q.; Li, W.-M. Identifying important nodes for temporal networks based on the ASAM model. *Phys. A Stat. Mech. Its Appl.* **2022**, *586*, 126455. [CrossRef]
19. Liu, L.; Zhang, S.; Zhang, L.; Pan, G.; Yu, J. Multi-UUV Maneuvering Counter-Game for Dynamic Target Scenario Based on Fractional-Order Recurrent Neural Network. *IEEE Trans. Cybern.* **2022**, *53*, 4015–4028. [CrossRef]
20. Nagy, M.; Vasarhelyi, G.; Pettit, B.; Roberts-Mariani, I.; Vicsek, T.; Biro, D. Context-dependent hierarchies in pigeons. *Proc. Natl. Acad. Sci. USA* **2013**, *110*, 13049–13054. [CrossRef]
21. Zhang, W.; Wang, N.; Wei, S.; Du, X.; Yan, Z. Overview of unmanned underwater vehicle swarm development status and key technologies. *J. Harbin Eng. Univ.* **2020**, *41*, 289–297.
22. Yang, P.; Xu, C.; Chen, H. Multi-attribute ranking method for identifying key nodes in complex networks based on GRA. *Int. J. Mod. Phys. B. Condens. Matter Phys. Stat. Phys. Appl. Phys.* **2018**, *32*, 1850363. [CrossRef]

23. Tyrsin, A.N. Entropy Modeling of Network Structures. *Autom. Remote Control* **2022**, *83*, 1608–1618. [CrossRef]
24. Zhang, Q.; Li, M.; Deng, Y. A betweenness structure entropy of complex networks. *Chaos Solitons Fractals* **2022**, *161*, 112264. [CrossRef]

Journal of
Marine Science and Engineering

MDPI

Article

Analysis of the Steady-Stream Active Flow Control for the Blended-Winged-Body Underwater Glider

Xiaoxu Du *, Xin Liu and Yani Song

School of Marine Science and Technology, Northwestern Polytechnical University, Xi'an 710072, China; lxnwpu1938@163.com (X.L.); syn_1103@163.com (Y.S.)
* Correspondence: nwpudu@163.com

Abstract: The BWB-UG is a glider with a smooth and integrated fuselage and wing. Its lift-to-drag properties are some of the most significant factors affecting its performance. In order to improve its hydrodynamic characteristics, the method of steady-stream active flow control (SS-AFC) is proposed. The computational fluid dynamics method is used to numerically investigate the SS-AFC of the BWB-UG. The mechanism of the SS-AFC effect on the lift-to-drag characteristics is revealed from the flow field aspect. The flow field of the BWB-UG before and after installing the SS-AFC was simulated using FLUENT. The results show that the SS-AFC can effectively optimise the hydrodynamic characteristics of the BWB-UG and can optimise the structure of the flow field around the BWB-UG. The steady-suction AFC can increase the lift-to-drag ratio of the BWB-UG by up to 45.01%. With the steady-jet AFC, the lift-to-drag ratio of the BWB-UG can be increased by as much as 93.17%.

Keywords: underwater glider; active flow control; hydrodynamic characteristics

Citation: Du, X.; Liu, X.; Song, Y. Analysis of the Steady-Stream Active Flow Control for the Blended-Winged-Body Underwater Glider. *J. Mar. Sci. Eng.* **2023**, *11*, 1344. https://doi.org/10.3390/jmse11071344

Academic Editor: Alessandro Ridolfi

Received: 18 May 2023
Revised: 7 June 2023
Accepted: 24 June 2023
Published: 30 June 2023

1. Introduction

The Blended-Wing-Body Underwater Glider (BWB-UG) is a tailless underwater vehicle with a full-wing configuration. It can dramatically improve the lift-to-drag ratio by increasing the area of the airfoil and providing internal space that overcomes many of the shortcomings of the traditional UG. Since American oceanographer Stommel first proposed the concept of the UG in 1989, researchers in many countries have gradually researched related technologies of the UG. The United States Naval Research Laboratory began to develop Liberdade-class large-scale BWB-UG in 2003, mainly including XRay, XRay2, and ZRay [1]. Researchers conducted a sea trial of the XRay in California. The horizontal velocity of the XRay was about, and the lift-to-drag ratio was 17. Northwestern Polytechnical University has also conducted relevant research on the BWB-UG. Du established the flow field calculation model for the BWB-UG, carried out the hydrodynamic simulation, established the 6-DOF dynamic model, and analysed the glide motion performance [2–4]. Li used free-form deformation to optimise the framework of geometric parameterization. An optimised BWB-UG was considered as the initial shape and four shape optimisation cases were performed for different design purposes using the proposed framework [5]. Sun carried out a shape optimisation design of the BWB-UG based on a global optimisation method with a maximum lift-to-drag ratio and maximum range as optimisation objectives [6–8]. However, existing studies show an irreconcilable contradiction between the optimisation of hydrodynamic performance and the maximization of piggyback space in the shape optimisation design process. At the same time, further improvement of the lift-to-drag ratio of the BWB-UG is limited by the occurrence of flow separation. In summary, the critical issue for further development in the field of overall BWB-UG design is whether a method can be found to optimise the hydrodynamic characteristics of the BWB-UG without modification of its shape.

Active flow control (AFC) is an emerging drag reduction technology to control the local flow field of the object by actively inputting appropriate disturbances to the local flow field of the object. AFC can improve the whole-field fluid structure by inputting or releasing energy at the critical point. At the same time, because of its initiative, this technology can effectively carry out accurate phase control in complex dynamic systems. The control method of steady-jet AFC is simple, and the device is easy to realize. GANESH N et al. used the CFD method to numerically study the flow field structure of NACA63 (4) -021 wing with steady-suction AFC. The results show that the aerodynamic performance can be improved by 28% by equipping suction at 0.3 times the chord length from the wing's leading edge [9]. Fatahian et al. conducted numerical studies on the steady-suction AFC of the NACA0012 flap. The results show that the steady-suction perpendicular to the airfoil surface can effectively improve the aerodynamic performance of the airfoil [10]. AFC technology was widely used in wind turbine lift drag reduction [11,12], ground vehicle energy-saving drag reduction [13–15], aircraft airfoil aerodynamic performance improvement [16–19], and other fields. Overall, AFC technology has broad application prospects in many fields. It has shown considerable potential and is likely to become a future breakthrough technology in fluid mechanics. The AFC has a good improvement effect on the hydrodynamic performance, and the structure is simple. Du et al. achieved excellent hydrodynamic optimisation results by applying the electromagnetic AFC technique to the overall design of the BWB-UG [20]. The AFC has reasonable practicability for the BWB-UG. However, from the current public research results, there are relatively few theoretical and experimental results on the AFC of the BWB-UG. The research on SS-AFC of the BWB-UG is still in the preliminary exploration stage. Therefore, studying the SS-AFC of the BWB-UG is of great theoretical significance and application value.

In this paper, the effect of the SS-AFC on the BWB-UG is analysed using the computational fluid dynamics method. Firstly, an accurate and effective numerical computational model is established. Secondly, the disturbance phenomenon of the BWB-UG equipped with the steady-jet AFC and the steady-suction AFC is numerically calculated by analysing the flow field structure characteristics and hydrodynamic trend characteristics of the BWB-UG before and after being fitted with the AFC. The mechanism of the influence of the SS-AFC on the BWB-UG is obtained.

2. Materials and Methods

2.1. Geometric Model and Reference Definition

The SS-AFC technology is considered equipped for the BWB-UG, as shown in Figure 1. The AFC system of the BWB-UG mainly includes nozzles, a small pump, and a simple pipeline. Since the AFC has a good improvement effect on the hydrodynamic performance of the underwater glider and its structure is simple, the AFC has reasonable practicability for the underwater glider. There are 20 nozzles on the BWB-UG. The jet or suction flow rate is V_{stream} and its direction angle is θ_{stream}. The specific design parameters of the model are shown in Table 1. The hydrodynamic characteristics of the BWB-UG at attack angles of $0° \sim 12°$ are calculated. The hydrodynamic characteristics before and after the equipment of the AFC are compared.

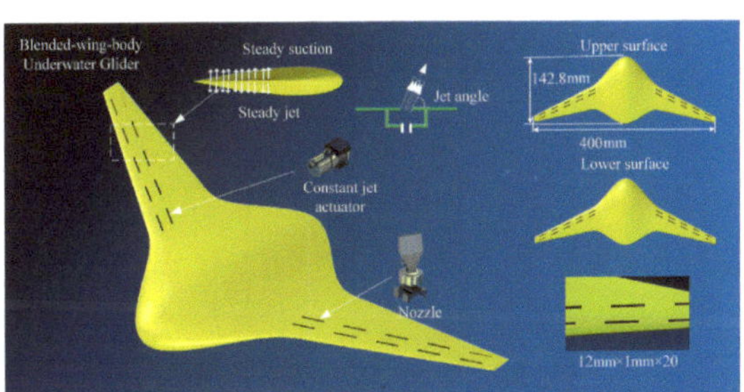

Figure 1. Steady-stream AFC of the BWB-UG.

Table 1. Model design parameters.

Name	Value
Spread length of UG	400 mm
Maximum chord length of UG	142.8 mm
Section of UG	NACA0015
Length of nozzle	12 mm
Width of nozzle	1 mm
Number of nozzles	20

The wet surface area S_h is used as the characteristic area to define the lift coefficient C_l, the drag coefficient C_d, and the lift-to-drag ratio R of the BWB-UG.

$$\begin{cases} C_l = \frac{2F_L}{\rho v_\infty^2 S_h} \\ C_d = \frac{2F_D}{\rho v_\infty^2 S_h} \\ R = \frac{C_l}{C_d} \end{cases} \tag{1}$$

where F_L is the lift force; F_D is the drag force; v_∞ is the flow velocity at infinity; and ρ is the density of sea water. Define the dimensionless expressions of velocity and pressure.

$$C_p = 2\frac{(P - P_0)}{\rho U^2} \tag{2}$$

$$C_v = \frac{V_{point}}{V_{stream}} \tag{3}$$

The Q-criterion is a fundamental law in fluid mechanics to characterize the change in flow length of a fluid after deflection.

$$\begin{aligned} Q = 0.5\times(&-ddx(u)^2 - ddy(v)^2 - ddz(w)^2 \\ &- 2 \times ddz(u) \times ddx(w) - 2 \times ddz(v) \times ddy(w) - 2 \times ddy(u) \times ddx(v)) \end{aligned} \tag{4}$$

2.2. Flow Field Calculation Model

The 3-D incompressible model is selected for the relevant research. The control equation is:

$$\frac{\partial(\rho u)}{\partial x} + \frac{\partial(\rho v)}{\partial y} + \frac{\partial(\rho w)}{\partial z} = 0 \tag{5}$$

$$\begin{aligned} \rho\left(\bar{u}\frac{\partial \bar{u}}{\partial x} + \bar{v}\frac{\partial \bar{u}}{\partial y} + \bar{w}\frac{\partial \bar{u}}{\partial z}\right) =& \mu\left(\frac{\partial^2 \bar{u}}{\partial x^2} + \frac{\partial^2 \bar{u}}{\partial y^2} + \frac{\partial^2 \bar{u}}{\partial z^2}\right) \\ &+ \left(\frac{\partial\left(-\rho\overline{u'u'}\right)}{\partial x} + \frac{\partial\left(-\rho\overline{u'v'}\right)}{\partial y} + \frac{\partial\left(-\rho\overline{u'w'}\right)}{\partial z}\right) - \frac{\partial p}{\partial x} + S_{fx} \end{aligned} \tag{6}$$

$$\begin{aligned} \rho\left(\bar{u}\frac{\partial \bar{v}}{\partial x} + \bar{v}\frac{\partial \bar{v}}{\partial y} + \bar{w}\frac{\partial \bar{v}}{\partial z}\right) =& \mu\left(\frac{\partial^2 \bar{v}}{\partial x^2} + \frac{\partial^2 \bar{v}}{\partial y^2} + \frac{\partial^2 \bar{v}}{\partial z^2}\right) \\ &+ \left(\frac{\partial\left(-\rho\overline{v'u'}\right)}{\partial x} + \frac{\partial\left(-\rho\overline{v'v'}\right)}{\partial y} + \frac{\partial\left(-\rho\overline{v'w'}\right)}{\partial z}\right) - \frac{\partial p}{\partial y} + S_{fy} \end{aligned} \tag{7}$$

$$\rho\left(\bar{u}\frac{\partial\bar{w}}{\partial x}+\bar{v}\frac{\partial\bar{w}}{\partial y}+\bar{w}\frac{\partial\bar{w}}{\partial z}\right)=\mu\left(\frac{\partial^2\bar{w}}{\partial x^2}+\frac{\partial^2\bar{w}}{\partial y^2}+\frac{\partial^2\bar{w}}{\partial z^2}\right)+\left(\frac{\partial\left(-\rho\overline{w'u'}\right)}{\partial x}+\frac{\partial\left(-\rho\overline{w'v'}\right)}{\partial y}+\frac{\partial\left(-\rho\overline{w'w'}\right)}{\partial z}\right)-\frac{\partial p}{\partial z}+S_{fz} \quad (8)$$

where p is fluid pressure; μ is the dynamic viscosity coefficient of fluid; $\bar{u}$,$\bar{v}$,$\bar{w}$ are the time-averaged velocity components of the fluid in x, y, z directions; u',v',w' are the pulsating velocity components of the fluid in x, y, z directions; S_{fx},S_{fy},S_{fz} are generalized source terms of momentum equations in x, y, z directions.

Taking into account the need for accuracy and efficiency in this numerical investigation, the Reynolds averaging method is considered. The shear stress transport (SST) $k-\omega$ model has excellent simulation accuracy for models such as inverse pressure gradient flow, airfoil winding flow, and jet flow in the near-wall region. Therefore, the $k-\omega$ two-equation model is used to close the basic governing equations. The transport equations of k and ω are expressed as:

$$\begin{cases} \frac{\partial(\rho k)}{\partial t}+\frac{\partial}{\partial x_j}\left(\rho u_j k-(\mu+\sigma^*\mu_t)\frac{\partial k}{\partial x_j}\right)=\tau_{tij}S_{ij}-\beta^*\rho\omega k \\ \frac{\partial(\rho\omega)}{\partial t}+\frac{\partial}{\partial x_j}\left(\rho u_j\omega-(\mu+\sigma\mu_t)\frac{\partial\omega}{\partial x_j}\right)=\alpha\frac{\omega}{k}\tau_{tij}S_{ij}-\beta\rho\omega^2 \end{cases} \quad (9)$$

2.3. Computational Domain and Boundary Conditions

The BWB-UG equipped with the SS-AFC is divided into rectangular computational domains. The fixed semimodal numerical calculation method is used. The model is 4 L from the velocity inlet and 3 L from the upper/lower wall, as shown in Figure 2. The boundary conditions, including symmetry surfaces, velocity inlets, pressure outlets, sliding walls, and no sliding walls are set. The flow field calculation domain satisfies the far-field boundary conditions.

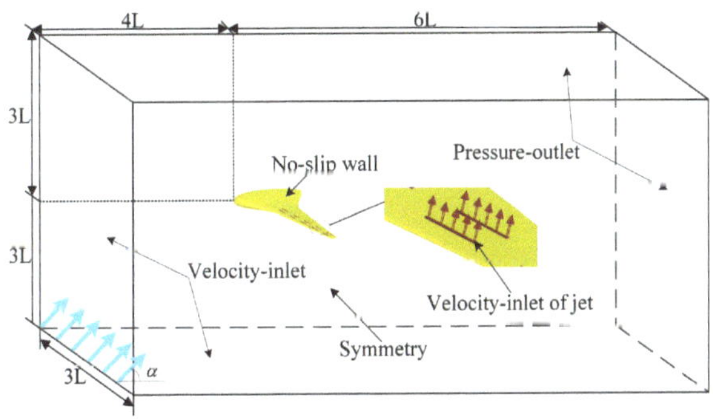

Figure 2. Calculation domain and related boundary conditions of the BWB-UG.

The computational domain of the BWB-UG is divided into high-quality hexahedral structural grids, as show in Figure 3. The mesh near the boundary layer of the BWB-UG wall and the steady-stream nozzles is encrypted to ensure that the Y^+ is less than 1. The number of grid units is about 5.13 million.

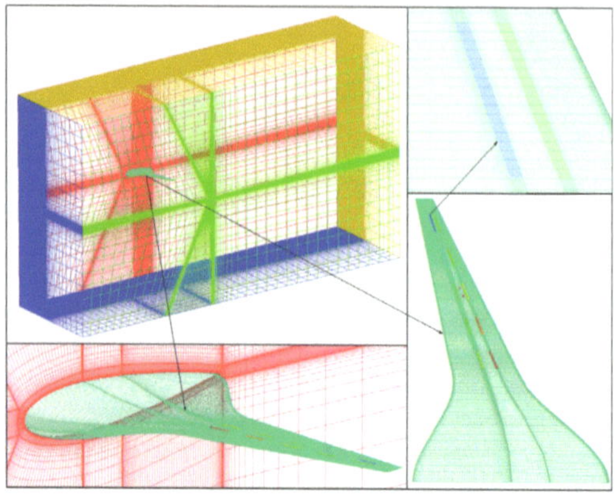

Figure 3. Grid meshing of the BWB-UG equipped with SS-AFC.

2.4. Validation of Numerical Calculation Method

Since the airfoil of the BWB-UG usually adopts the modified NACA airfoil, the NACA0012 standard airfoil model is used to validate the numerical calculation method. The chord length of the wing is c and the maximum thickness is $0.12c$.

The lift-to-drag characteristics of NACA0012 are solved using the established numerical flow field calculation method. By changing the angle of attack of the airfoil, the C_l and C_d of the airfoil at 0∼12° angle of attack are monitored, and the curve of the C_l and C_d changing with the angle of attack is plotted.

The numerical calculation results are compared with the experimental results in the literature [10]. The lift-to-drag ratio curve of the NACA0012 airfoil obtained by numerical calculation is consistent with the experimental results, as shown in Figure 4. This shows the feasibility of the established flow field numerical calculation method in the numerical calculation of the hydrofoil flow field and verifies that the established flow field mathematical model is reliable and accurate.

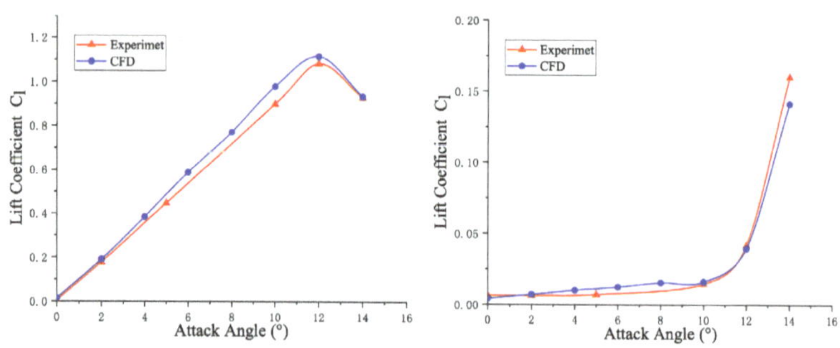

(**a**) The lift coefficient (**b**) The drag coefficient

Figure 4. Comparison between CFD numerical calculation results and experimental results.

2.5. Independence Validation of Flow Field Calculation Method

Grid convergence is a necessary condition to ensure the credibility of numerical computation. The convergence of the grid is evaluated in this work by the grid convergence index (GCI). The GCI rating method was proposed by P. J. Roache [21,22]. This method requires that the computational results must satisfy the monotonic convergence condition:

$$0 < (f_3 - f_2)/(f_2 - f_1) < 1 \tag{10}$$

where f_1, f_2, and f_3 are the numerically computed discrete solutions obtained through three grids ranging from rarefaction to denseness, respectively. The refinement rate of the grid is defined as

$$r_{k,k+1} = \frac{h_k}{h_{k+1}} \tag{11}$$

where h is the grid feature size. In general, the value of r is less than 2. The calculation formula of GCI is:

$$\begin{cases} GCI_k = F_s \left| \frac{\varepsilon_{k.k+1}}{r^p_{k,k+1}-1} \right| \\ GCI_{k+1} = F_s \left| \frac{r^p_{k,k+1}\varepsilon_{k.k+1}}{r^p_{k,k+1}-1} \right| \\ \varepsilon_{k.k+1} = \frac{f_{k+1}-f_k}{f_{k+1}} \end{cases} \tag{12}$$

where F_s is the grid convergence security factor; F is the numerically computed discrete solutions; $r_{k,k+1}$ is the mesh refinement ratio, which is usually required to be less than 1.2; and p is the convergence accuracy.

By adjustment of the distribution of the mesh nodes, four groups of meshes with different dispositions are calculated. The above four groups of grids are used to calculate the SS-AFC of the BWB-UG under 12° angle of attack. Meanwhile, the C_l calculated by different grids are monitored and the corresponding GCIs are calculated; the results are shown in Figure 5. When the number of grids exceeds 5130000, the C_l gradually stabilizes and the GCI is less than 1%. Therefore, 5.13 million grids are chosen to simulate the SS-AFC of the BWB-UG.

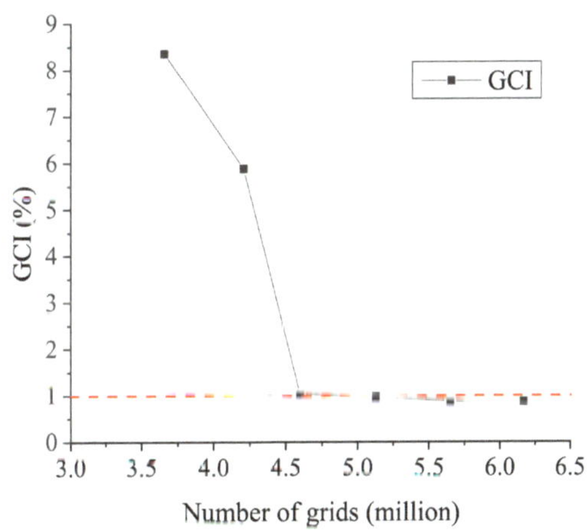

Figure 5. The curves of GCI with the number of grids.

3. Results and Discussion of Flow Field Calculation

3.1. Study on Steady-Suction AFC

The C_L and C_L of the BWB-UG before and after the installation of the steady-suction AFC are shown in Figure 6. The steady-state AFC can significantly enhance the C_L of the BWB-UG and increase the stall angle from 6° to 8° with the angle of attack in the range of 0~12°. The steady-suction AFC has a drag-reducing effect on the BWB-UG when the angle of attack is in the range of 4~12°. When the angle of attack is 8°, the drag reduction effect is best. The drag reduction rate can reach 15.4%.

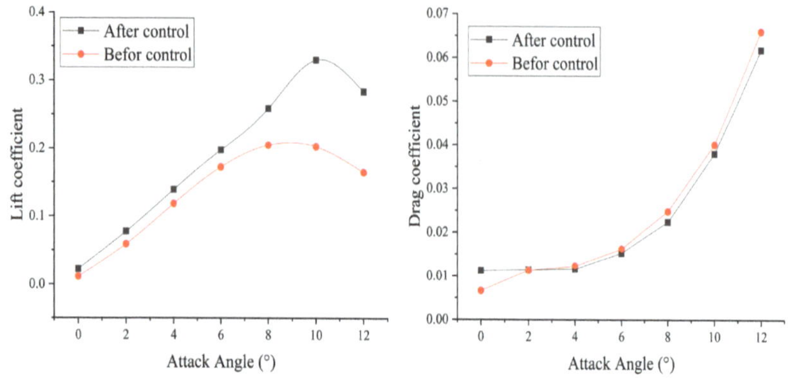

(a) The lift coefficient **(b)** The drag coefficient

Figure 6. Effect of steady suction on the BWB-UG.

Figure 7 compares the change in the lift-to-drag ratio of the BWB-UG before and after the installation of the steady-suction AFC. The steady-suction AFC can effectively increase the lift-to-drag ratio of the BWB-UG. The lift-to-drag ratio of the BWB-UG at 0~12° angle of attack is significantly improved by the addition of the steady-suction AFC. The lift amplitude of the lift-to-drag ratio reaches the maximum value of 3.78 when the angle of attack is $\alpha = 8°$. Under the condition of a large angle of attack range from 6~12°, the steady-suction AFC can increase the lift-to-drag ratio of the BWB-UG by up to 45.01%.

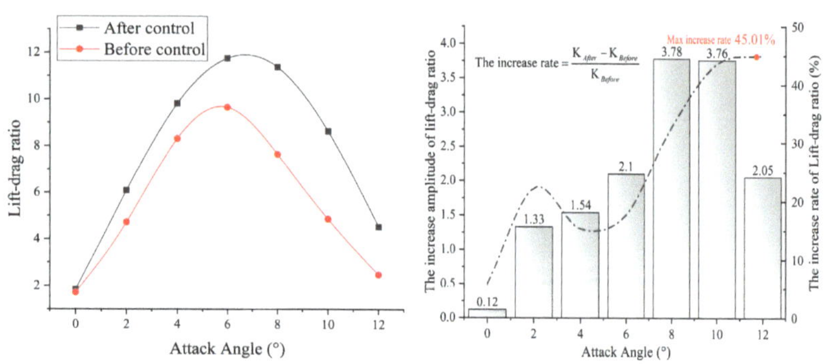

(a) lift-to-drag ratio **(b)** The increase amplitude and increase Rate

Figure 7. Effect of steady-suction on the lift-to-drag ratio of the BWB-UG.

From the perspective of flow structure and surface pressure distribution, the impact of the steady-suction AFC on the BWB-UG is further analysed. As illustrated in Figure 8, the pressure contours of the upper and lower surfaces of the BWB-UG before and after the installation of the steady-suction AFC at three typical angles of attack of 4°, 8°, and 12° are shown. Since the steady-suction nozzles are installed for the upper surface, the steady-suction has only a slight change on the pressure of the lower surface. The distribution of the pressure on the upper surface of the BWB-UG has been significantly changed. The range of the low-pressure area at the leading edge of the wing is extended for the three angles of attack mentioned above. This increases the pressure difference between the upper and lower surfaces. As a result, the lift of the BWB-UG is improved to varying degrees. Only at the angle of attack $\alpha = 8°$ does the pressure area of $0 \sim 20$ Pa on the lower surface increase slightly, and the pressure on the lower surface does not change significantly at other angles of attack.

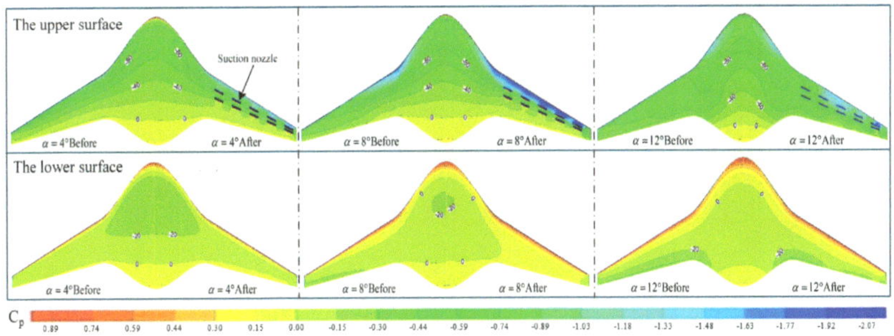

Figure 8. Pressure contours of the BWB-UG before and after the steady-suction AFC.

The flow control mechanism is further analysed by taking the flow field characteristics of the BWB-UG before and after the installation of the steady-suction AFC at an angle of attack of 8° as an example. It can be seen from Figure 9 that there is a region of fluid countercurrent motion on the upper surface of the BWB-UG prior to the installation of the steady-suction AFC. The steady-suction AFC realizes the boundary layer suction and the upper surface turbulent vorticity reduction. After the equipment of the steady-suction AFC, the separation is eliminated or weakened, the vorticity transported to the fluid is significantly reduced, the energy dissipation of the fluid is significantly weakened, and the purpose of drag reduction is achieved. According to Figure 9, the low-velocity fluid of the boundary layer is sucked into the steady-suction nozzles, the flow rate of the boundary layer fluid is increased, and therefore, the ability to resist fluid viscosity and adverse pressure gradients is enhanced. At this time, the wall adhesion is more robust, the fluid counter current motion area is significantly reduced, and the large separation vortex is effectively suppressed. Meanwhile, the coupling effect of steady-suction flow and incoming flow gives the fluid in the wing's leading edge higher flow velocity, so there is higher energy to overcome the flow separation. The flow separation vortex structure of the underwater glider is significantly reduced, thereby effectively improving its lift-to-drag characteristics.

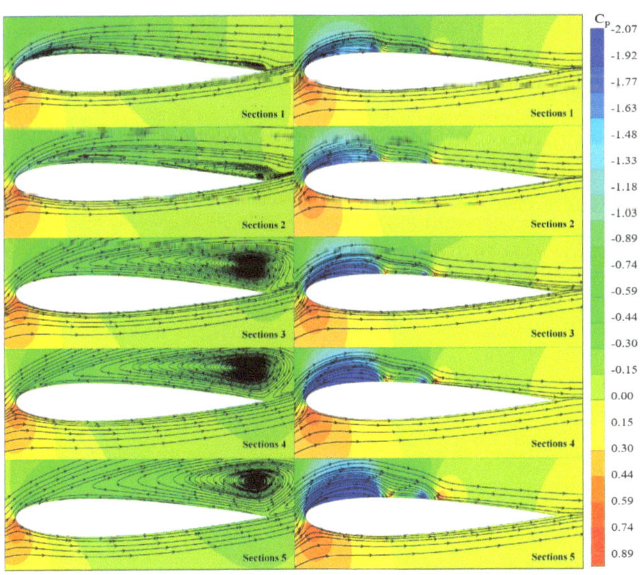

Figure 9. Pressure contours and streamline of the BWB-UG sections.

As shown in Figure 10, we have carefully analysed the physical field characteristics at the steady-suction AFC nozzle. The steady-suction side is the downward flow surface. The steady-suction AFC has little effect on the pressure and velocity distribution in the local flow field. The nature of the steady-suction AFC is that it will draw energy from the flow field and weaken the energy of the local flow field. As a result, the vortex structure created by the steady-suction AFC is very small.

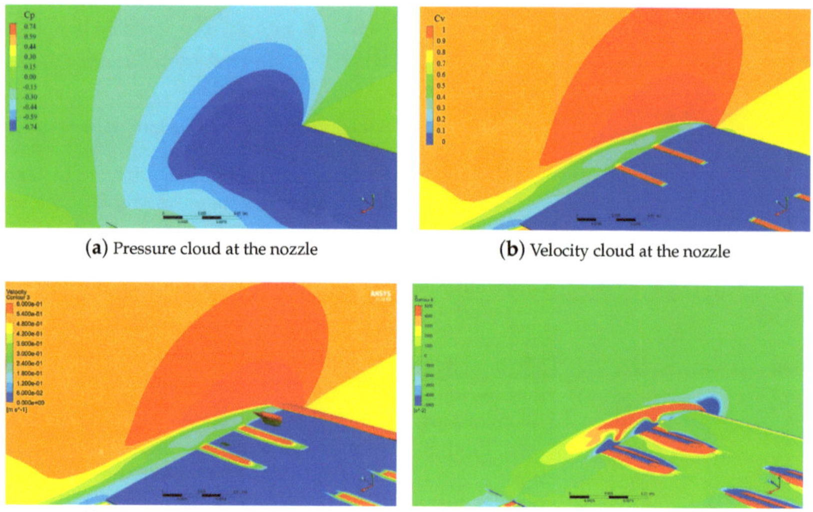

(**a**) Pressure cloud at the nozzle

(**b**) Velocity cloud at the nozzle

(**c**) Vortex structure at the nozzle

(**d**) Q criterion at the nozzle

Figure 10. Detail at the nozzle of steady-suction AFC.

In summary, the hydrodynamic performance of the BWB-UG is effectively improved by being equipped with the steady-suction AFC, and the lift-to-drag ratio can be increased by up to 45.1%. Through the analysis of the flow field structure, it can be seen that the steady-suction AFC can effectively improve the flow field structure and flow field stability. This has noticeable beneficial effects on increasing the glide ratio, improving the motion performance, enhancing the detection ability, and reducing the energy consumption of the BWB-UG.

3.2. *Study on Steady-Jet AFC*

The lift-to-drag ratio of the BWB-UG before and after the equipment of the steady-jet AFC is obtained in Figure 11. It can be seen that in the attack angle range of $0{\sim}6^\circ$, the steady-jet AFC can significantly improve the lift of the BWB-UG. It is also found that when the attack angle is greater than 12°, the steady-jet AFC can significantly improve the lift. Through the analysis of the flow field structure, it is found that when the angle of attack is greater than 12°, the vortex shedding phenomenon occurs in the flow field of the BWB-UG, and the steady-jet AFC can effectively suppress this phenomenon. However, the attack angle of the BWB-UG does not exceed 12° in regular operation. Therefore, it is no longer necessary to conduct in-depth research on the total effect of the steady-jet AFC when the attack angle is above 12°. The steady-jet AFC has little effect on the resistance of the underwater glider. When the attack angle is greater than 8°, the drag reduction effect is pronounced, and the drag reduction effect is maintained at about 4%.

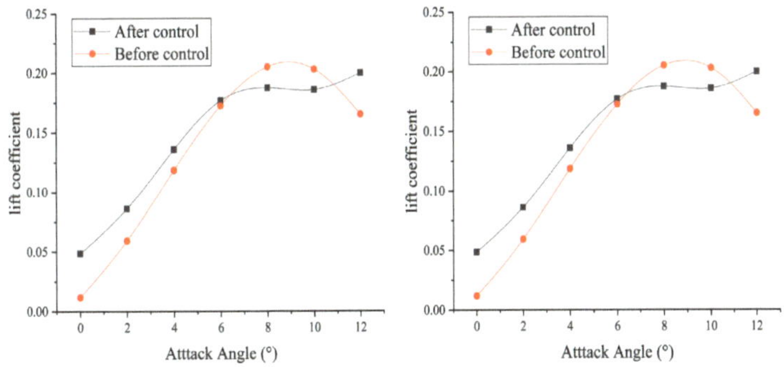

(a) The lift coefficient (b) The drag coefficient

Figure 11. Effect of steady-jet AFC on the BWB-UG.

As shown in Figure 12a, the steady-jet AFC significantly influences the lift-to-drag ratio. Especially in the case of low angles of attack, the improvement effect of the lift-to-drag ratio is pronounced. When the attack angle does not exceed 5°, the steady-jet AFC can effectively improve the lift-to-drag ratio of the BWB-UG. It can be seen from Figure 12b that the increased amplitude of the lift-to-drag ratio gradually decreases with the increase of attack angle, and the maximum increase amplitude of the lift-to-drag ratio is 1.56. When the attack angle exceeds 5°, the steady-jet AFC will slightly reduce the lift-to-drag ratio of the BWB-UG. The steady-jet AFC can improve the lift-to-drag ratio of the BWB-UG by 93.17% at most. The improved hydrodynamic performance is extensively valued for its glide motion stability, real-time maneuverability, and equipment safety.

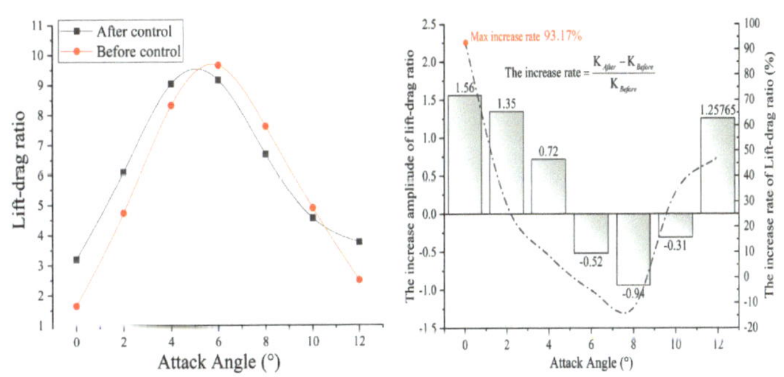

(a) lift-to-drag ratio (b) The increase amplitude and increase Rate

Figure 12. Effect of steady-jet AFC on the lift-to-drag ratio of the BWB-UG.

From the perspective of flow field structure and surface pressure distribution, the mechanism of steady-jet AFC on the BWB-UG is further analysed. Figure 13 shows the pressure contours of the surface of the BWB-UG at typical angles of attack before and after the installation of the steady-jet AFC. At angles of attack 4°, 8°, and 12° the effect of the steady-jet AFC on the lower surface pressure distribution is mainly reflected around the steady-jet nozzles on the wing and has little effect on the upper surface pressure distribution. The influence on the lower surface pressure distribution is mainly reflected around the steady-jet nozzles. The pressure from the steady-jet nozzles towards the leading edge is increased, while the pressure from the forward steady-jet nozzles towards the trailing edge of the wing is decreased. As the angle of attack increases, the area of high pressure on the underside gradually increases. Therefore, combined with the increase in the pressure area above 40 Pa and the change in the range of the same pressure area from 20 Pa to 0 Pa, it can be seen that the range of the pressure area increases most when the angle of attack is

4°. After equipping the steady-jet AFC at angles of attack of 8° and 12°, the depression between the pressure surface and the suction surface is slightly lower than that of the ground state due to the smaller range of the pressurization region and the larger range of the decompression region. The macroscopic manifestation is that the lift of the BWB-UG is lower than before. This is in agreement with the results of the previous studies on the lift and the drag in this work.

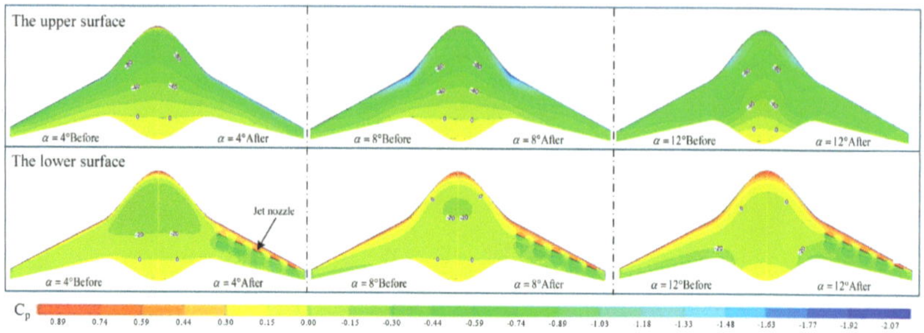

Figure 13. Pressure contours of the BWB-UG before and after the steady-jet AFC.

The lift and drag calculation results show that the steady-jet AFC, unlike the steady-suction, cannot optimise the high angle of attack hydrodynamical performance. From the perspective of the fluid-structure at the 8° angle of attack, the reason why the steady-jet AFC cannot increase the lift and reduce the drag at high angles of attack is analysed. It can be seen from Figure 14 that after the steady-jet AFC is installed on the underside of the wing, the range of the counter flow motion area above the wing of the BWB-UG is extended to varying degrees compared to that before the control. This phenomenon is most obvious in Sections 2 and 3.

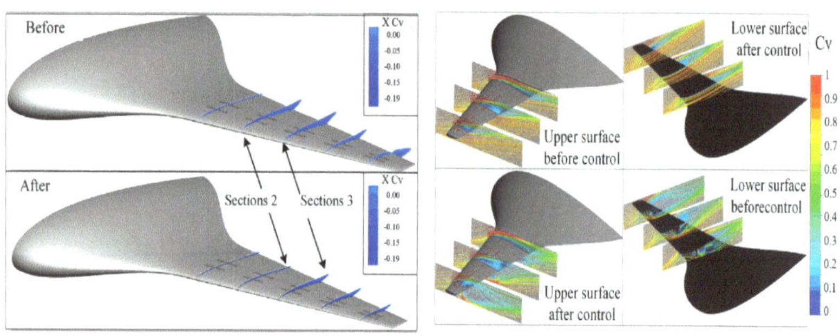

(**a**) Velocity contours on the BWB-UG sections (**b**) Streamline on the BWB-UG sections

Figure 14. The velocity distribution after the equipment of steady-jet AFC.

As can be seen in Figure 15, after the installation of the steady-jet AFC, the high-pressure area at the leading edge of the wing increases significantly and the low-pressure area at the trailing edge decreases significantly. The pressure drag increases as the pressure differential increases. In addition, since the differential pressure resistance is greater than the frictional resistance in the total resistance of the underwater glider at the angle of attack of 8°, the total resistance of the underwater glider after the installation of the steady-jet AFC is greater than that before the application of the control. The steady-jet AFC has little effect on the velocity distribution near the upper surface of the hydroplane. However, it has a coupling effect with the incoming flow. A huge flow separation vortex is formed near the trailing edge of the wing. The separation vortex bypasses the trailing edge of the wing to interfere with the flow separation region on the upper surface of the wing, which reduces the ability of the fluid near the upper wall to resist the fluid viscosity and the inverse pressure gradient. This increases the instability of the fluid flow and expands the flow separation area. As a result, at high angles of attack, the flow field characteristics of the submersible are worse than before the control.

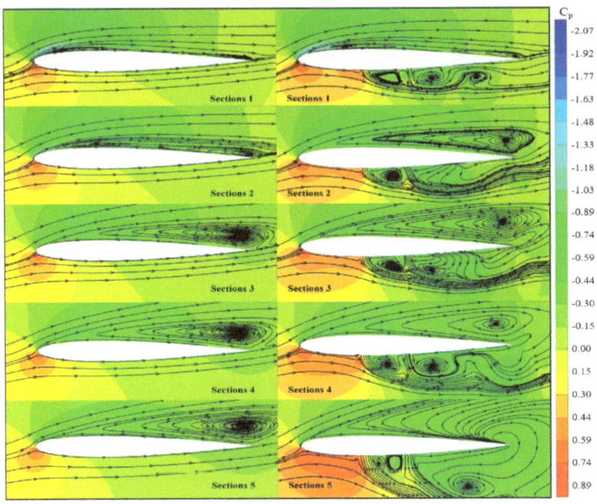

Figure 15. Pressure contours and streamline of the BWB-UG sections.

The physical field characteristics at the nozzle have been carefully analysed, as shown in Figure 16. The steady-jet side is the upward flow surface. The pressure and velocity distribution in the local flow field are significantly affected by the steady-jet AFC. The nature of the steady-jet AFC is to inject the energy into the flow field and perturb the local flow field. Therefore, the vortex structure created by the steady-jet AFC is very pronounced.

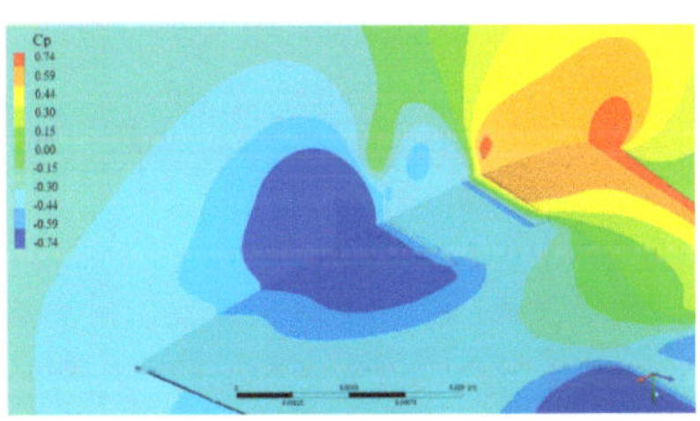

(**a**) Pressure cloud at the nozzle

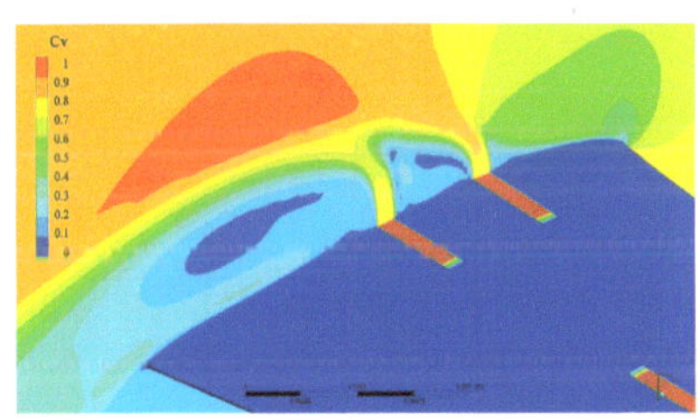

(**b**) Velocity cloud at the nozzle

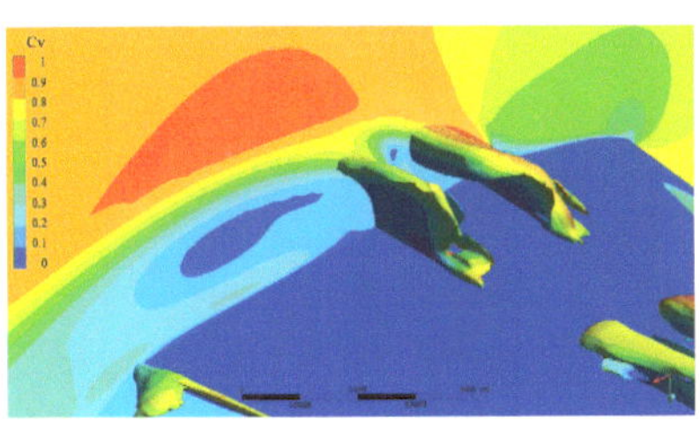

(**c**) Vortex structure at the nozzle

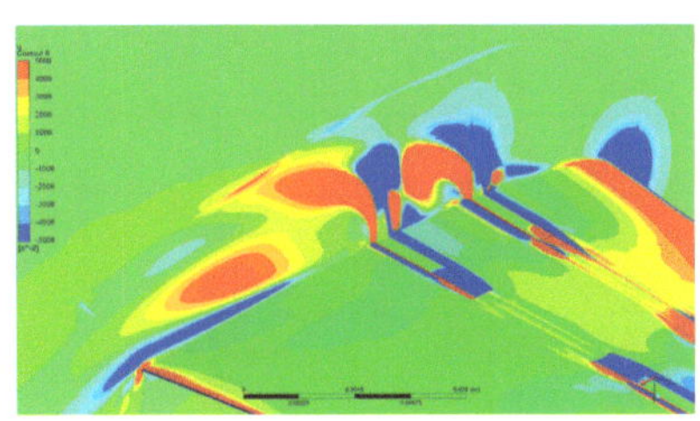

(**d**) Q criterion at the nozzle

Figure 16. Detail at the nozzle of the steady-jet AFC.

4. Conclusions

This paper proposes the steady-stream active control method for the underwater glider. The steady-suction and steady-jet AFC methods are investigated separately by numerical calculation methods. The effects of the steady-stream active control methods on the hydrodynamic and flow field structure of the glider are analysed. It is found that both steady-flow active control methods can significantly improve the lift-to-drag ratio of the underwater glider without changing its shape. The main conclusions are as follows:

(1) The lift-to-drag ratio of the BWB is greatly improved after the steady-suction AFC is installed. Especially at a large angle of attack, the steady-suction AFC can increase the lift-to-drag ratio of the BWB-UG by a maximum of 45.01%. In the angle of attack range 4~12°, the steady-suction AFC will reduce the drag of the BWB-UG. When the angle of attack is 8°, the drag reduction effect is at its best, and the drag reduction rate can be as high as 15.4%.

(2) The steady-jet AFC has a significant effect on the lift-to-drag ratio. The effect of improving the lift-to-drag ratio is particularly pronounced at low angles of attack. The steady-jet AFC can improve the lift-to-drag ratio of the BWB-UG by a maximum of 93.17%. The improved hydrodynamic performance of the BWB-UG is of great value for its glide stability, real-time maneuverability, and equipment safety.

Author Contributions: Conceptualization, X.D. and X.L.; software and formal analysis, X.L.; writing, Y.S. All authors have read and agreed to the published version of the manuscript.

Funding: The author(s) disclosed receipt of the following financial support for the research, authorship, and publication of this article. This work is supported by the National Natural Science Foundation of China (No. 51979227).

Institutional Review Board Statement: Not applicable.

Informed Consent Statement: Not applicable.

Data Availability Statement: The statement does not create new data.

Conflicts of Interest: The authors declare no conflict of interest.

Nomenclatures and Abbreviations

V_{stream}	Jet /suction flow rate
θ_{stream}	Direction angle
C_l	Lift coefficient
C_d	Drag coefficient
C_p	Pressure coefficient
$p_{surface}$	Surface pressure
p_0	Reference pressure
C_v	Velocity coefficient
V_{point}	Local fluid flow velocity
R	Lift-to-drag ratio
S_h	Wetted surface area
F_L	Lift force
F_D	Drag force
v_∞	Flow velocity at infinity
ρ	Density of seawater
p	Fluid pressure
μ	Dynamic viscosity coefficient
$\bar{u}$,$\bar{v}$,$\bar{w}$	Time-averaged velocity component
u',v',w'	Pulsating velocity components
S_{fx},S_{fy},S_{fz}	Generalized source terms
L	Spread length of BWB-UG
f	Discrete solution of the numerical calculation;
h	Grid feature size
r	Refinement ratio of the grid
F_s	Grid convergence security factor

References

1. D'Spain, G.L.; Jenkins, S.A.; Zimmerman, R.; Luby, J.C.; Thode, A.M. Underwater acoustic measurements with the Liberdade/X-Ray flying wing glider. *J. Acoust. Soc. Am.* **2005**, *117*, 2624. [CrossRef]
2. Du, X.; Song, B.w.; Pan, G. Dynamics and Simulation of Underwater Glide Vehicle. In Proceedings of the 2010 International Conference on Digital Manufacturing Automation, Changcha, China, 18–20 December 2010; Volume 1, pp. 597–599.
3. Du, X.; Ali, N. Numerical Computation of Wave Forces on Blended Winged-Body Underwater Glider using Panel Method. In Proceedings of the 2021 International Bhurban Conference on Applied Sciences and Technologies (IBCAST), Islamabad, Pakistan, 12–16 January 2021; pp. 924–931.
4. Du, X.; Zhang, L. Analysis on energy consumption of blended-wing-body underwater glider. *Int. J. Adv. Robot. Syst.* **2020**, *17*, 1729881420920534. [CrossRef]
5. Li, J.; Wang, P.; Chen, X.; Dong, H. Shape Optimization of Blended-Wing-Body Underwater Gliders Based on Free-Form Deformation. *J. Northwestern Polytech. Univ.* **2020**, *38*, 459–464.
6. Sun, C.; Song, B.; Wang, P. Parametric geometric model and shape optimization of an underwater glider with blended-wing-body. *Int. J. Nav. Archit. Ocean Eng.* **2015**, *7*, 995–1006. [CrossRef]
7. Sun, C.; Song, B.; Wang, P.; Wang, X. Shape optimization of blended-wing-body underwater glider by using gliding range as the optimization target. *Int. J. Nav. Archit. Ocean Eng.* **2017**, *9*, 693–704. [CrossRef]
8. Wang, X.; Song, B.; Wang, P.; Sun, C. Hydrofoil optimization of underwater glider using Free-Form Deformation and surrogate-based optimization. *Int. J. Nav. Archit. Ocean Eng.* **2018**, *10*, 730–740. [CrossRef]
9. Ganesh, N.; Arunvinthan, S.; Nadarajapillai, S. Effect of surface blowing on aerodynamic characteristics of tubercled straight wing. *Chin. J. Aeronaut.* **2019**, *32*, 1111–1120. [CrossRef]
10. Fatahian, E.; Nichkoohi, A.L.; Salarian, H.; Khaleghinia, J. Effects of the hinge position and suction on flow separation and aerodynamic performance of the NACA 0012 airfoil. *J. Braz. Soc. Mech. Sci. Eng.* **2020**, *42*, 86. [CrossRef]
11. Giorgi, M.G.D.; Luca, C.G.D.; Ficarella, A.; Marra, F.S. Comparison between synthetic jets and continuous jets for active flow control: Application on a NACA 0015 and a compressor stator cascade. *Aerosp. Sci. Technol.* **2015**, *43*, 256–280. [CrossRef]
12. Rezaeiha, A.; Montazeri, H.H.; Blocken, B. Active flow control for power enhancement of vertical axis wind turbines: Leading-edge slot suction. *Energy* **2019**, *189*, 116131. [CrossRef]
13. Lienhart, H.; Becker, S. Flow and Turbulence Structure in the Wake of a Simplified Car Model. *SAE Trans.* **2003**, *112*, 785–796.
14. McNally, J.; Fernandez, E.J.; Robertson, G.; Kumar, R.; Taira, K.; Alvi, F.S.; Yamaguchi, Y.; Murayama, K. Drag reduction on a flat-back ground vehicle with active flow control. *J. Wind Eng. Ind. Aerodyn.* **2015**, *145*, 292–303. [CrossRef]
15. Rouméas, M.; Gilliéron, P.; Kourta, A. Drag reduction by flow separation control on a car after body. *Int. J. Numer. Methods Fluids* **2009**, *60*, 1222–1240. [CrossRef]
16. Genç, M.S.; Kaynak, U.; Yapici, H. Performance of transition model for predicting low Re aerofoil flows without/with single and simultaneous blowing and suction. *Eur. J. Mech. B-Fluids* **2011**, *30*, 218–235. [CrossRef]
17. Huang, L.; Huang, P.G.; LeBeau, R.P.; Hauser, T. Numerical study of blowing and suction control mechanism on NACA0012 airfoil. *J. Aircr.* **2004**, *41*, 1005–1013. [CrossRef]
18. Xu, H.Y.; Dong, Q.; Qiao, C.L.; Ye, Z. Flow Control over the Blunt Trailing Edge of Wind Turbine Airfoils Using Circulation Control. *Energies* **2018**, *11*, 619. [CrossRef]
19. Yousefi, K.; Saleh, R.; Zahedi, P. Numerical study of blowing and suction slot geometry optimization on NACA 0012 airfoil. *J. Mech. Sci. Technol.* **2014**, *28*, 1297–1310. [CrossRef]
20. Du, X.; Liu, X.; Song, D. Coupled physics analysis of blended-wing-body underwater glider equipped with electromagnetic active flow control. *Ocean Eng.* **2023**, *278*, 114402. [CrossRef]
21. Baker, N.; Kelly, G.; O'Sullivan, P.D. A grid convergence index study of mesh style effect on the accuracy of the numerical results for an indoor airflow profile. *Int. J. Vent.* **2019**, *19*, 300–314. [CrossRef]
22. Roache, P.J. Error Bars for CFD. In Proceedings of the 41st Aerospace Sciences Meeting and Exhibit, Reno, NV, USA, 6–9 January 2003; p. 2003-408.

Journal of
Marine Science and Engineering

MDPI

Article

Underwater Small Target Detection Based on YOLOX Combined with MobileViT and Double Coordinate Attention

Yan Sun [1], Wenxi Zheng [1,*], Xue Du [2] and Zheping Yan [2]

1 College of Information and Communication Engineering, Harbin Engineering University, Harbin 150001, China
2 College of Intelligent Systems Science and Engineering, Harbin Engineering University, Harbin 150001, China; duxue@hrbeu.edu.cn (X.D.); yanzheping@hrbeu.edu.cn (Z.Y.)
* Correspondence: zhengwenxi@hrbeu.edu.cn

Abstract: The underwater imaging environment is complex, and the application of conventional target detection algorithms to the underwater environment has yet to provide satisfactory results. Therefore, underwater optical image target detection remains one of the most challenging tasks involved with neighborhood-based techniques in the field of computer vision. Small underwater targets, dispersion, and sources of distortion (such as sediment and particles) often render neighborhood-based techniques insufficient, as existing target detection algorithms primarily focus on improving detection accuracy and enhancing algorithm complexity and computing power. However, excessive extraction of deep-level features leads to the loss of small targets and decrease in detection accuracy. Moreover, most underwater optical image target detection is performed by underwater unmanned platforms, which have a high demand of algorithm lightweight requirements due to the limited computing power of the underwater unmanned platform with the mobile vision processing platform. In order to meet the lightweight requirements of the underwater unmanned platform without affecting the detection accuracy of the target, we propose an underwater target detection model based on mobile vision transformer (MobileViT) and YOLOX, and we design a new coordinate attention (CA) mechanism named a double CA (DCA) mechanism. This model utilizes MobileViT as the algorithm backbone network, improving the global feature extraction ability of the algorithm and reducing the amount of algorithm parameters. The double CA (DCA) mechanism can improve the extraction of shallow features as well as the detection accuracy, even for difficult targets, using a minimum of parameters. Research validated in the Underwater Robot Professional Contest 2020 (URPC2020) dataset revealed that this method has an average accuracy rate of 72.00%. In addition, YOLOX's ability to compress the model parameters by 49.6% efficiently achieves a balance between underwater optical image detection accuracy and parameter quantity. Compared with the existing algorithm, the proposed algorithm can carry on the underwater unmanned platform better.

Keywords: YOLOX; underwater target detection; MobileVIT; coordinate attention

Citation: Sun, Y.; Zheng, W.; Du, X.; Yan, Z. Underwater Small Target Detection Based on YOLOX Combined with MobileViT and Double Coordinate Attention. *J. Mar. Sci. Eng.* **2023**, *11*, 1178. https://doi.org/10.3390/jmse11061178

Academic Editor: Rafael Morales

Received: 26 April 2023
Revised: 25 May 2023
Accepted: 31 May 2023
Published: 5 June 2023

1. Introduction

The rapid advancement of science and technology has led to a shift in human behavior and spurred an interest in the exploitation of aquatic resources at the expense of land-based ones [1]. Since 70% of the planet is covered by water, there are plenty of underwater resources to drive future advances in science and technology, and, as the forerunner of subsurface resource utilization, underwater exploration has had ample time to make significant advances. Several researchers have devoted themselves to the study of underwater optical image detection, a cornerstone of modern underwater detection that has been used successfully in many areas of ocean exploration. However, underwater detection is limited by several issues that do not impede land detection. The complexity of the underwater imaging environment causes problems, such as location dispersion, color deviation, and blur, to name just a few. Furthermore, determining the target volume of an underwater object presents additional challenges. These factors make conventional target detection techniques incapable of meeting the demands of the modern industry. In addition to these issues, the limited use of underwater detection and underwater unmanned platforms transport has resulted in storage limitations for underwater equipment and vehicles, and the typically large models that land algorithms are based on are difficult to transport in underwater environments. Therefore, the development of an underwater optical image detection method with low parameters and high accuracy that is suited to the needs of underwater unmanned platforms is essential.

Existing target detection algorithms are mainly divided into two types: two-stage [2–4] and one-stage [5–8]. The former has stronger detection accuracy but a complex structure, while the latter has lower accuracy but a lightweight structure. In order to ensure the carrying on the underwater unmanned platform, we chose YOLOX [9] in the first-order algorithm. On the basis of inheriting CSPDarknet53 [10] and the feature pyramid network (FPN) [11] of YOLO series algorithms, YOLOX applies frameless detection in the YOLO algorithm for the first time, which reduces the computational complexity of YOLOX. However, due to the complex feature extraction of CSPDarknet, it is not friendly to the detection of underwater targets. In addition, the underwater unmanned platform has higher requirements for the storage of the algorithm, but the existing lightweight algorithm has insufficient feature extraction ability. Therefore, we chose the MobileVIT [12] lightweight model as the backbone network of the algorithm. In addition, in order to extract the shallow information of the target better, we proposed a new attention mechanism DCA based on CA [13] attention and applied it to YOLOX, so that the algorithm can obtain higher accuracy. Experiments show that the accuracy of URPC2020 data is up to 72.00% and the number of parameters is reduced by 49.6%.

With this background in mind, we present the main contributions of this paper:

(1) Within the mainstream of target detection methods, we chose YOLOX algorithms as the basic structure. By using MobileViT as the backbone network in YOLOX, we further improved the global feature extraction ability of the algorithm while reducing the number of parameters.

(2) To address problems posed by underwater targets characterized by small volumes, scattered distributions, and blurred imaging, we designed a DCA mechanism based on prior CA mechanisms. By improving the shallow feature extraction ability of the algorithm model, we enhanced its ability to extract the data of difficult targets.

(3) The results of our evaluation of the URPC2020 dataset show that our network model has better accuracy compared with the baseline method while reducing the number of parameters. Therefore, our method is not only feasible but also superior to the original baseline method.

2. Related Work

2.1. Object Detection

Researchers in the field of underwater target detection have applied convolutional neural networks (CNNs) extensively [14], and the existing target detection algorithms based on CNNs are mainly divided into two types: one-stage algorithms and two-stage algorithms. Two-stage algorithms, on the other hand, extract a series of candidate regions and classify them for target detection. Two-stage algorithms include the R-CNN [2], Fast R-CNN [3], and FasterR-CNN [4], among others. Two-stage algorithms have high accuracy, but their detection efficiency is lower than that of one-stage algorithms. One-stage algorithms use a first-order network to complete classification and location tasks, greatly improving detection efficiency and achieving a good balance between accuracy and algorithm volume. Some first-stage algorithms include the Single-Shot MultiBox Detector (SSD) [15] and the YOLO (You Only Look Once) [5–8] series. Although the YOLO series algorithm achieves a good balance in accuracy and complexity, the YOLO series algorithm has a poor detection effect on small targets and low recall rate, which affects the application of the YOLO algorithm underwater. Therefore, researchers have carried out a lot of studies based on YOLO series algorithms.

Chen et al. [16] proposed an underwater target recognition network based on improved YOLOv4. Lei et al. [17] applied YOLOv5 in underwater target detection. These tasks integrate the YOLO series into underwater target detection, Chen et al. [18]. Proposed an underwater target detection lightweight algorithm based on multi-scale feature fusion but do not take into account the hardware limitations of underwater detection.

Underwater targets are frequently dispersed, resulting in the loss of small targets, and the deep extraction of features often produces images that are blurred [19]. Therefore, there is a need to improve algorithms' abilities to extract shallow information, the low-level features and patterns typically captured by the initial layers of a neural network. Our proposed model extracts shallow information by utilizing an attention mechanism during the shallow information extraction stage.

In order to solve the problems surrounding feature extraction, we propose the introduction of a transformer [20]. Although transformers are still in their early stages, transformer-based models have already achieved excellent results in the field of natural language processing. In 2020, researchers applied the transformer model to computer vision [21] domain tasks and achieved promising results. Algorithms built on transformer-based models are better able to focus on feature extraction and, therefore, have stronger global feature extraction capabilities. Lei et al. [17] proposed YOLOv5 combined with a Swin Transformer. Chen et al. [22] proposed a lightweight underwater target detection algorithm based on dynamic sampling transformer and knowledge-distillation optimization, but they ignored the cost of the transformer.

Researchers are also starting to incorporate lightweight transformers into the algorithms they use for underwater detection. In 2021, MobileVIT combined a transformer with a lightweight algorithm for the first time, which can greatly reduce the amount of calculation, while guaranteeing the feature extraction capability, and can meet the needs of underwater unmanned platforms. Thus, we chose MobileVIT [12] to combined with YOLOX [9].

2.2. Lightweight Network and Attention

Considering the storage limitations of underwater unmanned platforms, underwater target detection algorithms normally face additional challenges due to large amount of algorithm parameters, which limits their speed and efficiency. To address this issue, lightweight algorithm research for target detection has attracted the attention of researchers in recent years and they have sought to develop algorithms optimized for resource-constrained devices. MobileNet (v1/v2) [23,24], SqueezeNet [25], ShuffleNet [26], GhostNet [27], and other lightweight CNN architectures have all been developed to provide deep learning algorithms that can extract features more efficiently by improving their convolution methods. Yeh et al. [28] proposed a lightweight deep neural network for joint learning of underwater object detection and color conversion. Wang et al. [29] applied a novel attention-based lightweight network for multiscale object detection in underwater images. The downside is that the use of lightweight methods inevitably leads to reductions in detection accuracy.

In underwater target detection, attention mechanisms are frequently used in feature extraction, and, in mobile networks, attention mechanisms have proven their usefulness in computer vision through their ability to achieve efficient feature extraction at a relatively low cost. The squeeze-and-excitation network (SENet) [30], convolutional block attention module (CBAM) [31], and CA mechanism [13] are among the definitive attention mechanisms noted for their high efficiency. SENet compresses and maps 2D features to prioritize informative channels; CBAM further improves on spatial information coding by applying a large-size kernel to the feature map and using convolutional layers. Zhang et al. [32] proposed underwater object detection based on YOLOv4 and multi-scale attentional feature fusion. Li et al. [33] applied YOLOv4 combined channel attention to detect underwater biology. In order to enhance the feature extraction capability of the algorithm, we designed a new attention mechanism named DCA. DCA can integrate feature information extracted by CNN and the transformer to further enhance the shallow information extraction of the algorithm.

The rest of this paper is organized as follows: in Section 3, we introduce the YOLOX algorithm, MobileViT algorithm, and CA attention mechanism based on which the algorithm in this paper is based; in Section 4, we introduce the new DCA attention mechanism proposed in this paper and the improved algorithm combining the MobileViT and YOLOX algorithm; in Section 5, we verify the effectiveness of the algorithm through experiments; and in the fifth chapter, we draw a conclusion. The flow chart of this paper is shown in Figure 1.

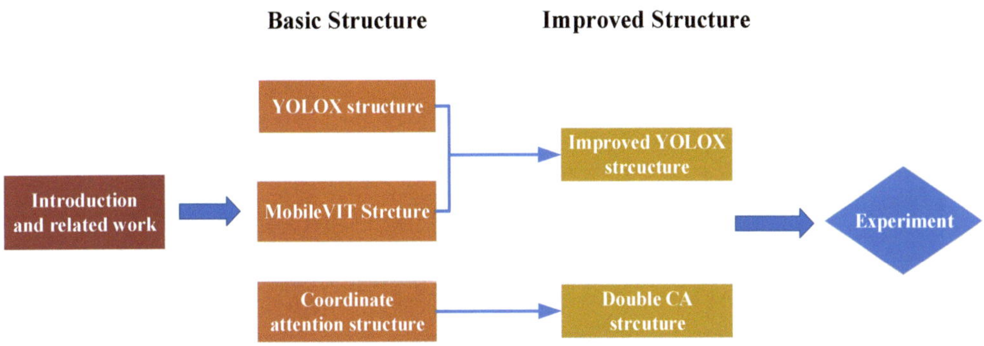

Figure 1. The flow chart of algorithm structure.

3. Basic Structure

3.1. YOLOX Structure

YOLOX improves on prior YOLO object detection algorithms, which are a series of single-shot detectors developed by Megvii Technology in 2021. A key innovation of YOLOX is the compartmentalization of tasks performed by separate components, such as the image pre-processor, the backbone network, the neck network, and the predictive head network. YOLOX uses mosaic data enhancement during the image pre-processing stage and selects four images from the dataset for stitching and testing, which can enrich image backgrounds. In the backbone network, YOLOX uses CSPDarknet CNNs for feature extraction, consistent with previous generations of YOLO algorithms. FPN and PANNet are used in the neck network to combine features from different layers, allowing shallow information to guide the deep information, thereby retaining the position information of the input image. A major innovation of YOLOX is its decoupled head architecture; the head network makes predictions about objects in the input image—such as target category, background, and coordinate information—before relegating the classification and regression tasks into separate modules. This improves the expression ability and accuracy of the algorithm and accelerates convergence. In addition, YOLOX optimizes the number of parameters and further improves the accuracy of the algorithm using anchor-free [34] detection and SimOTA [35] tag allocation. Despite its excellent detection accuracy, YOLOX operates at a low cost; therefore, we chose YOLOX as our baseline method. In order to extract special features of underwater targets, we optimized the original YOLOX by further improving the model's detection accuracy and reducing the number of parameters. The YOLOX structure is shown in Figure 2.

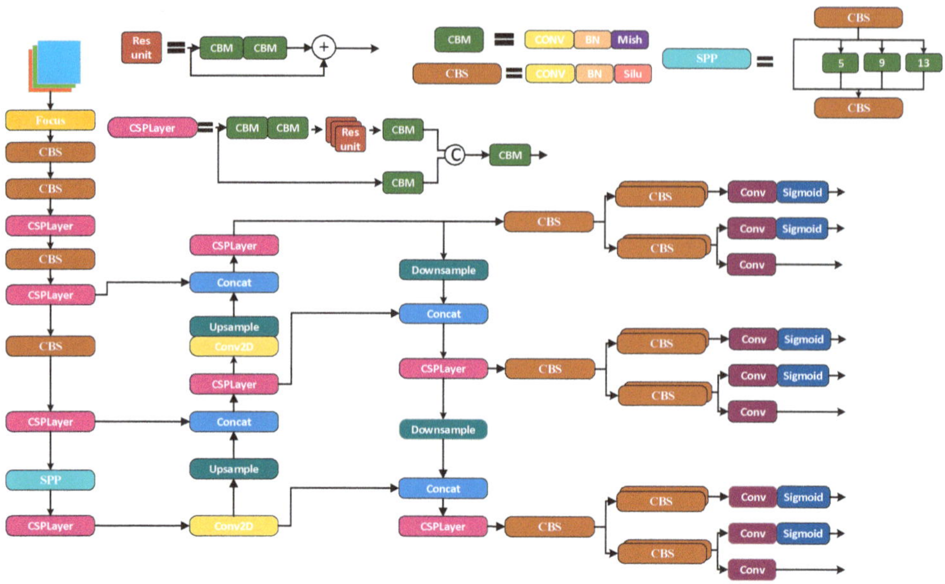

Figure 2. The network structure of YOLOX.

3.2. MobileVIT Structure

MobileViT is a computer vision model that combines mobile-friendly CNNs with vision transformers. CNNs focus on the extraction of local information, ignoring correlations within this information, and their use of excessive convolution leads to the loss of key information from the target. Compared with CNNs, transformer methods perform better in global feature extraction, and transformers are also better able to identify correlations between adjacent positions, improving how the shallow information of an image is saved. However, because the transformer is a heavyweight model lacking the inductive bias that would allow it to migrate directly to target detection, it often results in algorithms that operate poorly. MobileViT, on the other hand, combines CNN and transformer layers, resulting in a model featuring the efficient and lightweight aspects of CNNs as well as the strong overall vision capacity of the transformer method. The two core components of MobileViT are MobileViTBlock and MobileNetV2Block. MobileNetV2Block is the inverted residual block component in MobileNetV2. The Mobilenet v2 block is shown in Figure 3

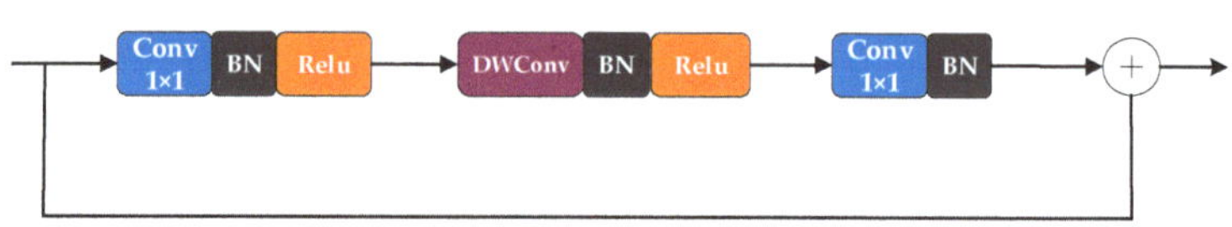

Figure 3. The network structure of MobileNet v2Block.

This component enables an algorithm to retain its CNN while significantly reducing the amount of calculation and number of parameters required, while additionally avoiding the need for the extensive transformer operations typically performed by MobileViT. MobileNetBlock, a building block in MobileNet, first adjusts the number of channels through 3×3 convolution and 1×1 convolution, and then extracts global features through the unfold, transformer, and fold technique. After the channel is adjusted via 1×1 convolution, it is concatenated with the original special diagnosis map by shortcut, and then feature fusion is performed by 3×3 convolution. The MobileVIT block is shown in Figure 4

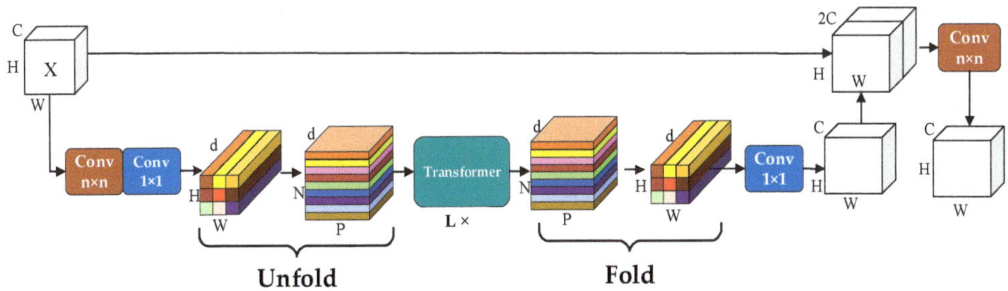

Figure 4. The network structure of MobileVIT block.

3.3. Coordinate Attention

Attention mechanisms have been proven to significantly improve the performance of neural networks. However, due to the large amount of computation required by self-attention modules (also known as self-attention mechanisms), they can only be used in large models and are not suitable for mobile networks. Most of the attention mechanisms applicable to mobile networks are based on global pooling, which reduces the spatial dimensions of a feature map to a single value per channel—ignoring the location information of the features. This lack of position information affects the structure of the object captured by target detection. In addition, CNNs can only capture local relationships; our proposed coordination attention mechanism solves this problem by decomposing channel attention into two one-dimensional feature coding processes, which are characterized in two directions. This allows for the capture of remote dependencies in one direction while retaining accurate location information in the other direction. The two directional feature maps are then combined with the input feature maps to enhance the direction and position information of the target. This method allows an algorithm to focus on the global area of a network at a low computational cost, while the use of two parallel one-dimensional feature decoding can reduce the feature information loss caused by global pooling.

We use spatial extents of pooling kernels $(H,1)$ or $(1,W)$ to encode each channel along the horizontal coordinate and the vertical coordinate separately. Thus, the output of the c-th channel at height h can be formulated as:

$$Z_c^w(w) = \frac{1}{W} \sum_{0 \leq i < W} x_c(h,i) \tag{1}$$

Similarly, the output of the c-th channel at width w can be formulated as:

$$Z_c^w(w) = \frac{1}{H} \sum_{0 \leq i < H} x_c(j,w) \tag{2}$$

Specifically. Equations (1) and (2) produce the aggregated feature maps, and then we start to concatenate and then send them to a shared 1×1 convolutional transformation function F_1,which can be formulated as:

$$\mathrm{f} = \delta\left(F_1([z^h, z^w])\right) \tag{3}$$

where $[.,.]$ denotes the splicing operations along the spatial dimension, δ denotes the non-linear activation function, and $\mathrm{f} \in \mathbb{R}^{C/r \times (H+W)}$ is the intermediate feature map encoding spatial information in both horizontal and vertical directions. In addition, r is the reduction ratio. We then split f along the spatial dimension into two separate tensors, $\mathrm{f}^w \in \mathbb{R}^{C/r \times W}$ and $\mathrm{f}^h \in \mathbb{R}^{C/r \times H}$. F_h and F_w are 1×1 convolutional transformations which are utilized to separately transform f^h and f^w to tensors with the same channel number as the input, which can be formulated as:

$$g^h = \sigma\left(F_h(f^h)\right) \tag{4}$$

$$g^w = \sigma(F_w(f^w)) \tag{5}$$

where σ denote the sigmoid function. To satisfy the demand of the lightweight model, we frequently reduce the channel number f with an appropriate reduction ratio r. The outputs g^h and g^w are expanded and used as attention weights separately. Finally, Figure 5 shows the structure of the CA mechanism, and the output of the coordinate attention block can be formulated as:

$$y_c(i,j) = x_c(i,j) \times g_c^h(i) \times g_c^w(j) \tag{6}$$

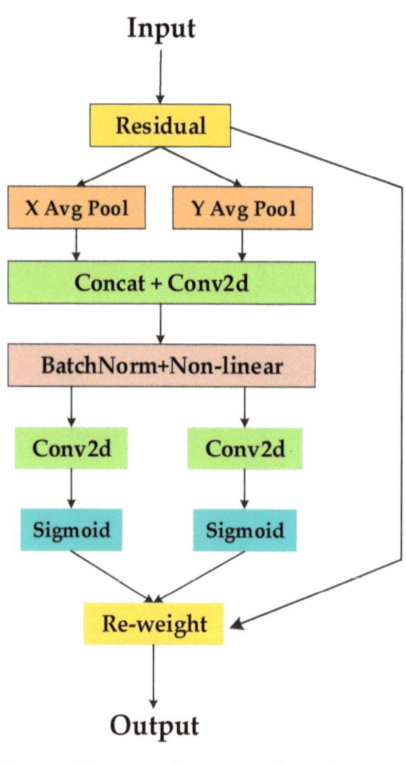

Figure 5. The network structure of coordinate attention.

4. Proposed Structure

4.1. Double Coordinate Attention

Underwater small target detection methods require improvements in shallow information feature extraction. To this end, we found that shallow feature extraction can be further strengthened by utilizing a CA mechanism and fusing the feature information in the backbone network before and after the block. Additionally, we aggregated the features in different directions, so that the algorithm can focus more attention on the location of shallow information. After obtaining the weight in each direction, it is combined with the input from both sides, allowing the shallow information of the output feature map to be better expressed. With this improved backbone network structure, we confirmed that MobileViT block does not change the structure of the input features when features are fused before or after this group of layers. Thus, we designed proposed coordinate attention named double coordinate attention (DCA). We tested this in consideration of how global features and local features are emphasized during feature extraction, and the results suggest distinct advantages for detecting the location of small targets.

We encoded the two input feature graphs Z_n and Z_v along two directions, respectively, and carried out feature convolution feature fusion on the feature codes in the same direction. Therefore, the horizontal output after convolution fusion Z^h can be formulated as:

$$Z^h = Z_n^h(h) \otimes Z_v^h(h) \tag{7}$$

Similarly, the vertical output Z^w can be formulated as:

$$Z^w = Z_n^w(w) \otimes Z_v^w(w) \tag{8}$$

where $\otimes$ represents the convolution operation. n and v represent the feature graphs of different inputs, respectively. Similar to the CA mechanism, we convolved the two input feature graphs and multiplied them with the attention weights of the two directions g^h and g^w to obtain the output Y, which can be formulated as:

$$y_c(i,j) = (x_n(i,j) \otimes x_v(i,j)) \times g_c^h(i) \times g_c^w(j) \tag{9}$$

By combining the two input weights and inputting the special detection map with attention into FPN, the multi-feature extraction can be further guided. The DCA attention mechanism extracted in this paper can not only retain the acquisition of global features in the transformer, but also take into account the capture of local features in CNN and enhance the feature extraction of shallow information. This can better satisfy the needs of underwater target detection. Figure 6 shows the structure of the DCA mechanism.

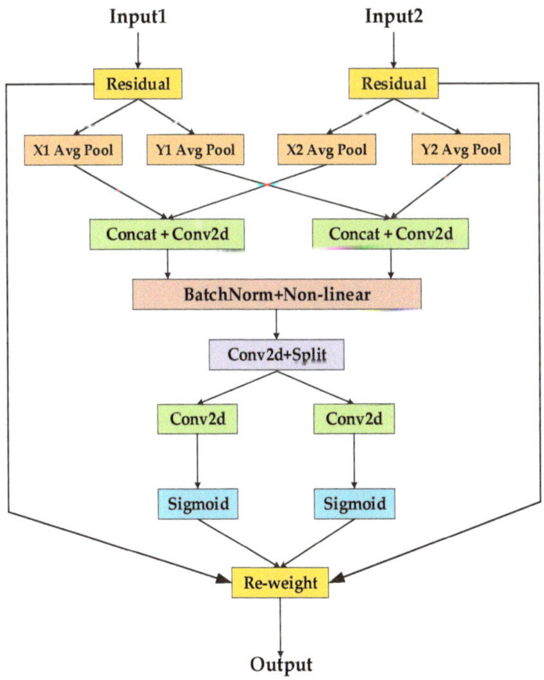

Figure 6. The network structure of double coordinate attention.

4.2. Improved Network Structure

Because YOLOX has advantages in image enhancement, target classification, and label classification in the YOLO series of algorithms, we optimized the backbone network and feature extraction. YOLOX follows the backbone network of the previous generation of YOLO algorithm, CSPDarknet53, which has excellent and different performance in feature extraction. However, because the extraction of shallow information for underwater small targets is very important, the multi-layer convolution of CSPDarknet will cause serious loss of shallow information, make the detection of small targets more difficult, and the multi-layer convolution layer will make the algorithm cumbersome. Therefore, we chose MobileVIT as the backbone network of the algorithm for small target feature extraction. MobileVIT combines the lightweight algorithm MobileNet and vision transformer. It not only retains the lightweight and efficient features of CNN, but also reflects the feature that the transformer focuses on global features. For the support of shallow information to small targets, we added the attention mechanism DCA when outputting the first two features, which can greatly enhance the extraction of shallow information in the algorithm and enhance the detection accuracy of underwater small targets. Figure 7 shows the structure of the proposed YOLOX.

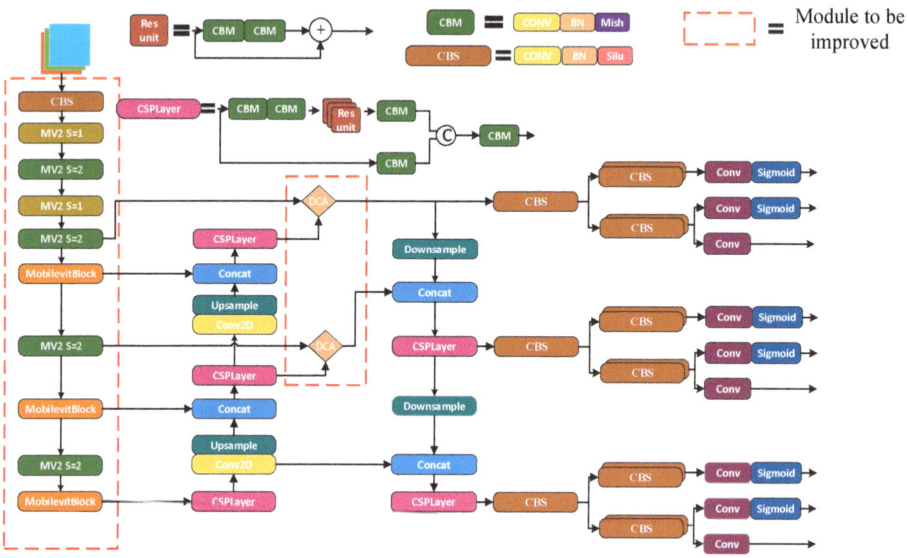

Figure 7. The network structure of improved YOLOX.

5. Experiment and Analysis

5.1. Datasets

This experiment uses the URPC2020 dataset for validation experiments, which includes five categories: starfish, scallops, waterweeds, echinus, and holothurian. It also included 5543 training images, of which aquatic plants were officially recognized as a negligible target, and it only contains 82 targets. However, in order to verify the algorithm's ability to detect small underwater targets, we still included aquatic plants in the detection category, and this paper ultimately uses 5543 images and five categories and, according to the ratio of 9:1, it is randomly divided into training set and validation set.

5.2. Experimental Enviroment

The experimental environment is shown in Table 1. The hardware used for this experiment was a Ryzen 7 3700x; NVIDIA RTX 3090 graphics card; Ubuntu 18.04

operating system; CUDA version 11.6; PyTorch version 1.11.0; and Python environment 3.9.12.

Table 1. The environment for the experiments.

Environment	Versions or Model Number
CPU	Ryzen 7 3700x
GPU	NVIDIA RTX 3090
OS	Ubuntu 16.04
CUDA	11.6
Python	V1.11.0
Pytorch	V3.9.12

5.3. Parameter Settings

We validated the improved YOLOX algorithm proposed in this paper on the URPC2020 dataset. The input resolution was uniformly resized to 608 × 608. For the task, the original YOLOX-S model was adopted as the basis for experimental comparison. For fair comparison, we trained all models across 300 epochs by using SGD with weight decay of 0.0005. The initial learning rate was set to 0.001 and the batch size was set to 16. All other parameters were kept the same as YOLOX-S.

5.4. Results and Analysis

We used the URPC dataset to train and test the model. mAP was the standard to measure the accuracy of the model in target detection. Since the model size was also one of the evaluation criteria in this paper, we also took the parameter size, Flops, as one of the measurement criteria. In order to verify the effectiveness of the experiment, this paper uses the ablation experiment, in which Model 1 uses the main framework of the YOLOX algorithm, and replaces the backbone network of YOLOX with MobileVIT. Among the models, Model 1 adopts the main framework of the YOLOX algorithm, while replacing the backbone network of YOLOX with MobileViT, and replacing the convolution in FPN with depthwise separable convolution. The rest of the network structure is consistent with YOLOX. Model 2 is based on Model 1 and adds CA attention mechanism to the two layers of shallow information output, with the rest remaining unchanged. Model 3 is based on Model 2 and replaces the shallowest layer's CA attention mechanism with the DCA attention mechanism, with the rest remaining unchanged. Model 4 is based on Model 2 and replaces both instances of the CA attention mechanism with the DCA attention mechanism, with the rest remaining unchanged.

The results in Table 2 show that using MobileVIT as the backbone network in this paper demonstrates the advantage of MobileVIT in reducing the number of parameters and improving the accuracy of target detection. Furthermore, the added attention mechanism proves that the coordinate attention mechanism can less sacrifice the parameter and Flops, and the DCA attention mechanism proposed in this paper can also better control the number of parameters. Model 4, proposed in this paper, can meet the carrying demand of underwater unmanned platforms.

Table 2. Parameter of ablation experiment.

Model	Method				Parameter(M)	Flops
	Baseline	Dw	CA	DCA		
YOLOX	√				8.94	26.64
Model 1		√			4.37	24.88
Model 2		√	√		4.39	24.92
Model 3		√	√	√	4.42	25.18
Model 4		√		√	4.51	25.35

The results in Table 3 show that Model 1 proposed in this paper demonstrates the advantages of the transformer algorithm in feature extraction, and Model 4, compared with Model 2 and Model 3, proves that the proposed DCA attention mechanism can perform better in shallow information extraction. Furthermore, the water seeds category listed in this paper has a significant increase in mAP compared with the baseline algorithm, proving that the DCA attention mechanism has a better prospect for detecting difficult and small targets.

Table 3. Results of ablation experiment on the URPC2020 dataset.

Model	mAP	Holothurian	Echinus	Starfish	Scallop	Waterweeds
YOLOX	66.92	67.00	87.09	79.49	83.05	17.96
Model 1	68.69	71.64	87.13	80.59	83.23	20.87
Model 2	71.01	73.24	86.93	80.42	82.81	31.66
Model 3	70.75	75.23	87.19	80.29	82.84	28.22
Model 4	72.00	73.42	87.37	79.40	82.97	36.87

The results in Table 4 show thatcompared with existing algorithms, the proposed algorithm achieves the best balance in precision and parameter quantity, and proves that the proposed algorithm can be carried on the underwater unmanned platform. Figure 8 shows the results of the proposed method for the detection.

Table 4. Result of different algorithms on the URPC2020 dataset.

Model	mAP	Holothurian	Echinus	Starfish	Scallop	Parameter (M)
YOLOv4	81.01	71.21	89.94	85.58	77.30	64.04
T-YOLOv4	68.69	54.09	80.43	77.94	58.87	5.96
YOLOX	79.16	67.00	87.09	79.49	83.05	8.94
Model 4	80.79	73.42	87.37	79.40	82.97	4.51

Note: Data YOLOv4 and T-YOLOv4 are quoted from the literature [32].

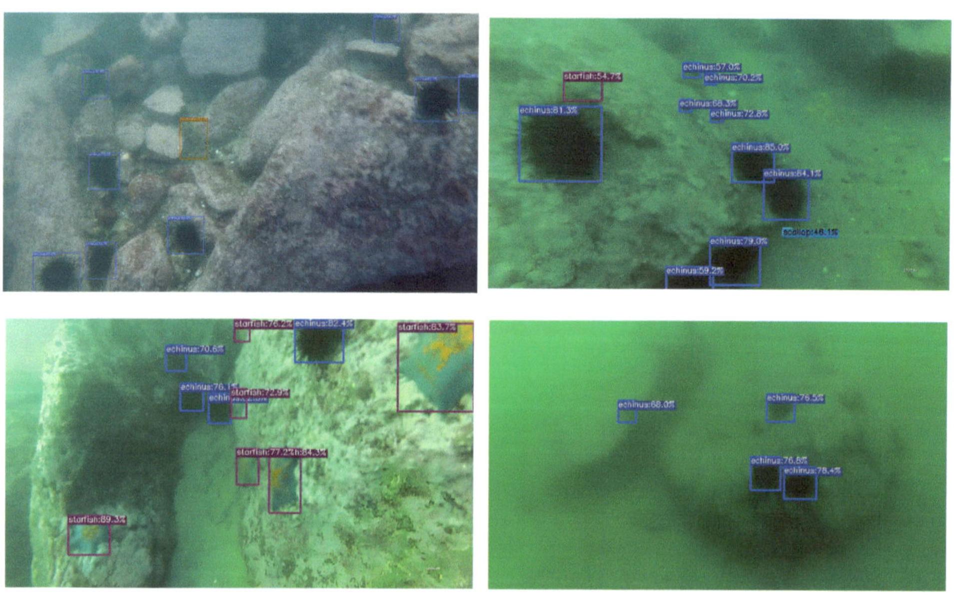

Figure 8. The detection results of our method in URPC2020.

6. Discussion

6.1. Underwater Target Detection Combined with Transformer

CNN-based object detection algorithms have always been a research hotspot and have been considered the foundation of object detection algorithms for many years. However, in underwater environments, the lack of global features and target loss caused by multiple convolutions limit the development of object detection algorithms. Previously, transformers have been widely used in the field of natural language processing. It was not until 2021 that transformers entered the field of image processing and achieved good results. Therefore, this paper focuses on combining MobileViT with YOLOX for object detection. Experiments show that this method can improve the accuracy of object detection while reducing the number of parameters. Thus, the transformer's ability to extract features from targets is more suitable for underwater imaging environments, reducing target loss and demonstrating better accuracy for detecting difficult targets. However, since transformers need to consider global features, the detection speed will face significant challenges. Therefore, the optimization of the network structure should also consider improving the detection speed. In addition, compared with the existing YOLO algorithm, CNN-based deep feature information extraction still has certain advantages in high-precision categories. Therefore, further experiments are needed to combine CNN with the transformer.

6.2. Challenges of Underwater Small Target Detection

Existing detection algorithms enhance object detection accuracy through data augmentation, multi-feature fusion, and attention mechanisms. Among them, the impact of feature maps at different scales in multi-feature fusion on small object detection varies. In the experiments, we added attention mechanisms at different scales to enhance feature extraction, but the experimental results show that the large scale has a much greater impact on shallow information extraction than the small scale, and adding attention at the small scale may even reduce algorithm accuracy. Therefore, for underwater small targets, there should be more improvements in feature fusion. In addition, attention mechanisms mostly act as removable components added to existing algorithms, and experimental results show that attention mechanisms perform well in underwater small object detection. However, attention mechanisms rely too heavily on the structure of existing algorithms, and the improvement of experimental results is uncertain. The DCA attention mechanism proposed in this paper is not universal in improving the accuracy of target detection, and some categories are difficult to be improved. Therefore, there is still considerable room for the development of attention mechanisms in the field of underwater object detection.

6.3. Future Research Focus on Underwater Small Target Detection

From the experimental results, we analyzed that the reason why sea urchins can obtain the best performance is that the color of the sea urchin is monotonous and has obvious color difference with the color of underwater imaging, and the underwater dataset is large. Compared with other targets with high accuracy, the sea cucumber has a similar color and smaller size to the underwater background. Moreover, compared with the YOLOX algorithm, the detection rate of sea-participating seagrass has been significantly improved after combining with MobileViT feature extraction. Therefore, we believe that shallow feature extraction is positive for the detection of small underwater targets. Therefore, in the detection of small underwater targets, it is necessary to strengthen the shallow features. In addition, since the detection category with color difference has higher accuracy, it may bring unexpected results to be added into the target detection algorithm, such as reducing underwater color bias and increasing contrast.

7. Conclusions

This paper proposes an underwater object detection model that provides a good balance between accuracy and memory. In our work process, in order to establish an underwater object detection algorithm that can be mounted on underwater unmanned platforms, we chose YOLOX as the base algorithm and used MobileViT as the backbone network to replace YOLOX's CSPDarknet53 for extracting global features from images. At the same time, depthwise separable convolution is used in the neck for reconstruction to control the number of parameters in the algorithm. We also improved the DCA attention mechanism based on the CA attention mechanism to enhance feature extraction capability between the backbone network and the neck, strengthening the global and shallow features of the algorithm. This is beneficial for underwater unmanned platforms to extract small and difficult targets in water. Experiments show that the proposed network is feasible. Our algorithm reduces the number of parameters by 49.6% compared with YOLOX, and the accuracy on the URPC dataset is still improved, especially in the detection of small targets in the dataset. The results show that the algorithm has certain advantages and performs better in terms of parameter count and accuracy compared with the listed algorithms.

In our future work, we will further advance the balance between parameter count and accuracy in underwater object detection. We will also consider incorporating algorithm speed into the algorithm evaluation. The following methods will become important directions for our upcoming work: model pruning, numerical acceleration techniques, and loss function improvements.

Author Contributions: Conceptualization, Y.S. and W.Z.; methodology, W.Z.; software, W.Z.; validation, Y.S., W.Z., X.D. and Z.Y.; formal analysis, Y.S.; investigation, W.Z. and X.D.; resources, Y.S.; data curation, Y.S.; writing—original draft preparation, W.Z.; writing—review and editing, Y.S. and X.D.; visualization, W.Z.; supervision, Y.S.; project administration, Y.S. and Z.Y.; funding acquisition, Y.S., X.D. and Z.Y. All authors have read and agreed to the published version of the manuscript.

Funding: This research was funded by National Natural Science Foundation of China, grant number 52171297.

Institutional Review Board Statement: Not applicable.

Informed Consent Statement: Not applicable.

Data Availability Statement: Not applicable.

Conflicts of Interest: The authors declare no conflict of interest.

References

1. Sahoo, A.; Dwivedy, S.K.; Robi, P.S. Advancements in the field of autonomous underwater vehicle. *Ocean. Eng.* **2019**, *181*, 145–160. [CrossRef]
2. Girshick, R.; Donahue, J.; Darrell, T.; Malik, J. Rich feature hierarchies for accurate object detection and semantic segmentation. In Proceedings of the IEEE Conference on Computer Vision and Pattern Recognition, Columbus, OH, USA, 23–28 June 2014.
3. Girshick, R. Fast R-CNN. In Proceedings of the IEEE International Conference on Computer Vision 2015, Santiago, Chile, 7–13 December 2015.
4. Ren, S.; He, K.; Girshick, R.; Sun, J. Faster R-CNN: Towards real-time object detection with region proposal networks. *Adv. Neural Info Rmation Process. Syst.* **2015**, *28*, 91–99. [CrossRef]
5. Redmon, J.; Divvala, S.; Girshick, R.; Farhadi, A. You Only Look Once: Unified, Real-Time Object Detection. In Proceedings of the CVPR, Las Vegas, NV, USA, 27–30 June 2016.
6. Redmon, J.; Farhadi, A. YOLO9000: Better, Faster, Stronger. In Proceedings of the IEEE Conference on Computer Vision & Pattern Recognition, Honolulu, HI, USA, 21–26 June 2017; pp. 6517–6525.
7. Redmon, J.; Farhadi, A. Yolov3: An incremental improvement. *arXiv* **2018**, arXiv:1804.02767.
8. Bochkovskiy, A.; Wang, C.Y.; Liao, H.Y.M. Yolov4: Optimal speed and accuracy of object detection. *arXiv* **2020**, arXiv:2004.10934.
9. Ge, Z.; Liu, S.; Wang, F.; Li, Z.; Sun, J. Yolox: Exceeding yolo series in 2021. *arXiv* **2021**, arXiv:2107.08430.
10. Wang, C.-Y.; Mark Liao, H.-Y.; Wu, Y.-H.; Chen, P.-Y.; Hsieh, J.-W.; Yeh, I.-H. CSPNet: A New Backbone That Can Enhance Learning Capability of CNN. In Proceedings of the 2020 IEEE/CVF Conference on Computer Vision and Pattern Recognition Workshops (CVPRW), Seattle, WA, USA, 14–19 June 2020; pp. 1571–1580.
11. Lin, T.-Y.; Dollar, P.; Girshick, R.; He, K.; Hariharan, B.; Belongie, S. Feature Pyramid Networks for Object Detection. In Proceedings of the 2017 IEEE Conference on Computer Vision and Pattern Recognition (CVPR), Honolulu, HI, USA, 21–26 July 2017; pp. 936–944.
12. Mehta, S.; Rastegari, M. Mobilevit: Light-weight, general-purpose, and mobile-friendly vision transformer. *arXiv* **2021**, arXiv:2110.02178.

13. Hou, Q.; Zhou, D.; Feng, J. Coordinate attention for efficient mobile network design. In Proceedings of the IEEE/CVF Conference on Computer Vision and Pattern Recognition, Nashville, TN, USA, 20–25 June 2021; pp. 13713–13722.
14. Lu, H.; Li, Y.; Zhang, Y.; Chen, M.; Serikawa, S.; Kim, H. Underwater Optical Image Processing: A Comprehensive Review. *Mobile Netw. Appl.* **2017**, *22*, 1204–1211. [CrossRef]
15. Liu, W.; Anguelov, D.; Erhan, D.; Szegedy, C.; Reed, S.; Fu, C.Y.; Berg, A.C. SSD: Single shot multibox detector. In Proceedings of the European Conference on Computer Vision, Amsterdam, The Netherlands, 11–14 October 2016.
16. Chen, L.; Zheng, M.; Duan, S.; Luo, W.; Yao, L. Underwater Target Recognition Based on Improved YOLOv4 Neural Network. *Electronics* **2021**, *10*, 1634. [CrossRef]
17. Lei, F.; Tang, F.; Li, S. Underwater Target Detection Algorithm Based on Improved YOLOv5. *J. Mar. Sci. Eng.* **2022**, *10*, 310. [CrossRef]
18. Chen, L.; Yang, Y.; Wang, Z.; Zhang, J.; Zhou, S.; Wu, L. Underwater Target Detection Lightweight Algorithm Based on Multi-Scale Feature Fusion. *J. Mar. Sci. Eng.* **2023**, *11*, 320. [CrossRef]
19. Liu, B. Research on Feature Extraction and Target Identification in Machine Vision Underwater and Surface Image. Ph.D. Thesis, Dalian University of Technology, Dalian, China, 2013.
20. Vaswani, A.; Shazeer, N.; Parmar, N.; Uszkoreit, J.; Jones, L.; Gomez, A.N.; Lukasz, K.; Polosukhin, I. Attention is all you need. *Adv. Neural Inf. Process. Syst.* **2017**, *30*.
21. Dosovitskiy, A.; Beyer, L.; Kolesnikov, A.; Weissenborn, D.; Zhai, X.; Unterthiner, T.; Houlsby, N. An image is worth 16×16 words: Transformers for image recognition at scale. *arXiv* **2020**, arXiv:2010.11929.
22. Chen, L.; Yang, Y.; Wang, Z.; Zhang, J.; Zhou, S.; Wu, L. Lightweight Underwater Target Detection Algorithm Based on Dynamic Sampling Transformer and Knowledge-Distillation Optimization. *J. Mar. Sci. Eng.* **2023**, *11*, 426. [CrossRef]
23. Howard, A.G.; Zhu, M.; Chen, B.; Kalenichenko, D.; Wang, W.; Weyand, T.; Andreetto, M.; Adam, H. Mobilenets: Efficient convolutional neural networks for mobile vision applications. *arXiv* **2017**, arXiv:1704.04861.
24. Howard, A.; Zhmoginov, A.; Chen, L.C.; Sandler, M.; Zhu, M. Inverted residuals and linear bottlenecks: Mobile networks for classification, detection and segmentation. *arXiv* **2018**, arXiv:1801.04381.
25. Iandola, F.N.; Han, S.; Moskewicz, M.W.; Ashraf, K.; Dally, W.J.; Keutzer, K. SqueezeNet: AlexNet-level accuracy with 50x fewer parameters and <0.5 MB model size. *arXiv* **2016**, arXiv:1602.07360.
26. Zhang, X.; Zhou, X.; Lin, M.; Sun, J. Shufflenet: An extremely efficient convolutional neural network for mobile devices. In Proceedings of the IEEE Conference on Computer Vision and Pattern Recognition, Salt Lake City, UT, USA, 18–23 June 2018; pp. 6848–6856.
27. Han, K.; Wang, Y.; Tian, Q.; Guo, J.; Xu, C.; Xu, C. Ghostnet: More features from cheap operations. In Proceedings of the IEEE/CVF Conference on Computer Vision and Pattern Recognition, Seattle, WA, USA, 13–19 June 2020; pp. 1580–1589.
28. Yeh, C.H.; Lin, C.H.; Kang, L.W.; Huang, C.H.; Lin, M.H.; Chang, C.Y.; Wang, C.C. Lightweight deep neural network for joint learning of underwater object detection and color conversion. *IEEE Trans. Neural Netw. Learn. Syst.* **2021**, *33*, 6129–6143. [CrossRef]
29. Wang, J.; He, X.; Shao, F.; Lu, G.; Jiang, Q.; Hu, R.; Li, J. A Novel Attention-Based Lightweight Network for Multiscale Object Detection in Underwater Images. *J. Sens.* **2022**, *2022*. [CrossRef]
30. Hu, J.; Shen, L.; Sun, G. Squeeze-and-excitation networks. In Proceedings of the IEEE Conference on Computer Vision and Pattern Recognition, Salt Lake City, UT, USA, 18–23 June 2018; pp. 7132–7141.
31. Woo, S.; Park, J.; Lee, J.Y.; Kweon, I.S. Cbam: Convolutional block attention module. In Proceedings of the European Conference on Computer Vision (ECCV), Munich, Germany, 8–14 September 2018; pp. 3–19.
32. Zhang, M.; Xu, S.; Song, W.; He, Q.; Wei, Q. Lightweight Underwater Object Detection Based on YOLO v4 and Multi-Scale Attentional Feature Fusion. *Remote Sens.* **2021**, *13*, 4706. [CrossRef]
33. Li, A.; Yu, L.; Tian, S. Underwater Biological Detection Based on YOLOv4 Combined with Channel Attention. *J. Mar. Sci. Eng.* **2022**, *10*, 469. [CrossRef]
34. Tian, Z.; Shen, C.; Chen, H.; He, T. Fcos: Fully convolutional one-stage object detection. In Proceedings of the ICCV, Seoul, Korea, 27 October–2 November 2019.
35. Ge, Z.; Liu, S.; Li, Z.; Yoshie, O.; Sun, J. Ota: Optimal transport assignment for object detection. In Proceedings of the CVPR, Montreal, BC, Canada, 11–17 October 2021.

Journal of
Marine Science and Engineering

MDPI

Article

Physical Consistent Path Planning for Unmanned Surface Vehicles under Complex Marine Environment

Fang Wang [1,2], Yong Bai [3] and Liang Zhao [3,*]

1 School of Information and Electrical Engineering, Hangzhou City University, Hangzhou 310015, China; wangfang_hz@163.com
2 Science and Technology on Underwater Vehicle Technology Laboratory, Harbin Engineering University, Harbin 150001, China
3 College of Civil Engineering and Architecture, Zhejiang University, Hangzhou 310058, China; baiyong@zju.edu.cn
* Correspondence: liamzhao13@zju.edu.cn

Abstract: The increasing demand for safe and efficient maritime transportation has underscored the necessity of developing effective path-planning algorithms for Unmanned Surface Vehicles (USVs). However, the inherent complexities of the ocean environment and the non-holonomic properties of the physical system have posed significant challenges to designing feasible paths for USVs. To address these issues, a novel path planning framework is elaborately designed, which consists of an optimization model, a meta-heuristic solver, and a Clothoid-based path connector. First, by encapsulating the intricate nature of the ocean environment and ship dynamics, a multi-objective path planning problem is designed, providing a comprehensive and in-depth portrayal of the underlying mechanism. By integrating the principles of the candidate set random testing initialization and adaptive probability set, an enhanced genetic algorithm is devised to fully exploit the underlying optimization problem in constrained space, contributing to the global searching ability. Accounting for the non-holonomic constraints, the fast-discrete Clothoid curve is capable of maintaining and improving the continuity of the path curve, thereby promoting strong coordination between the planning and control modules. A thorough series of simulations and comparisons conducted in diverse ocean scenarios has conclusively demonstrated the effectiveness and superiority of the proposed path planning framework.

Keywords: path planning; unmanned surface vehicles; path smoothing; multi-objective; genetic algorithm

Citation: Wang, F.; Bai, Y.; Zhao, L. Physical Consistent Path Planning for Unmanned Surface Vehicles under Complex Marine Environment. *J. Mar. Sci. Eng.* **2023**, *11*, 1164. https://doi.org/10.3390/jmse11061164

Academic Editor: Sergei Chernyi

Received: 11 May 2023
Revised: 23 May 2023
Accepted: 30 May 2023
Published: 1 June 2023

1. Introduction

With artificial intelligence at the helm, the advent of unmanned surface vehicles (USVs) has garnered significant attention, fueled by their potential to revolutionize maritime operations by enhancing safety and efficiency [1–6]. However, the successful deployment of USVs depends on the development of autonomous technology, which refers to the ability of these vehicles to plan and execute their missions in complex environments without human intervention, thereby enabling safe and efficient navigation [7,8]. Generally speaking, the most common approaches that contribute to the autonomous level of USVs are perception, localization and mapping, path planning and decision-making, and control system design. Central to achieving autonomy in USVs is the challenge of path planning, which involves determining an optimal path for the vehicle to traverse in order to accomplish its mission objectives while adhering to a set of predetermined rules and regulations. Compared to other types of autonomous vehicles, such as unmanned ground vehicles (UGVs), USV path planning may incorporate specialized techniques to handle challenges, such as wave prediction models, collision avoidance strategies for vessels, or algorithms, that account for hydrodynamic effects on the vehicle's motion. This task is particularly challenging due to the dynamic nature of maritime environments, which are subject to constantly changing weather conditions, currents, and other environmental factors that can impact navigation [9]. Achieving this goal requires the development of sophisticated path planning algorithms that enable these vehicles to navigate complex environments with minimal human intervention, paving the way for a future of safe and efficient maritime navigation [10].

The field of path planning for USVs has been an active area of research in recent years, with numerous studies investigating the development of effective planning strategies for USVs. In general, two primary categories of path planning algorithms have been proposed: global approaches and local approaches [7]. Global approaches involve the generation of a complete path for the USV based on prior knowledge of the environment, usually represented as a map. Such methods typically employ high-level planning techniques that treat the USV as a point object, neglecting its maneuverability and physical constraints. These methods are, therefore, more suitable for planning routes for long-distance voyages, where the emphasis is on efficient and safe navigation over extended periods. In contrast, local approaches generate a path by utilizing local information collected during the mission, enabling the USV to adapt to unexpected obstacles or changes in the environment. These methods fully consider the physical bounds of the USV's mechanical system, leading to more precise tracking performance for the low-level controller. Although the design of such methods is generally more complex, as it requires the integration of high-level planning and low-level control techniques to ensure effective operation, it is more applicable in practice.

Presently, there is a growing affinity for deterministic approaches in path planning, with various methods, such as A* and D* lite, basking in the limelight of scientific popularity. In particular, Yu and Wang [11] have put forward a hybrid algorithm that fuses artificial potential field (APF) and D* lite to navigate complex environments. This approach not only minimizes time cost but also enhances path safety through the APF. Nonetheless, it overlooks disturbances and energy consumption. Similarly, Yu et al. [12] have proposed an improved D* lite that reduces expanded nodes, validated via simulations. However, the simulation fails to consider ship dynamics, smoothness, and safety. Meanwhile, Song et al. [13] have utilized various smoothing techniques to mitigate the jagged effect in A*, which has been demonstrated to be effective through both experiments and simulations, making it a practical choice. Furthermore, Shah and Gupta [14] have presented a quadtree representation of the marine environment, which accelerates the A* algorithm without significantly sacrificing solution optimality, as shown in simulations. To facilitate path planning for working ships in offshore wind farms, Xie et al. [15] devised a multi-direction A* algorithm modified by an artificial potential field. Compared with the real-case trajectory, the minimum distance to the wind turbines has increased, and the path length outside the wind farm decreased dramatically. To solve the path planning problem under changing environments with multiple dynamic obstacles, Yao et al. [16] proposed an Improved D* lite algorithm, which has demonstrated its efficacy in real-time path planning through simulation results. Although deterministic approaches have emerged as popular and reliable methods for path planning, these methods can be computationally expensive, particularly when operating in large and complex environments. This can have a significant impact on their performance, making them less suitable for real-time applications.

As a result, the meta-heuristic algorithms, including ant colony optimization (ACO) [17], particle swarm optimization (PSO) [18], and genetic algorithms (GA) [19], have emerged as promising alternatives for path planning in marine robotics. These algorithms offer a set of high-level strategies to search for solutions, allowing them to optimize paths while considering multiple objectives with a comparatively low computational burden. Considering the effects of currents, Krell et al. [20] devised an improved PSO method implemented in visibility graphs. For the safe navigation of ships, a quasi-reflection-based PSO was proposed by Xue [21]. Incorporating the environmental loads, a hierarchical path planning framework based on GA is developed by Wang and Xu [10]. For rapid path generation, a leader-vertex ant colony optimization algorithm is proposed by Liang et al. [22], which ensures a leader of the ant colony and optimizes the route by vertex method. For both global and local path planning, a series of studies on artificial fish swarm algorithms have been conducted by Zhao et al. [7,8,23]. Under different current distribution, a comprehensive study has been conducted by Ma et al. [24] using multi-objective dynamic augmented PSO. Organically bridging the planning and tracking, Wang et al. [25] devised an elite-duplication GA (EGA) strategy to optimally generate sparse waypoints in a constrained space. However, meta-heuristic algorithms commonly encounter a significant hurdle with regard to their global searching capabilities, as they are prone to be trapped in local minima or suboptimum, thereby impeding the identification of the global optimum necessary for producing high-quality trajectories. Additionally, the computational efficiency of existing methods is not satisfactory enough to facilitate efficient path generation within high-dimensional configuration spaces [26]. Therefore, there is an imperative need for innovative techniques that can enhance the global searching capability and convergence rate of meta-heuristic algorithms for path planning.

In addition, a noteworthy limitation of most existing methods is the neglect of the non-holonomic constraints of the vehicle, which can lead to paths that are potentially infeasible. Specifically, the USV, being a non-holonomic robot, often functions as an underactuated system during its missions, resulting in limited maneuverability and motion flexibility [27,28]. Analogous to unmanned ground vehicles (UGVs), this restricts the USV's motion to forward velocity and manipulation of the heading angle to attain its desired position, thus precluding lateral movement. Consequently, it is of paramount importance to ensure smooth and continuous transitions of yaw and curvatures at turning points in order to devise an effective trajectory for a USV. For instance, sharp turns may be deemed unfeasible for a USV due to the significant sideslip that ensues, deviating from the planned path. Thus, the motion dynamics of the USV should be meticulously accounted for in the path planning process [10].

Inspired by the aforementioned literature review, this paper proposes a novel GA-variant meta-heuristic algorithm in combination with a fast-discrete Clothoid curve to optimize path generation. The main contributions of this paper are illustrated as follows:

- By capturing the non-holonomic nature of USVs and the intricate ocean dynamics, a sophisticated optimization model is carefully devised for the path planning problem, whereby the effects of currents, increments of curvatures, and constraints of physical system are addressed jointly.
- Introducing the random testing initialization algorithm and the adaptive design in the selection procedure, the proposed GA-variant facilitates strong global searching capabilities and a fast convergence rate, thereby contributing to the optimal generation of waypoint sequence.
- Accommodating the non-holonomic constraints, the fast-discrete Clothoid curve is able to preserve and enhance the continuity of the path curve, resulting in robust coordination between the planning and control module.

This paper is organized in the following structures: Section 2 presents the detailed modeling of the environment, USV, and the optimization model. Section 3 introduces the methodology. Illustrative simulation results are shown in Section 4. The conclusion is drawn in Section 5.

2. Problem Formulation

2.1. Environment Model

In this research, we consider the marine surface area represented as $\mathscr{M}$ in the Euclidean space $\mathbb{R}^2$. $\mathscr{M}$ is divided into obstacle area $\mathscr{M}_o$ and obstacle-free motion area $\mathscr{M}_f$, respectively. The relationship between these two grids is illustrated in Equation (1). The path P of the USV is defined as a sequence of connected elementary waypoints, denoted by $\boldsymbol{p}_i (i = 1, 2, 3, \ldots, m)$. By following the path P, the USV moves from the initial position $\boldsymbol{p}_S(x_S, y_S)$ to the final position $\boldsymbol{p}_E(x_E, y_E)$ while avoiding numerous obstacle areas $\mathscr{M}_o$ ($\mathscr{M}_o = \{O_1, O_2, \ldots, O_k\}$, where k denotes the number of obstacles).

$$\mathcal{M}_f + \mathcal{M}_o = \mathcal{M} \tag{1}$$

Accordingly, to guarantee the safety, the generated path should be restricted to $\mathcal{M}_f$ domain which is given as:

$$P = \cup_{i=1}^{m} \boldsymbol{p}_i \subset \mathcal{M}_f \tag{2}$$

2.2. Currents Model

According to previous research [20], ocean currents have a significant impact on the energy consumption of USVs. Therefore, when carrying out activities, USVs tend to choose a path that allows them to take advantage of the currents. Two types of currents exist: fixed and time-varying. Fixed currents are common in inland water voyages while time-varying currents are present in large-scale ocean environments. Assuming that the velocity of the USV in the body frame is $\boldsymbol{v} = [u,\ v,\ r]^T$ and the current velocity in the body frame is $\boldsymbol{v}_c$, the USV velocity, taking the effects of the currents into account, can be expressed as:

$$\boldsymbol{v}_r = \boldsymbol{v} + \boldsymbol{v}_c \tag{3}$$

2.3. USV Model

Typically, the motion of the USV can be regarded as a rigid body motion on the horizontal plane, as shown in Figure 1a, with three degrees of freedom: surge, sway, and yaw. Consequently, the state-space model for the USV, accounting for the impact of the current, can be derived as follows:

$$\begin{aligned} \dot{\boldsymbol{\eta}} &= \boldsymbol{R}(\psi)\boldsymbol{v_r} \\ \boldsymbol{M}\dot{\boldsymbol{v}}_{\boldsymbol{r}} + \boldsymbol{C}(\dot{\boldsymbol{v}}_{\boldsymbol{r}})\dot{\boldsymbol{v}}_{\boldsymbol{r}} + \boldsymbol{D}\boldsymbol{v_r} &= \boldsymbol{\tau} \end{aligned} \tag{4}$$

where the position and yaw angle in the earth-fixed inertial frame {n} are represented by the vector $\boldsymbol{\eta} = [x,\ y,\ \psi]^T$, while the relative velocities in the body-fixed frame {b} are denoted by $\boldsymbol{v_r} = [u_r,\ v_r,\ r]^T$, and the control signals are represented by the vector $\boldsymbol{\tau} = [\tau_u,\ 0,\ \tau_r]^T$. This paper considers the underactuated configuration of the USV, which means that the surge force and yaw moment are the only control forces. With these assumptions, the rotation matrix $\boldsymbol{R}(\psi)$, mass matrix $\boldsymbol{M}$, Coriolis matrix $\boldsymbol{C}$, and damping matrix $\boldsymbol{D}$ can be expressed as:

$$\begin{aligned}
\boldsymbol{R}(\psi) &= \begin{bmatrix} cos(\psi) & -sin(\psi) & 0 \\ sin(\psi) & cos(\psi) & 0 \\ 0 & 0 & 1 \end{bmatrix} \\
\boldsymbol{M} &= \begin{bmatrix} m - X_{\dot{u}} & 0 & 0 \\ 0 & m - Y_{\dot{v}} & 0 \\ 0 & 0 & I_z - N_{\dot{r}} \end{bmatrix} = \begin{bmatrix} m_{11} & 0 & 0 \\ 0 & m_{22} & 0 \\ 0 & 0 & m_{33} \end{bmatrix} \\
\boldsymbol{C} &= \begin{bmatrix} 0 & 0 & -(m - Y_{\dot{v}})v \\ 0 & 0 & (m - X_{\dot{u}})u \\ (m - Y_{\dot{v}})v & -(m - X_{\dot{u}})u & 0 \end{bmatrix} \\
&= \begin{bmatrix} 0 & 0 & -m_{22}v \\ 0 & 0 & m_{11}u \\ m_{22}v & -m_{11}u & 0 \end{bmatrix} \\
\boldsymbol{D} &= \begin{bmatrix} d_{11} & 0 & 0 \\ 0 & d_{22} & 0 \\ 0 & 0 & d_{33} \end{bmatrix}
\end{aligned} \tag{5}$$

To represent the USV's orientation relative to the earth-fixed inertial frame, we use a rotation matrix $\boldsymbol{R}(\psi)$ that transforms the body-fixed frame. The mass matrix $\boldsymbol{M} = \boldsymbol{M}^{\boldsymbol{T}} > 0$ takes into account the USV's inertial properties and hydrodynamic added mass. The matrix $\boldsymbol{D}$ incorporates damping coefficients, while the Coriolis matrix $\boldsymbol{C}$ captures the Coriolis and centripetal effects and can be obtained from $\boldsymbol{M}$.

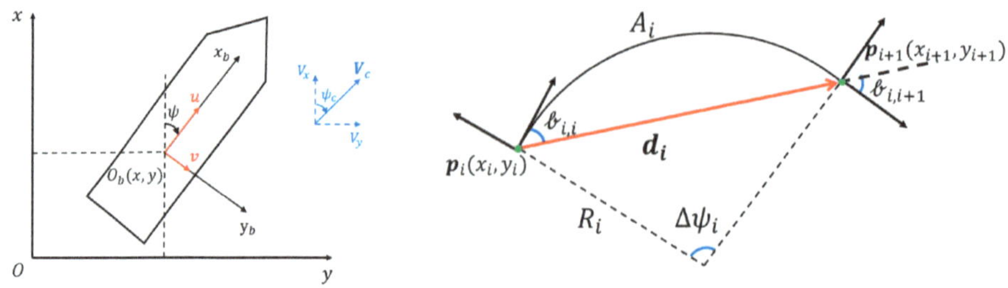

Figure 1. (**a**) Coordinate system; (**b**) Definition of a path curve.

Due to its non-holonomic nature, a USV typically has limited maneuverability and motion flexibility during most operations [28]. Similar to UGVs, the non-holonomic constraint restricts its lateral motion, meaning that the USV can only use its forward velocity while adjusting the heading angle to reach a desired position.

In Figure 1b, let $\boldsymbol{d}_i = [x_{i+1} - x_i, y_{i+1} - y_i, 0]^T$ denote the position vector between two points, and $\boldsymbol{p}_i$ and $\boldsymbol{p}_{i+1}$ represent the waypoints. The angles between $\boldsymbol{d}_i$ and $\boldsymbol{p}_i$ and $\boldsymbol{p}_{i+1}$ are defined as $\vartheta_{i,\,i}$ and $\vartheta_{i,\,i+1}$, respectively. In order to ensure a continuous path, the straight line and turning motions require that two consecutive positions $\boldsymbol{p}_i$ and $\boldsymbol{p}_{i+1}$ lie on a common arc of constant curvature, which can be expressed as:

$$\vartheta_{i,\,i} = \vartheta_{i,\,i+1} \tag{6}$$

Moreover, the turning angle at any point on the path must be restricted in the dynamic bounds, which gives:

$$\Delta\psi_i \leq \Delta\psi_{max} \tag{7}$$

2.4. Optimization Terms

2.4.1. Cruising Time

Since the path length and energy consumption objectives are interdependent, modifying the design variables affects both objectives equally. To jointly represent these objectives, we have adopted the cruising time t as the first objective. Let $\boldsymbol{v}_i$ be the velocity of the USV at $\boldsymbol{p}_i$ and $\boldsymbol{v}_c$ be the current velocity. The resultant velocity is denoted by $\boldsymbol{v}_r = \boldsymbol{v}_i + \boldsymbol{v}_c$. The cruising time t_i between $\boldsymbol{p}_i$ and $\boldsymbol{p}_{i+1}$ can be calculated as follows:

$$t_i = \frac{||\boldsymbol{p}_{i+1} - \boldsymbol{p}_i||}{\boldsymbol{v}_{i,c}} \tag{8}$$

Then, the total cruising time is calculated by:

$$\mathscr{T} = \sum_{i=1}^{m-1} t_i \tag{9}$$

The calculation of $\boldsymbol{v}_c$ is performed using the current distribution function. In most cases, $\boldsymbol{v}_i$ is considered to be a constant value in the same direction as $\boldsymbol{p}_i$. Therefore, t_i represents the nominal cost of travel time and does not accurately reflect the actual travel time.

2.4.2. Smoothness and Continuity

The additional cost incurred due to yaw is closely linked to the motion control performance of the USV. In order to enhance the smoothness of the trajectory, an objective function is introduced. To achieve this, constraint (9) is added as a quadratic penalty term to the objective function. The turning angle $\Delta\psi_i$ between waypoints $\boldsymbol{p}_i$ and $\boldsymbol{p}_{i-1}$ in the path P is calculated as $\Delta\psi_i = |\psi_i - \psi_{i-1}|$, where $\psi_i = atan((y_i - y_{i-1})/(x_i - x_{i-1}))$ and $\psi_{i-1} = atan((y_{i-1} - y_{i-2})/(x_{i-1} - x_{i-2}))$. The smoothest path is achieved when the changes in $\Delta\psi_i$ $(i = 2, 3, \ldots, m)$ and curvature $|\mathscr{b}_{i-1,\,i-1} - \mathscr{b}_{i-1,\,i}|$ are minimized. Hence, the objective function for achieving the smoothest path is defined as:

$$\vartheta = \sum_{i=2}^{m} \Delta\psi_i + \sum_{i=2}^{m} |\mathscr{b}_{i-1,\,i-1} - \mathscr{b}_{i-1,\,i}| \tag{10}$$

2.4.3. Path Safety

To ensure the safe movement of the USV, it is crucial to find a collision-free path that also maintains a safe distance from the obstacles. In addition to satisfying the conditions outlined in Equation (2), the minimum clearance from obstacles, denoted as d_i, is used to determine the safety of the solution. Specifically, we define two circular areas with radii d_{min} and d_{max} around each path waypoint $\boldsymbol{p}_i$. The distance between each path waypoint $\boldsymbol{p}_i$ and its closest obstacle O_i $(O_i \subset \mathscr{M}_o)$ is represented by $d_i = ||\boldsymbol{p}_i, O_i||, (i = 1, 2, 3, \ldots, m)$. The safety of each point along the path can then be evaluated by:

$$\mathscr{D}_i = \begin{cases} 0, & d_i \geq d_{max} \\ \frac{d_{max} - d_i}{d_{max} - d_{min}}, & d_{min} < d_i < d_{max} \\ 1, & d_i \leq d_{min} \end{cases}, \; i = 1, 2, 3, \ldots, m \tag{11}$$

Therefore, to ensure the safety of the path, the third objective is to minimize the minimum value of $\mathscr{D}_i$, see the following:

$$\mathscr{D} = \text{argmin}\,\{\mathscr{D}_1, \mathscr{D}_2, \ldots, \mathscr{D}_i\}, \; i = 1, 2, 3, \ldots, m \tag{12}$$

2.5. Problem Statement

The path planning model for the problem is formed as:

$$\min \mathscr{J} = \mathscr{T} + \vartheta + \mathscr{D} \tag{13}$$

s.t.

$$\begin{aligned}
&\mathscr{M}_f = \mathscr{M} - \mathscr{M}_o \\
&P = \cup_{i=1}^{m} \boldsymbol{p}_i \subset \mathscr{M}_f, \; i = 1, 2, 3, \ldots, m \\
&\boldsymbol{p}_1 = (x_1, y_1) = (x_S, y_S) \\
&\boldsymbol{p}_m = (x_m, y_m) = (x_E, y_E) \\
&\boldsymbol{v} + \boldsymbol{v}_c \geq 0 \\
&\boldsymbol{v}_r = \boldsymbol{v} + \boldsymbol{v}_c, \; i = 1, 2, 3, \ldots, m \\
&\mathscr{b}_{i,\,i} = \mathscr{b}_{i,\,i+1}, \; i = 1, 2, 3, \ldots, m-1 \\
&t_i = \frac{||\boldsymbol{p}_{i+1} - \boldsymbol{p}_i||}{\boldsymbol{v}_{i,c}}, \; i = 1, 2, 3, \ldots, m-1 \\
&\mathscr{T} = \sum_{i=1}^{m-1} t_i, \; i = 1, 2, 3, \ldots, m \\
&\psi_i = atan((y_i - y_{i-1})/(x_i - x_{i-1})), \; i = 2, 3, \ldots, m \\
&\Delta\psi_i = |\psi_i - \psi_{i-1}|, \; i = 2, 3, \ldots, m \\
&\Delta\psi_i \leq \Delta\psi_{max} \\
&\vartheta = \sum_{i=2}^{m} \Delta\psi_i + \sum_{i=2}^{m} |\mathscr{b}_{i-1,\,i-1} - \mathscr{b}_{i-1,\,i}|, \; i = 2, 3, \ldots, m \\
&d_i = ||\boldsymbol{p}_i, O_i||, (i = 1, 2, 3, \ldots, m) \\
&\mathscr{D}_i = \begin{cases} 0, & d_i \geq d_{max} \\ \frac{d_{max} - d_i}{d_{max} - d_{min}}, & d_{min} < d_i < d_{max} \\ 1, & d_i \leq d_{min} \end{cases}, \; i = 1, 2, 3, \ldots, m \\
&\mathscr{D} = argmin\,\{\mathscr{D}_1, \mathscr{D}_2, \ldots, \mathscr{D}_i\}, \; i = 1, 2, 3, \ldots, m
\end{aligned} \tag{14}$$

3. Solver Design

3.1. Adaptive-Elite Genetic Algorithm

The genetic algorithm (GA) was initially proposed by Professor J. Holland in 1973 as a meta-heuristic optimization method. By simulating the evolutionary process of an artificial population, the GA manipulates each individual in the population through genetic operations, such as selection, crossover, and mutation. The process generates a new population with the best-performing individuals from the previous generation as the parents. The population evolves through several generations, and the individuals with the best fitness values are selected as the optimal solutions.

The GA's strength lies in its ability to search a large solution space using stochastic searches and evolutionary operations, such as crossover and mutation, making it effective in handling non-linear and non-convex optimization problems. Moreover, the GA's population size enables it to mitigate the impact of hyperparameter selection by allowing the algorithm to sample from a diverse set of solutions. Given that the optimization problem presented in this paper is an NP-hard nonlinear problem, we choose the GA as the primary framework.

3.1.1. Chromosome Representation

In evolutionary algorithms, chromosomes can be represented in various ways, such as binary-coded, real-coded, and decimal-coded. In our paper, we utilized the real-coded chromosome to directly represent the USV's path. Specifically, we use a sequence of points that begins at an origin position and ends at a destination point. Each point, denoted as $p_i = (x_i, y_i)$, is saved along with its x and y coordinates and a pointer to the next point in the path. Figure 2 illustrates this representation.

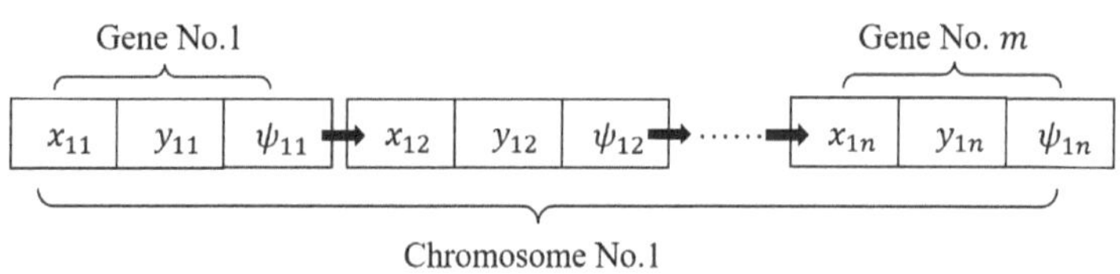

Figure 2. Definition of chromosome.

3.1.2. Initialization

The original population for the algorithm is obtained through population initialization. In the case of the original GA, a certain number of solutions are randomly generated in the solution space without the use of a heuristic function. This can lead to a random and unfocused solving process, resulting in a high proportion of poor solutions and low-quality genes in the population. This, in turn, requires a long convergence time during subsequent evolution and makes the solving process prone to being trapped in a local optimum.

To address this issue, we have developed a modified initialization method for GA inspired by failure analysis techniques used in software systems. Specifically, we have incorporated a candidate set adaptive random testing (ART) approach to improve the diversity of the initial population. By enhancing the initial diversity, the ART-based initialization method allows the GA to explore a broader range of potential solutions. This exploration can improve the algorithm's ability to escape local optima and discover better solutions in the search space. Consequently, it enhances the chances of finding high-quality solutions and can potentially accelerate the convergence toward optimal or near-optimal solutions. In summary, compared to standard random initialization, the ART-based initialization method in the GA offers the advantage of generating an initial population that is more diverse and better distributed throughout the search space. This increased diversity can facilitate improved exploration of the solution space and potentially lead to better overall performance and convergence in the GA.

The main steps of the initialization process are illustrated as follows:

Step 1: m candidate individuals $C = c_1, c_2, \ldots, c_m$ are randomly generated.

Step 2: The objective distances between each candidate individual and the current individuals in the population set $P = p_1, p_2, \ldots, p_n$ are calculated.

Step 3: The shortest distance between each candidate individual and the population set is identified.

Step 4: The candidate individual with the maximum distance value is selected and added to the population set P.

3.1.3. Selection Operator

In a genetic algorithm, the selection operator is responsible for choosing individuals from a population for the crossover operator. This selection process is carried out based on a predefined regulation called the Roulette Wheel Selection (RWS) method. To perform Roulette Wheel Selection (RWS) in a genetic algorithm (GA), first, we compute the fitness values for each individual in the population and normalize them to obtain probabilities. Then, calculate the cumulative probabilities by summing up the normalized fitness values. After that, generate a random number between 0 and 1, and select individuals whose cumulative probability exceeds this random number. Repeat the selection process as needed to obtain the desired number of parents. Use the selected parents for crossover or recombination to generate offspring for the next generation. This iterative process allows individuals with higher fitness to have a greater chance of being selected, promoting the propagation of favorable traits in the GA. The RWS method ensures that the fitter individuals have a higher chance of being selected for the crossover, thus improving the overall quality of the population in the subsequent generation. The selection probability of each individual can be expressed as follows:

$$P^i_{select} = F(x_i) / \sum_{i=1}^{n} F(x_i) \tag{15}$$

where x_i denotes the individual and $f(x_i)$ is the corresponding fitness value. As can be seen from Equation (15), better individuals have more chances to be selected by RWS, which leads to better solutions.

3.1.4. Hybrid Crossover

Crossover operators are utilized to combine two solutions and generate a new offspring with better performance in terms of a predefined objective. These operators can be applied to solutions with the same or different number of waypoints. The first crossover operator involves calculating the mean of the two parent solutions to produce a new offspring.

$$x^f = \frac{x_1^p + x_2^p}{2}, \quad y^f = \frac{y_1^p + y_2^p}{2} \tag{16}$$

where two parents have gene coordinates denoted by x_1^p, y_1^p, and x_2^p, y_2^p, respectively. The gene coordinates of the offspring are represented by x^f and y^f. First, two parent chromosomes are selected according to the selection operator. Second, we select one of the parents as a reference chromosome. In this procedure, if the number of waypoints in the parents are the same, we choose the reference randomly. Otherwise, the one with smaller waypoint number is chosen. Then, waypoints of the offspring $\left(x^f, y^f\right)$ are calculated by taking the mean of each waypoint of the reference chromosome and the nearest waypoint of the other parent. To determine the gene coordinates of the offspring $\left(x^f, y^f\right)$, the average of each gene in the selected parent and the closest gene in the other parent are computed.

To enhance the variability of the population and explore the entire available space, the second crossover operator is utilized in which the two parents are randomly merged:

$$x^f = kx_1^p + (1-k)x_2^p, \quad y^f = ky_1^p + (1-k)y_2^p \tag{17}$$

where vector k consists of random numbers ranging from -1 to 1. When the number of genes differs between the parents, a similar approach to the first operator is used to combine genes with minimal distance. The primary aim of the first operator is to escape local optima, while the second one explores the environment randomly, preventing premature convergence.

3.1.5. Mutation Operator

The mutation operator in a Genetic Algorithm has a critical function in maintaining diversity within the population. The primary goal of the mutation operator is to randomly modify the value(s) of one or more genes within an individual's chromosome. By introducing such changes, the mutation operator assists in preventing the GA from becoming trapped in a local optimum, which would hinder the search for the global optimum. In the absence of the mutation operator, the GA may converge to a suboptimal solution that is in close proximity to the initial population. Hence, mutation serves as a crucial component of GA by promoting exploration of the search space and preventing premature convergence to suboptimal solutions. This paper introduces two mutation operators to facilitate the genetic process:

The first operator is a random mutation that selects one position on chromosomes and changes the value in the free space as shown in the following figure:

A different mutation operator is utilized to enhance the path's smoothness and length by adjusting the position of a gene. The operator integrates the present position ($\boldsymbol{p}_i$) of a gene with the directions towards the genes located on either side, $\boldsymbol{p}_{i-1}$ and $\boldsymbol{p}_{i+1}$, using the subsequent expressions:

$$\begin{aligned} x_i^f &= x_i^p + m\left(x_{i-1}^p - x_i^p\right) + n\left(x_{i+1}^p - x_i^p\right) \\ y_i^f &= y_i^p + m\left(y_{i-1}^p - y_i^p\right) + n\left(y_{i+1}^p - y_i^p\right) \end{aligned} \quad (18)$$

where m and n are random positive coefficients from 0 to 1. As illustrated in Figure 3b, this mutation operator leads to the creation of paths with shorter lengths and better smoothness. Combining both mutation operators result in a powerful tool that enhances both the searching and convergence capabilities of the algorithm.

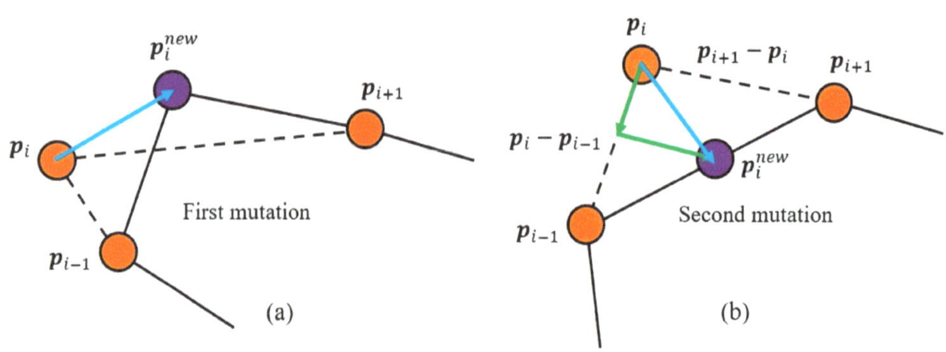

Figure 3. Mutation operators. (**a**) first mutation, (**b**) second mutation.

3.1.6. Fitness Design

In this paper, a multi-objective fitness function is devised. For this purpose, a weighted linear combination of the mentioned objectives is considered:

$$F = c_1 w_t \mathscr{T} + c_2 w_\theta \vartheta + c_3 w_d \mathscr{D} \quad (19)$$

The formula involves several parameters, including $\mathscr{T}$ (cruising time), ϑ (smoothness objective), and $\mathscr{D}$ (safety level), where w_t, w_θ, and w_d represent weight values, and their sum is equal to 1. To maintain consistency in the indicators' magnitudes, coefficients c_1, c_2, and c_3 have been set to 0.1, 1, and 100, respectively, as shown in the equation.

Selecting appropriate weight values is a vital aspect of the algorithm's performance. However, relying solely on empirical methods can be subjective. In an effort to achieve more balanced results, the Delphi weighting method [19] was employed to determine the weight of each indicator. As a result, the weight coefficients for cruising time, smoothness, and safety are 0.395, 0.275, and 0.330, respectively.

3.1.7. Determination of p_c and p_m

The conventional Genetic Algorithm for USV path planning relies on two essential parameters, namely, the crossover rate and mutation rate, to regulate p_c and p_m of individuals in each iteration. However, using fixed values for these parameters may pose certain challenges. For instance, employing a large crossover and mutation probability can make it difficult to retain the best individuals, slow down population convergence, and consequently, delay the generation of the inspection path, thereby impacting operational efficiency. Conversely, a small p_c and p_m can negatively affect the searching process, leading to the local optimum. This, in turn, causes the USV to travel longer distances, reducing its efficiency.

To tackle the aforementioned challenges, a modified approach is suggested. This approach involves adjusting p_c and p_m during the algorithm execution. Specifically, in the early stages of the algorithm, p_c and p_m are increased to improve the global search ability, while in the final stage, the probabilities are decreased to facilitate good convergence. Adaptive probabilities allow the GA to dynamically adjust the rates of crossover and mutation based on the progress of the algorithm. Initially, higher probabilities promote exploration by encouraging diverse offspring. As the GA progresses, the probabilities can be reduced, shifting the focus towards exploitation of promising solutions. This balance between exploration and exploitation helps the GA efficiently search for optimal or near-optimal solutions.

To achieve this, adaptive functions are formulated as follows:

$$p_c = p_{c0} e^{-\frac{a}{F}} \tag{20}$$

where p_{c0} represents the initial p_c, a is the scaling coefficient, and F here is the average fitness of the population. Similarly, the mutation probability can be obtained with same structure as follows:

$$p_m = p_{m0} e^{-\frac{b}{F}} \tag{21}$$

where p_{m0} represents the initial p_m and b is the scaling coefficient. These functions dynamically adjust the crossover probability based on the mean fitness degree of the population at each generation. As a result, the USV can achieve a balance between exploration and exploitation, leading to faster convergence and better results.

3.2. *Fast-Discrete Clothoid Curve*

To ensure real-time performance and accommodate the USV's kinematic constraints, we introduce a Fast-Discrete Clothoid Path (FDCP) to construct and connect the path. The FDCP employs a sequence of control points, referred to as waypoints, which are linked together using Clothoid segments. However, accurately generating Clothoids can be difficult due to their non-linear nature and multiple solutions. Thus, instead of directly computing the parameters of the Clothoid segments, our algorithm utilizes a variational approach that produces a polyline with linear discrete curvature, which approximates the Clothoid segment. This approach allows for efficient and precise path planning for the USV, while taking into account the vehicle's non-holonomic features.

To determine the position of intersection points, the following conditions must be met when inserting or updating point C between neighboring points B and D, as shown in Figure 4. To simplify the calculations, a normalized configuration is used where point B is located at $(-1, 0)$ and point D at $(1, 0)$. For each of the five control points denoted as

P, its left and right neighbors are identified as P_l and P_r, respectively. The insertion point C must satisfy the following conditions to approximate the Clothoids accurately between these control points:

- C must lie on the perpendicular bisector between B and D.
- The curvature at each point should vary linearly.

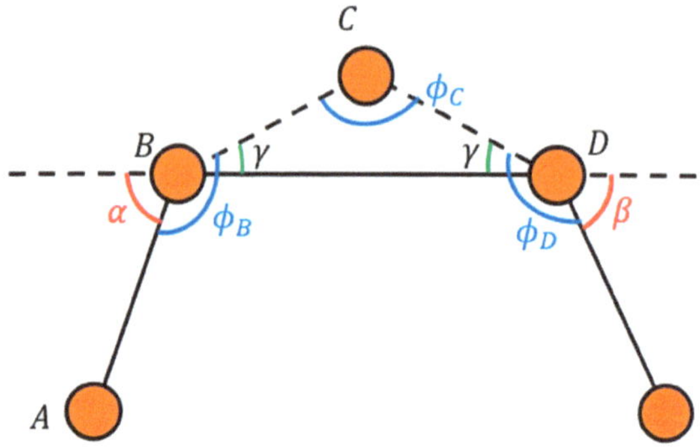

Figure 4. Locating point C.

From the abovementioned conditions, we have the following equation:

$$\rho_C = \frac{1}{2}(\rho_B + \rho_D) \tag{22}$$

where ρ denote the curvature at each point, it can be approximated by:

$$\rho_P = 2\frac{\pi - \phi_P}{|P_l P| + |PP_l|} \tag{23}$$

where ϕ_P is the angle between $|P_l P|$ and $|PP_l|$, and $|P_l P|$ is the scalar value of the length between P_l and P. According to the geometric relations in Figure 4, we obtain:

$$\begin{aligned} \phi_B &= \pi - \alpha + \gamma \\ \phi_C &= \pi - 2\gamma \\ \phi_D &= \pi - \beta + \gamma \end{aligned} \tag{24}$$

Substituting ϕ_P in previous equation using Equation (16), we have:

$$\begin{aligned} \rho_B &= 2\tfrac{\alpha-\gamma}{|AB|+|BC|} \\ \rho_C &= 2\tfrac{2\gamma}{|BC|+|CD|} \\ \rho_D &= 2\tfrac{\beta-\gamma}{|CD|+|DE|} \end{aligned} \tag{25}$$

As more and more points are inserted, the polyline gets refined and the angles between segments approach π. Therefore, with a large number of sample points, we can approximate $|BC| = |CD| = 1$. Solving Equation (14), γ can be obtained by:

$$\gamma = \frac{\beta(|AB| + 1) + \alpha(|DE| + 1)}{2|AB||DE| + 3(|AB| + |DE|) + 4} \tag{26}$$

Point C is now inserted on the perpendicular bisector between B and D in distance $|CD| \tan \gamma$. By iteratively inserting the intersection points (such as point C), we can approximate the Clothoid path with satisfactory computational performance.

Clothoid curves provide a continuous change in curvature, resulting in a smooth transition between straight segments and curved segments of a path. This helps reduce abrupt changes in the path and improves the vehicle's stability and comfort. Moreover, by gradually changing the curvature, Clothoid curves minimize lateral acceleration during turns. This reduces the forces acting on the USV, enhancing safety and stability during maneuvering. Additionally, Clothoid curves enable more precise and controlled maneuvering. They allow for gradual changes in heading angle, facilitating smooth turns, and transitions between different paths or waypoints.

The flowchart of the methodology is illustrated by Figure 5.

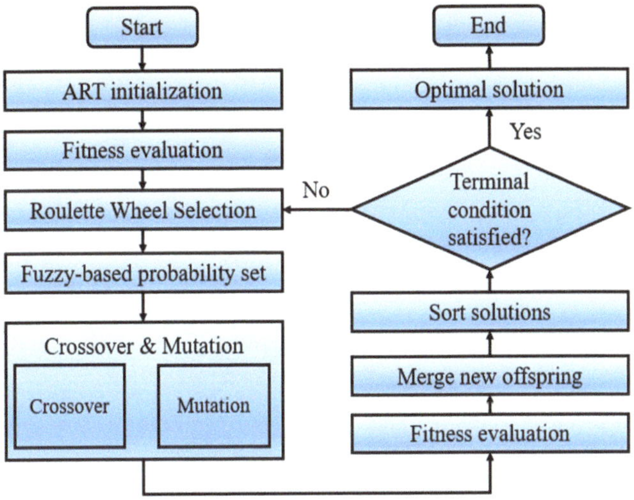

Figure 5. Flowchart of the proposed algorithm.

4. Results and Discussion

In this section, illustrative simulations have been carried out to evaluate the performance of our proposed method progressively through convergence test and simulation under a time-varying environment. The simulations are conducted via MATLAB 2021a environment with a PC configured with Intel (R) Core (TM) i7-13400 CPU and 16-GB RAM.

4.1. Convergence and Quality Test

In this section, simulations have been carried out to analyze the convergence characteristic of our proposed method. We have selected some other state-of-the-art methods from existing reliable references for comparison, including conventional genetic algorithm, D* lite [16], Hybrid A* [29], and RRT* [30]. The selected environment maps are presented in Figure 6, the start and goal points are marked as blue and red dots, respectively. It is worth noting that since we only test the convergence behavior and solution quality, the effects of time-varying currents are not considered. To maintain the efficiency and without loss of solution quality, we set the maximum number of waypoints in each path is 20 according to [31]. Table 1 shows the dimensions and coordinates of the given points.

Table 1. Environment setting.

	Map Size	Start	Destination
Scenario 1	300×400	(150, 300)	(200, 90)
Scenario 2	300×400	(50, 330)	(250, 250)
Scenario 3	400×400	(290, 350)	(390, 60)
Scenario 4	400×400	(150, 240)	(350, 120)

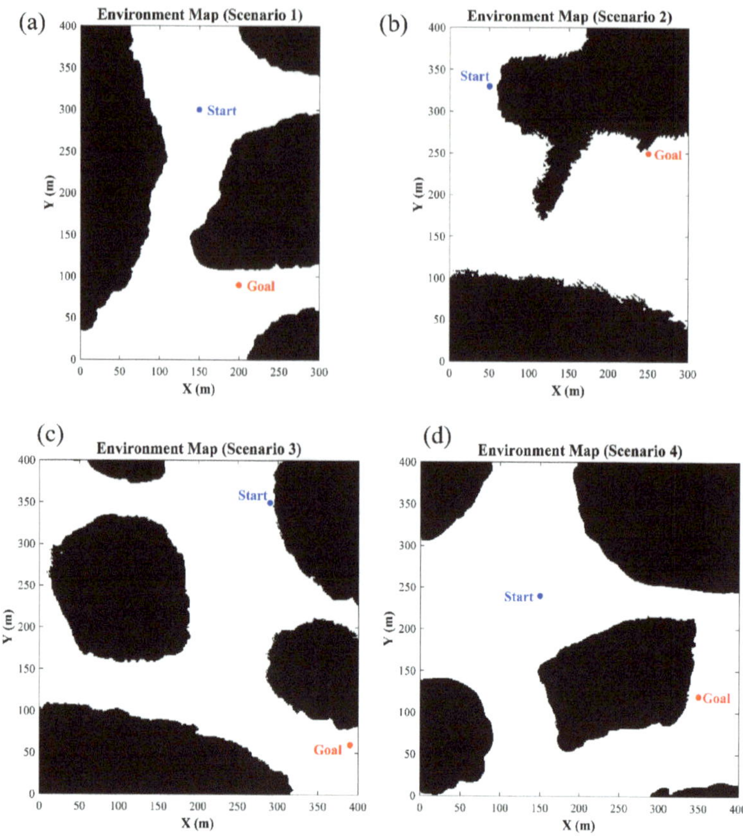

Figure 6. Environment map (**a**) Scenario 1; (**b**) Scenario 2; (**c**) Scenario 3; (**d**) Scenario 4.

The parameters are set as follows:

- In this case, the ocean current is fixed with velocity of 0.3 m/s and direction of $-70°$.
- Proposed: Population size = 100; generation = 200; $p_{c_0} = 0.8$; $p_{m_0} = 0.05$; $d_{min} = 5$ m; $d_{max} = 30$ m; $w_t = 0.395$; $w_\theta = 0.275$; $w_d = 0.330$; $a = 150$; and $b = 50$.
- Traditional GA: Population size = 100; generation = 200; $p_c = 0.8$; and $p_m = 0.05$.
- D* lite: search directions = 8.
- Improved hybrid A*: Minimum turning radius = 4 m; Motion primitive length = 4 m.
- RRT*: Max-iteration = 2500; Max-Connection distance = 1 pixel.

The tabulated data presented in Table 2 provides quantitative results for the proposed algorithm. The results reveal that the algorithm exhibits a significant reduction in time cost with 0.36 s, 0.613 s, 0.484 s, and 0.391 s for the four scenarios, respectively. This represents a considerable improvement of over 60% when compared to the traditional GA. The algorithm's increased speed is primarily due to the new initialization process. Additionally, the algorithm's robustness is evaluated through the standard deviation (SD) of the time cost, which is 0.012 s, 0.022 s, 0.022 s, and 0.008 s for each case, respectively. Furthermore, the proposed algorithm is shown to provide a satisfactory minimum path length of 253.5 pixels, 352.0 pixels, 356.0 pixels, and 298.0 pixels for each scenario, respectively. Although other methods may produce slightly smaller values in some cases, the proposed algorithm provides more practical and reasonable solutions. It is important to note that the relatively low path lengths produced by D* lite and Hybrid A* are due to their reliance on optimal search based on A*, which aims for the shortest path. However, this approach often results in paths that are too close to obstacles.

Table 2. Statistical results.

Methods	Performance	Scenario 1	Scenario 2	Scenario 3	Scenario 4
Proposed	Time (s)	0.360	0.613	0.484	0.391
	Time SD (s)	0.012	0.022	0.022	0.008
	AVG length (pixels)	253.880	354.020	358.460	300.210
	Minimum length (pixels)	253.500	352.000	356.000	298.000
GA	Time (s)	0.905	1.113	1.132	1.179
	Time SD (s)	0.068	0.074	0.102	0.115
	AVG length (pixels)	273.590	373.000	368.490	395.340
	Minimum length (pixels)	251.000	348.000	353.500	296.500
D* lite	Time (s)	3.789	14.946	2.589	3.683
	Time SD (s)	0.155	0.603	0.037	0.064
	AVG length (pixels)	253.296	356.718	353.480	299.841
	Minimum length (pixels)	253.296	356.718	353.480	299.841
Hybrid A*	Time (s)	4.731	1.093	9.127	3.612
	Time SD (s)	0.096	0.035	0.376	0.065
	AVG length (pixels)	253.676	351.758	369.485	303.112
	Minimum length (pixels)	253.676	351.758	369.485	303.112
RRT*	Time (s)	0.449	1.906	1.571	3.346
	Time SD (s)	0.171	0.457	0.552	1.309
	AVG length (pixels)	324.637	487.845	497.646	458.019
	Minimum length (pixels)	287.810	403.188	413.889	358.982

The visualized results of the simulations are presented in Figure 7. The red curve depicts the smoothed path generated by the proposed method. The results demonstrate that the curve is smooth without any abrupt turns. On the other hand, the results generated by GA, D* lite, Hybrid A*, and RRT* exhibit relatively large angle changes, particularly RRT*, which exhibits the poorest performance in terms of smoothness. Furthermore, the proposed method produces a safer path than the other methods. This is primarily due to the inclusion of a safety distance term in the cost function, which forces the path to remain at a safe distance from its nearest obstacle. In contrast, the results produced by other methods often remain too close to the obstacles in some sections of the path.

Table 3 presents a comparative study of path quality, focusing on two key features: minimum clearance d from obstacles and path smoothness. The minimum clearance from obstacles measures the safety level of the results by calculating the distance between each path segment and its nearest obstacle. It is important to note that the safety distance utilized in the simulation is set at 5 m. Path smoothness measures the degree of smoothness of the path. The results presented in Table 3 reveal that the proposed method produces the safest path with the minimum distance from obstacles of 12.649 m, 11.663 m, 10.557 m, and 5.0 m for each scenario, respectively. In contrast, traditional GA and RRT* fail to satisfy the safety requirement in most cases. Moreover, the methods given by D* lite and Hybrid A* exhibit the worst performance in terms of safety, failing in all scenarios. Therefore, they are not suitable for real-world applications. In terms of path smoothness, the proposed method produces the smoothest paths (as seen in Figure 7) with values of 174.547, 149.454, 129.088, and 211.538 for each case, respectively, significantly outperforming the other methods. For instance, the proposed method's path smoothness value is 5–6 times smaller than that of Hybrid A* and D* lite.

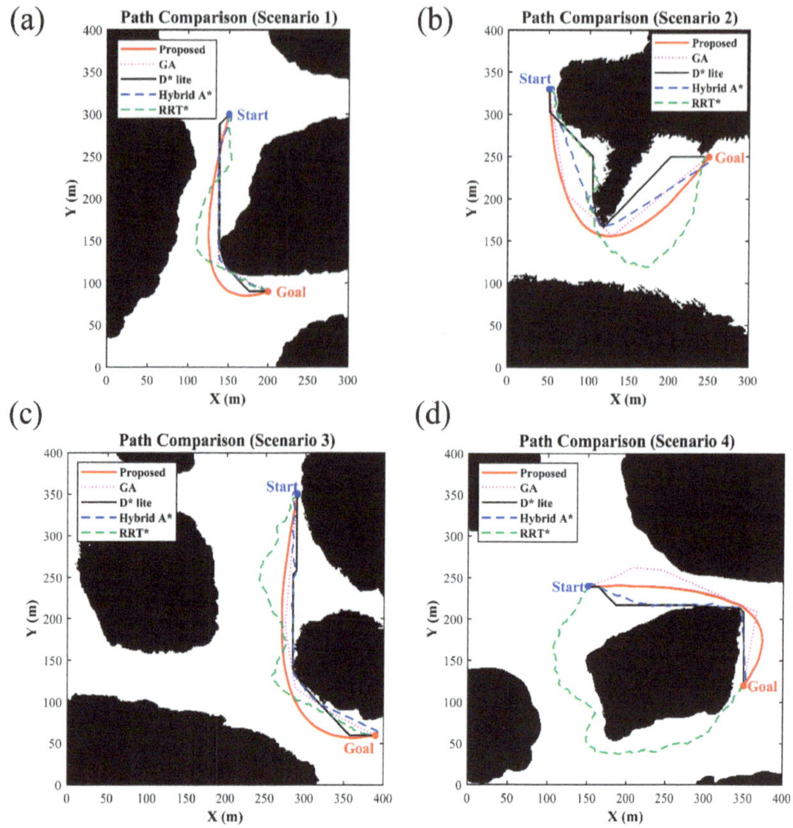

Figure 7. Visualized results (**a**) Scenario 1; (**b**) Scenario 2; (**c**) Scenario 3; (**d**) Scenario 4.

Table 3. Comparison of path quality.

Methods	Performance	Scenario 1	Scenario 2	Scenario 3	Scenario 4
Proposed	d [1] (m)	12.649 (○)	11.663 (○)	10.557 (○)	5.000 (○)
	Smoothness (deg)	174.547	149.454	129.088	211.538
GA	d (m)	1.000 (X)	1.414 (X)	1.414 (X)	5.099 (○)
	Smoothness (deg)	590.291	343.702	528.768	227.737
D* lite	d (m)	1.000 (X)	1.000 (X)	1.000 (X)	0 (X)
	Smoothness (deg)	945	720	720	450
Hybrid A*	d (m)	1.000 (X)	1.000 (X)	0 (X)	1.000 (X)
	Smoothness (deg)	1223	1404	1912	1263
RRT*	d (m)	1.414 (X)	1.000 (X)	5.831(○)	2.236 (X)
	Smoothness (deg)	303.395	555.388	426.283	762.342

[1] A safety distance of 5 m is established. If the minimum clearance requirement is met, a symbol of ○ is displayed within the bracket, while X is used when it is not met.

4.2. Testing in USV Model

In this subsection, simulation studies and comprehensive comparisons are provided by conducting experiments on a prototype USV Otter (see Figure 8, www.maritimerobotics.com (accessed on 1 May 2023), Table 4 shows the maneuvering derivatives). It is worth mentioning that the paths generated by D* lite and Hybrid A* are not suitable for real application because they would collide with the obstacles. Therefore, the paths given by GA and RRT* are selected for the simulation.

Figure 8. USV model.

Table 4. Maneuvering derivatives of Otter.

Inertial Related	Value	Damping Related	Value
m_{11}	85.28	d_{11}	−77.55
m_{22}	162.50	d_{22}	−0.02
m_{33}	41.45	d_{33}	−41.45

In simulations, Figure 9 demonstrates the alterations in course angle and speed for scenario 1 and scenario 2. The proposed method generates a path with gentle and steady changes in course and velocities, as depicted in the figures. The small deviation between the actual and desired signals suggests the feasibility of the proposed method in conjunction with the USV control system. However, the path created by GA exhibits sudden changes in both signals, resulting in a significant deviation at the point of the sudden course alteration. On the other hand, RRT* generated the poorest solution, with jagged changes in course angle and a substantial deviation between the reference and actual signals. Additionally, the simulation shows unstable velocity. Similar results are obtained by scenario 3 and scenario 4.

In Table 5, the energy and time cost of the simulations are displayed. The proposed method exhibits the lowest energy and time cost among all cases, as demonstrated in the table. This outcome is mainly attributed to the novel cost function incorporated into the proposed method, which directs the USV to move along the current direction. Conversely, the results produced by RRT* display the poorest performance with the highest energy consumption and computational time. This inferior performance is primarily due to the numerous abrupt changes in course angle and unnecessary turns.

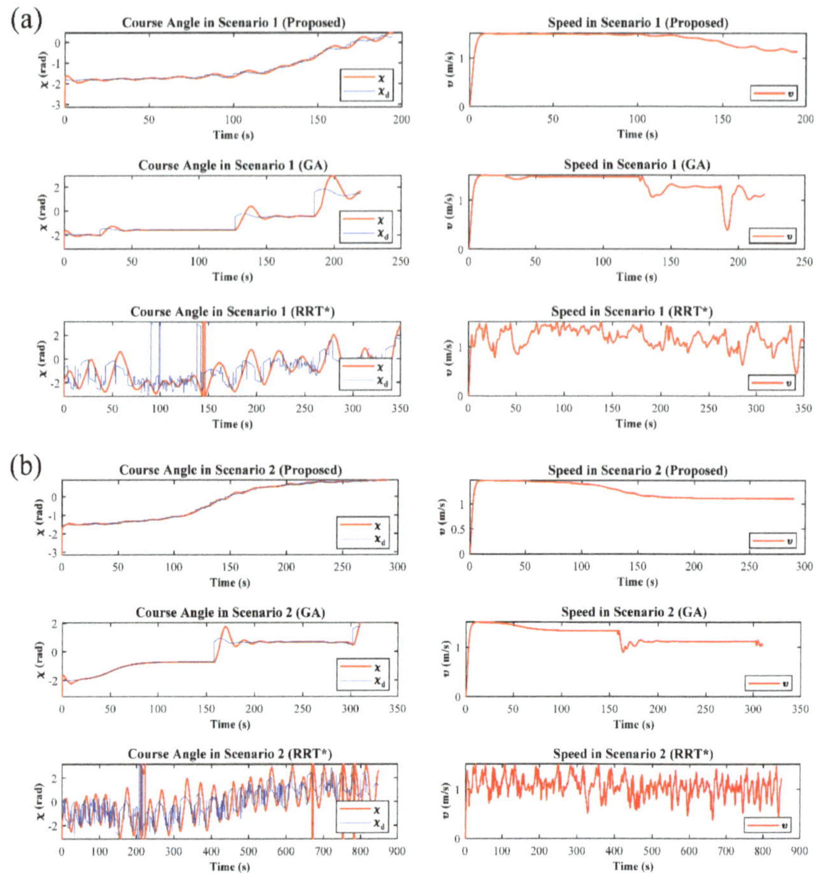

Figure 9. Simulation results for proposed method, GA, and RRT*: (**a**) Scenario 1; (**b**) Scenario 2.

Table 5. Energy cost and time cost during the mission.

	Proposed		GA		RRT*	
	Energy cost (kJ)	Time cost (s)	Energy cost (kJ)	Time cost (s)	Energy cost (kJ)	Time cost (s)
Scenario 1	27.4	194.8	29.8	219.9	42.7	349.9
Scenario 2	36.8	298.9	39.1	310.0	94.0	849.6
Scenario 3	37.4	266.6	39.6	289.8	65.8	530.0
Scenario 4	31.7	246.9	35.0	285.0	82.4	708.9

4.3. Simulation in Time-Vary Ocean Environments

In this section, we will evaluate the method under time-varying ocean currents. We have selected some state-of-the-art methods from existing reliable references for comparison, including improved artificial fish swarm algorithm [8] and multi-objective enhanced GA (MOEGA) [32]. The ocean current model used in this paper is based on the numerical solution of water jet structure [33]:

$$\phi(x,y) = 1 - \tanh\left(\frac{y - B(t)\cos(k(x-ct))}{(1 + k^2 B(t)^2 \sin^2(k(x-ct)))^{\frac{1}{2}}}\right) \quad (27)$$

where $B(t)$ and k are the properly adimensionalized amplitude and wavenumber of the undulation in the stream function. The specific expression for $B(t)$ is:

$$B(t) = B_o + \epsilon\cos(\omega t + \theta) \quad (28)$$

with $B_o = 1.2$, $c = 0.12$, $k = 0.84$, $\omega = 0.4$, $\epsilon = 0.3$ and $\theta = \pi/2$. The velocity field is obtained by the following expression:

$$U(x,y,t) = -\frac{\partial\phi}{\partial y} \quad V(x,y,t) = \frac{\partial\phi}{\partial x} \quad (29)$$

where $U(x,y,t)$ and $V(x,y,t)$ are the x– and y- components of the velocity vector at time in the location (x, y).

The parameters are set as follows:

- Environment: map size: 500 * 500, Start = (80, 150), Goal = (480, 330), and the ocean current is set as Equation (27) for Case 1 while we multiply -1 on the y component for Case 2.
- Proposed: Population size = 100, generation = 200, $p_{c_0} = 0.8$, $p_{m_0} = 0.05$, $d_{min} = 5$ m, $d_{max} = 30$ m, $w_t = 0.395$, $w_\theta = 0.275$, $w_d = 0.330$, $a = 150$, and $b = 50$.
- IAFSA: Population size = 100, $\delta = 0.618$, $step = 1$; $visual = 10$, and $try\_number = 8$.
- MOEGA: Population size = 100, generation = 200, $p_c = 0.8$, $p_m = 0.05$, $d_{min} = 5$ m, $d_{max} = 30$ m, $w_t = 0.395$, $w_\theta = 0.275$, and $w_d = 0.330$.

Table 6 presents the quantitative outcomes, including path distance, cruising time ($\mathscr{T}$), smoothness (ϑ), and minimum distance to obstacles (d), for both Case 1 and Case 2. The visualized solutions for each case are also depicted in Figures 10 and 11. According to the results in Table 6, the proposed algorithm delivers solutions with higher quality paths (as highlighted in bold in the table), outperforming other methods in terms of cruising time, smoothness, and safety in most scenarios. However, it is noteworthy that the proposed algorithm results in the lowest safety value (8 m to the obstacle) due to the significant impact of energy consumption on optimization. It should be noted that an 8 m safety distance is acceptable in real application [34]. Moreover, the IAFSA method yields the worst outcomes, which may be attributed to its random behavior during the algorithm process, leading to abrupt points along the path. As demonstrated in Figures 10 and 11, our proposed model leverages the currents to reduce energy consumption by selecting intersection points that align with the current direction. Overall, the presented results indicate that the proposed method exhibits superior performance to the other two algorithms.

Table 6. Statistical measurements of the paths.

	Indicators	Proposed	IAFSA	MOEGA
Case 1	Distance (m)	458.166	471.691	483.598
	$\mathscr{T}$ (s)	209.198	214.430	221.039
	ϑ (deg)	123.130	212.378	145.872
	d (m)	18.682	5.000	16.279
Case 2	Distance (m)	540.065	543.593	566.971
	$\mathscr{T}$ (s)	245.489	254.095	257.724
	ϑ (deg)	155.034	164.404	228.242
	d (m)	8.062	18.934	17.029

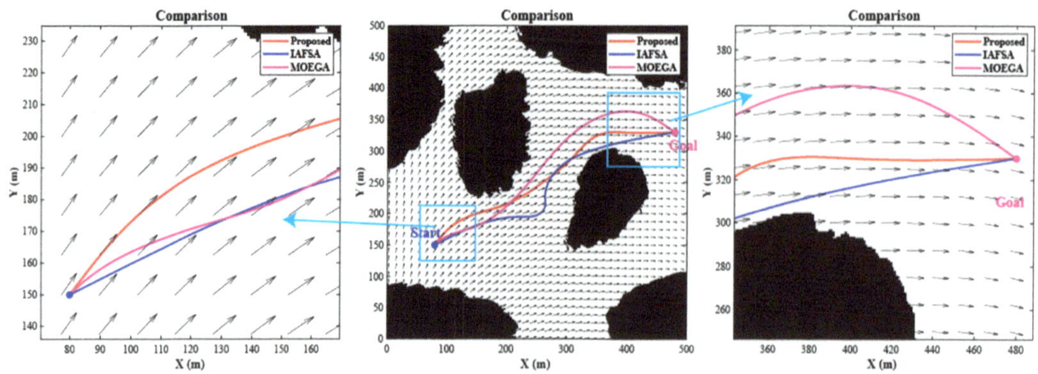

Figure 10. Visualized results of Case 1.

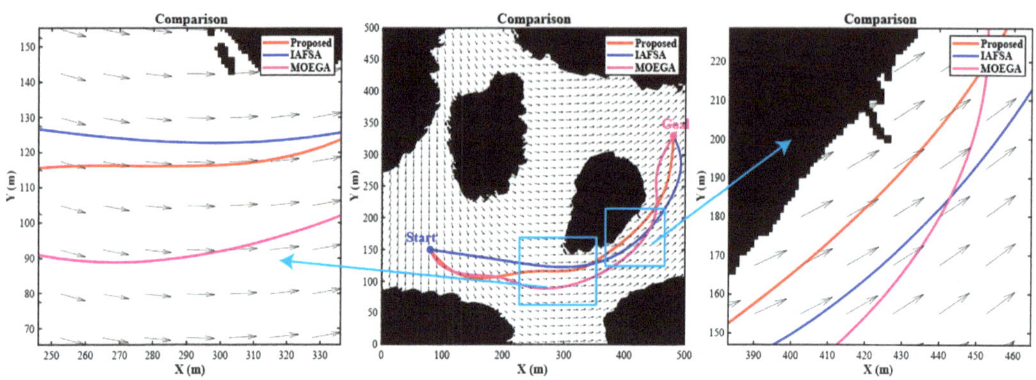

Figure 11. Visualized results of Case 2.

5. Conclusions

This paper presents a thorough investigation into the path planning problem for USVs. The proposed algorithm generates a path that is both optimally safe and quickly convergent, exhibiting strong adaptability to complex environments. In comparison to the existing literature, our method outperforms other algorithms across all problem variations. Additionally, the fast-discrete Clothoid curve is utilized to maintain path curve continuity and ensure reliable coordination between the planning and control modules, while also accommodating non-holonomic constraints. Simulation studies and comprehensive comparisons in various ocean scenarios have been conducted to illustrate the effectiveness and superiority of the proposed path planning framework.

Author Contributions: Conceptualization, F.W. and L.Z.; methodology, F.W.; software, F.W.; validation, F.W. and L.Z.; formal analysis, Y.B; writing—original draft preparation, F.W. and L.Z.; writing—review and editing, F.W. and L.Z.; visualization, F.W.; supervision, Y.B.; project administration, Y.B. and L.Z.; funding acquisition, F.W. All authors have read and agreed to the published version of the manuscript.

Funding: This research was funded by the Stable Supporting Fund of Science and Technology on Underwater Vehicle Technology (grant number: JCKYS2022SXJQR-01).

Institutional Review Board Statement: Not applicable.

Informed Consent Statement: Not applicable.

Data Availability Statement: No data was used in this paper.

Acknowledgments: The authors thank the editor-in-chief, the associate editor, and the anonymous referees for their comments and suggestions.

Conflicts of Interest: The authors declare no conflict of interest.

References

1. Zhou, C.; Gu, S.; Wen, Y.; Du, Z.; Xiao, C.; Huang, L.; Zhu, M. The Review Unmanned Surface Vehicle Path Planning: Based on Multi-Modality Constraint. *Ocean Eng.* **2020**, *200*, 107043. [CrossRef]
2. Öztürk, Ü.; Akdağ, M.; Ayabakan, T. A Review of Path Planning Algorithms in Maritime Autonomous Surface Ships: Navigation Safety Perspective. *Ocean Eng.* **2022**, *251*, 111010. [CrossRef]
3. Ai, B.; Jia, M.; Xu, H.; Xu, J.; Wen, Z.; Li, B.; Zhang, D. Coverage Path Planning for Maritime Search and Rescue Using Reinforcement Learning. *Ocean Eng.* **2021**, *241*, 110098. [CrossRef]
4. Fiskin, R.; Atik, O.; Kisi, H.; Nasibov, E.; Johansen, T.A. Fuzzy Domain and Meta-Heuristic Algorithm-Based Collision Avoidance Control for Ships: Experimental Validation in Virtual and Real Environment. *Ocean Eng.* **2021**, *220*, 108502. [CrossRef]
5. Krichen, M. Improving Formal Verification and Testing Techniques for Internet of Things and Smart Cities. *Mobile Netw. Appl.* **2019**. [CrossRef]
6. Khan, M.A. A Formal Method for Privacy-preservation in Cognitive Smart Cities. *Expert Syst.* **2022**, *39*, e12855. [CrossRef]
7. Zhao, L.; Wang, F.; Bai, Y. Route Planning for Autonomous Vessels Based on Improved Artificial Fish Swarm Algorithm. *Ships Offshore Struct.* **2022**, *18*, 897–906. [CrossRef]
8. Zhao, L.; Bai, Y.; Wang, F.; Bai, J. Path Planning for Autonomous Surface Vessels Based on Improved Artificial Fish Swarm Algorithm: A Further Study. *Ships Offshore Struct.* **2022**. [CrossRef]
9. MahmoudZadeh, S.; Abbasi, A.; Yazdani, A.; Wang, H.; Liu, Y. Uninterrupted Path Planning System for Multi-USV Sampling Mission in a Cluttered Ocean Environment. *Ocean Eng.* **2022**, *254*, 111328. [CrossRef]
10. Wang, N.; Xu, H. Dynamics-Constrained Global-Local Hybrid Path Planning of an Autonomous Surface Vehicle. *IEEE Trans. Veh. Technol.* **2020**, *69*, 6928–6942. [CrossRef]
11. Yu, X.; Wang, Y. A Time Dimension-Added Multiple Obstacles Avoidance Approach for Unmanned Surface Vehicles. *Ocean Eng.* **2022**, *252*, 111201. [CrossRef]
12. Yu, J.; Liu, G.; Xu, J.; Zhao, Z.; Chen, Z.; Yang, M.; Wang, X.; Bai, Y. A Hybrid Multi-Target Path Planning Algorithm for Unmanned Cruise Ship in an Unknown Obstacle Environment. *Sensors* **2022**, *22*, 2429. [CrossRef] [PubMed]
13. Song, R.; Liu, Y.; Bucknall, R. Smoothed A* Algorithm for Practical Unmanned Surface Vehicle Path Planning. *Appl. Ocean Res.* **2019**, *83*, 9–20. [CrossRef]
14. Shah, B.C.; Gupta, S.K. Long-Distance Path Planning for Unmanned Surface Vehicles in Complex Marine Environment. *IEEE J. Ocean. Eng.* **2020**, *45*, 813–830. [CrossRef]
15. Xie, L.; Xue, S.; Zhang, J.; Zhang, M.; Tian, W.; Haugen, S. A Path Planning Approach Based on Multi-Direction A* Algorithm for Ships Navigating within Wind Farm Waters. *Ocean Eng.* **2019**, *184*, 311–322. [CrossRef]
16. Yao, Y.; Liang, X.; Li, M.; Yu, K.; Chen, Z.; Ni, C.; Teng, Y. Path Planning Method Based on D* Lite Algorithm for Unmanned Surface Vehicles in Complex Environments. *China Ocean Eng.* **2021**, *35*, 372–383. [CrossRef]
17. Lyridis, D.V. An Improved Ant Colony Optimization Algorithm for Unmanned Surface Vehicle Local Path Planning with Multi-Modality Constraints. *Ocean Eng.* **2021**, *241*, 109890. [CrossRef]
18. Guo, X.; Ji, M.; Zhao, Z.; Wen, D.; Zhang, W. Global Path Planning and Multi-Objective Path Control for Unmanned Surface Vehicle Based on Modified Particle Swarm Optimization (PSO) Algorithm. *Ocean Eng.* **2020**, *216*, 107693. [CrossRef]
19. Hao, K.; Zhao, J.; Li, Z.; Liu, Y.; Zhao, L. Dynamic Path Planning of a Three-Dimensional Underwater AUV Based on an Adaptive Genetic Algorithm. *Ocean Eng.* **2022**, *263*, 112421. [CrossRef]
20. Krell, E.; King, S.A.; Garcia Carrillo, L.R. Autonomous Surface Vehicle Energy-Efficient and Reward-Based Path Planning Using Particle Swarm Optimization and Visibility Graphs. *Appl. Ocean Res.* **2022**, *122*, 103125. [CrossRef]
21. Xue, H. A Quasi-Reflection Based SC-PSO for Ship Path Planning with Grounding Avoidance. *Ocean Eng.* **2022**, *247*, 110772. [CrossRef]
22. Liang, C.; Zhang, X.; Han, X. Route Planning and Track Keeping Control for Ships Based on the Leader-Vertex Ant Colony and Nonlinear Feedback Algorithms. *Appl. Ocean Res.* **2020**, *101*, 102239. [CrossRef]
23. Wang, F.; Zhao, L.; Bai, Y. Path Planning For Unmanned Surface Vehicles Based On Modified Artificial Fish Swarm Algorithm With Local Optimizer. *Math. Probl. Eng.* **2022**, *2022*, 1283374. [CrossRef]
24. Ma, Y.; Hu, M.; Yan, X. Multi-Objective Path Planning for Unmanned Surface Vehicle with Currents Effects. *ISA Trans.* **2018**, *75*, 137–156. [CrossRef]
25. Wang, N.; Zhang, Y.; Ahn, C.K.; Xu, Q. Autonomous Pilot of Unmanned Surface Vehicles: Bridging Path Planning and Tracking. *IEEE Trans. Veh. Technol.* **2022**, *71*, 2358–2374. [CrossRef]
26. Meng, J.; Liu, Y.; Bucknall, R.; Guo, W.; Ji, Z. Anisotropic GPMP2: A Fast Continuous-Time Gaussian Processes Based Motion Planner for Unmanned Surface Vehicles in Environments With Ocean Currents. *IEEE Trans. Automat. Sci. Eng.* **2022**, *19*, 3914–3931. [CrossRef]

27. Liu, Y.; Bucknall, R.; Zhang, X. The Fast Marching Method Based Intelligent Navigation of an Unmanned Surface Vehicle. *Ocean Eng.* **2017**, *142*, 363–376. [CrossRef]
28. Liu, Y.; Bucknall, R. The Angle Guidance Path Planning Algorithms for Unmanned Surface Vehicle Formations by Using the Fast Marching Method. *Appl. Ocean Res.* **2016**, *59*, 327–344. [CrossRef]
29. Petereit, J.; Emter, T.; Frey, C.W.; Kopfstedt, T.; Beutel, A. Application of Hybrid A* to an Autonomous Mobile Robot for Path Planning in Unstructured Outdoor Environments. In Proceedings of the ROBOTIK 2012; 7th German Conference on Robotics, Munich, Germany, 21–22 May 2012; pp. 1–6.
30. Karaman, S.; Frazzoli, E. Sampling-Based Algorithms for Optimal Motion Planning. *Int. J. Robot. Res.* **2011**, *30*, 846–894. [CrossRef]
31. Liu, J.; Anavatti, S.; Garratt, M.; Abbass, H.A. Modified Continuous Ant Colony Optimisation for Multiple Unmanned Ground Vehicle Path Planning. *Expert Syst. Appl.* **2022**, *196*, 116605. [CrossRef]
32. Nazarahari, M.; Khanmirza, E.; Doostie, S. Multi-Objective Multi-Robot Path Planning in Continuous Environment Using an Enhanced Genetic Algorithm. *Expert Syst. Appl.* **2019**, *115*, 106–120. [CrossRef]
33. Alvarez, A.; Caiti, A.; Onken, R. Evolutionary Path Planning for Autonomous Underwater Vehicles in a Variable Ocean. *IEEE J. Ocean. Eng.* **2004**, *29*, 418–429. [CrossRef]
34. Zhao, L.; Bai, Y.; Paik, J.K. Global-Local Hierarchical Path Planning Scheme for Unmanned Surface Vehicles under Dynamically Unforeseen Environments. *Ocean. Eng.* **2023**, *280*, 114750. [CrossRef]

Journal of
Marine Science and Engineering

MDPI

Article

Automatic Alignment Method of Underwater Charging Platform Based on Monocular Vision Recognition

Aidi Yu, Yujia Wang *, Haoyuan Li and Boyang Qiu

College of Mechanical and Electrical Engineering, Harbin Engineering University, Harbin 150001, China; yuaidi@hrbeu.edu.cn (A.Y.); 13163663608@163.com (H.L.); boyangqiu@hrbeu.edu.cn (B.Q.)
* Correspondence: wangyujia@hrbeu.edu.cn

Abstract: To enhance the crypticity and operational efficiency of unmanned underwater vehicle (UUV) charging, we propose an automatic alignment method for an underwater charging platform based on monocular vision recognition. This method accurately identifies the UUV number and guides the charging stake to smoothly insert into the charging port of the UUV through target recognition. To decode the UUV's identity information, even in challenging imaging conditions, an encryption encoding method containing redundant information and an ArUco code reconstruction method are proposed. To address the challenge of underwater target location determination, a target location determination method was proposed based on deep learning and the law of refraction. The method can determine the two-dimensional coordinates of the target location underwater using the UUV target spray position. To meet the real-time control requirements and the harsh underwater imaging environment, we proposed a target recognition algorithm to guide the charging platform towards the target direction. The practical underwater alignment experiments demonstrate the method's strong real-time performance and its adaptability to underwater environments. The final alignment error is approximately 0.5548 mm, meeting the required alignment accuracy and ensuring successful alignment.

Keywords: real-time control; underwater image restoration; thresholding; visual servoing; target recognition; monocular vision; alignment

Citation: Yu, A.; Wang, Y.; Li, H.; Qiu, B. Automatic Alignment Method of Underwater Charging Platform Based on Monocular Vision Recognition. *J. Mar. Sci. Eng.* **2023**, *11*, 1140. https://doi.org/10.3390/jmse11061140

Academic Editor: Rafael Morales

Received: 13 April 2023
Revised: 17 May 2023
Accepted: 27 May 2023
Published: 29 May 2023

1. Introduction

Unmanned underwater vehicles (UUVs) play an irreplaceable role in various fields, serving as oceanic equipment suitable for underwater tasks. They are widely utilized for tasks including seafood fishing, subsea pipeline tracking, seafloor mapping, submarine cable laying, and marine resource exploration. However, as the scope of UUV missions continues to expand, the issue of their endurance has become a focal point. Due to the limited energy carried by UUVs, frequent charging becomes necessary. However, surface charging not only reduces the operational efficiency of UUVs and increases costs but also compromises their stealth capabilities during mission execution [1,2]. To address this challenge, underwater charging platforms have emerged, enabling UUVs to recharge without the need to surface. Currently, there is a wealth of research on UUV docking. Researchers have utilized navigation systems such as acoustics [3,4], optics [5], and electromagnetics [6,7] to guide UUVs into docking station (DS). However, to the best of our knowledge, there are few methods available for guiding the charging stake to accurately insert into the UUV's charging port after docking. Achieving automatic alignment of underwater charging platforms is a current trend in the development of underwater equipment technology and holds significant research and practical value.

Although there is limited research on automatic alignment of underwater charging platforms, the process of inserting the charging stake into the UUV's charging port can be conceptualized as a peg-in-hole assembly. The solution to this problem can be broadly categorized into contact-based and non-contact-based methods. Contact-based methods [8] typically involve the end of the shaft contacting the plane where the hole is located, followed by using a force sensor to search for the hole's position on the plane. This method has low safety and can potentially damage the outer surface of the UUVs. Non-contact-based methods can be divided into methods based on laser alignment instruments, acoustic sensors, and vision sensors. The core components of a laser alignment instrument are a semiconductor laser that emits laser beams and a photoelectric semiconductor position detector that collects information about the position of the laser spot [9]. Therefore, precise alignment can be achieved by installing laser alignment instruments on the hole axis. However, underwater small planktonic organisms cause light scattering, affecting the alignment accuracy, and suspended particles in the water obstruct the laser, rendering the alignment process unable to continue. Due to the slow attenuation of sound waves underwater, acoustic sensors are widely used in various underwater positioning and navigation tasks [4], but their alignment accuracy is lower in short-distance scenarios. Vision-based alignment, on the other hand, corrects alignment deviations through visual feedback, providing positioning and guidance for fragile or easily disturbed objects without physical contact [10]. It exhibits robustness in underwater environments and meets alignment requirements in terms of accuracy. Therefore, we adopt a vision-based non-contact alignment approach to guide the charging stake into the UUV's charging port.

Alignment operations using vision sensors have been widely studied in various fields. For instance, Fan et al. [11] proposed a laser-vision-sensor-based method for initial point alignment of narrow weld seams by utilizing the relationship between laser streak feature points and initial points. They obtain a high signal-to-noise image of the narrow weld seam using the laser vision sensor, and then calculate the 3D coordinates of the final image feature point and the initial point based on the alignment model. Finally, they control the actuator to achieve the initial point alignment. In another study, Chen et al. [12] developed an automatic alignment method for tracking the antenna of an unmanned aerial vehicle (UAV) using computer vision technology. The antenna angle is adjusted using the relative position between the center of the UAV image and the center of the camera image fixed on the antenna. The two image centers overlap during antenna alignment. Similarly, Jongwon et al. [13] designed a vision system that uses three cameras to locate the wafer's position for wafer alignment.

Underwater image processing techniques are of great importance in the field of underwater charging platform alignment due to the problems of low contrast, blurred edges with blue-green tones, and other problems. Traditional underwater image processing techniques can be divided into two categories: image enhancement and image restoration. Image enhancement algorithms [14] include histogram equalization, white balance, Retinex, wavelet transform, etc. These algorithms can render underwater objects clearer by enhancing image contrast and denoising. Image restoration techniques recover images by solving two unknown variables in the Jaffe–McGlamery [15,16] underwater imaging model: the transmission map and background light. For example, the dark channel prior algorithm (DCP) proposed by He et al. [17,18], which simplifies the JM model by introducing a priori knowledge that the dark channel value of a clear, fog-free image is close to zero. Variants of the DCP algorithm [19–22] have been developed and optimized over time, achieving better results. In recent years, convolutional neural networks have made remarkable achievements in multiple fields such as image classification [23], object detection [24], and instance segmentation [25]. Increasingly, many networks are being used to process underwater images. Some of these networks are end-to-end [26–28], which output the recovered image directly after inputting the original image, while others use deep learning to derive some of the physical parameters of the underwater imaging model and then perform image restoration [29]. This method has good performance and strong robustness but is not suitable for situations with limited hardware resources.

Accurate pose estimation is a primary prerequisite for successful alignment. Pose estimation technology recovers the position and orientation of an object by observing the correspondence between its image and features [30], which can be divided into three types: corners, lines, and ellipses (circles). Luckett et al. [31] compared the performance of these three features and found that the accuracy and precision of corner and line features increase as the distance decreases, but in high-noise environments, ellipse features have the strongest robustness. To address the issue of ellipse detection accuracy, Zhang et al. [32] improved the circle-based ellipse detection method and designed a sub-pixel edge-based ellipse detection method. This improved the accuracy of ellipse detection, especially in cases where the ellipse is incomplete. It was the first to prove that improving the accuracy of ellipse edges helps to improve the detection accuracy of ellipses. Huang et al. [33] proposed a universal circle and point fusion framework that can solve pose estimation problems with various feature combinations, combining the advantages of both features with high accuracy and robustness. Meng et al. [30] proposed a perspective circle and line (PCL) method that uses the perspective view of a single circle and line to recover the position and orientation of an object, which solves the duality and restores the roll angle.

We proposed an automatic alignment method for an underwater charging platform based on monocular vision recognition. After the UUV enters the underwater charging platform, the method accurately identifies its number and guides the charging platform to move towards the target direction using target recognition. This is achieved by calculating the deviation between the current position of the target keypoints and the target position, which aligns the charging stake with the UUV's charging port. The main contributions of this paper are as follows:

1. A single-camera visual-recognition-based UUV underwater alignment method is proposed that includes an encoding and decoding method for encrypted graphic targets, a method for determining the two-dimensional coordinates of the target location, and a target recognition algorithm, which can guide the charging stake on the charging platform to smoothly insert into the UUV's charging port.
2. The method can adapt to underwater environments and has certain robustness to partial occlusion. Additionally, this method requires less computational resources, lower hardware requirements, shorter processing times, and satisfies real-time control requirements. Moreover, the detection accuracy of this method meets the requirements for smooth alignment.

The rest of this paper is organized as follows. Section 2 describes the proposed single-camera visual-recognition-based UUV underwater alignment method in detail. Section 3 presents the experimental results of the proposed method. In Section 4, we analyze the experimental results from Section 3 and describe the shortcomings of our method. Finally, Section 5 presents our conclusions.

2. Methods

The structure of the charging platform utilized in this method is illustrated in Figure 1, where both the camera and charging stake are attached to the axial sliding table, which is fixed on the circumferential turntable. Meanwhile, the UUV is secured onto the alignment platform. Due to the positioning of the alignment platform, the camera's distance from the UUV target at the target position is a known, fixed value. Consequently, the two-dimensional coordinates of the target's keypoints on the UUV remain unchanged at the target position. By comparing the two-dimensional coordinates of the target keypoints at the current position and the target position, the direction of motion of the charging stake can be determined.

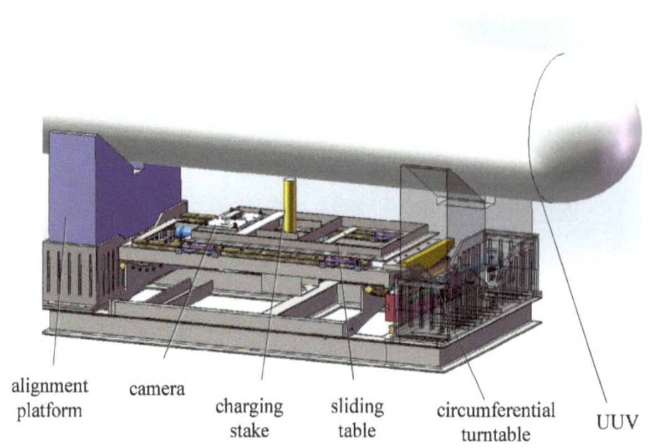

Figure 1. Structure of the underwater charging platform used in this method.

The proposed method comprises three stages, as shown in Figure 2. In the first stage, the UUV's identity information is decoded to obtain its number, which serves as an index to retrieve the registered information of the UUV within the charging platform. This information includes the UUV's charging voltage, size, and target spray position. Subsequently, the UUV is firmly clamped onto the alignment platform by the clamping device of the underwater charging platform. In the second stage, the UUV's size and target spray position are obtained based on the retrieved information, and the target position for docking is determined. This target position is the two-dimensional coordinate on the camera imaging plane where the keypoints of the UUV's target are located when the charging stake on the underwater charging platform can be inserted into the UUV's charging port. In the third stage, the keypoints of the UUV's target are recognized, and the charging stake is guided to move towards the target position by calculating the deviation between the current position and the target position. The stake is first aligned circumferentially and then aligned axially until the distance between the current position and the target position is within the allowable error range, as shown in Figure 3.

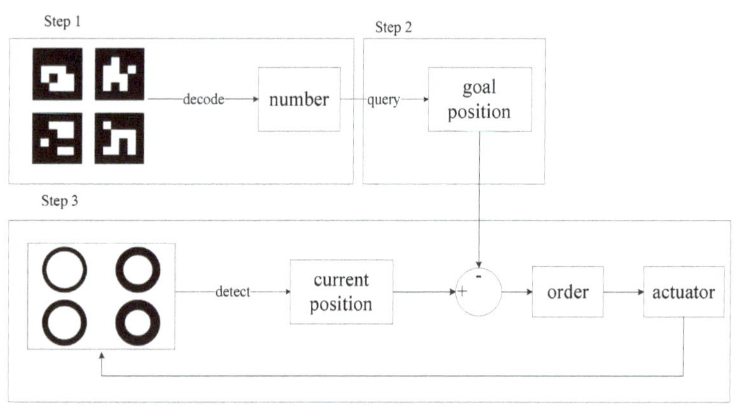

Figure 2. Schematic diagram of the proposed method.

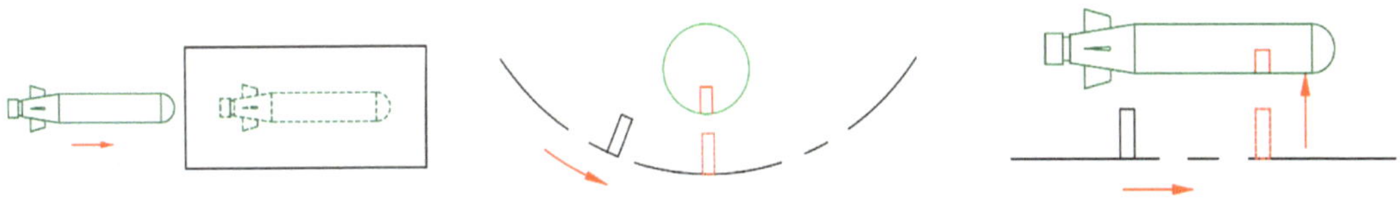

(a) (b) (c)

Figure 3. Schematic diagram of charging process. (**a**) UUV enters the underwater charging platform. In the figure, the arrow represents the direction of UUV movement. (**b**) Circumferential alignment. In the figure, the arrow represents the rotational direction of the stake during circumferential alignment. (**c**) Axial alignment. In the figure, the arrow represents the translational direction of the stake during axial alignment.

2.1. Encoding and Decoding

2.1.1. Encoding

Before UUVs can charge or exchange information with underwater charging platforms, their identity information should be determined to identify their model and charging voltage, and to confirm their mission type and ensure secure information exchange. Therefore, a UUV identity information encryption and coding method is necessary to ensure information security. Firstly, the UUV number is expanded to three digits, with leading zeros added if necessary. The UUV number is denoted as $A_1A_2A_3(A_1, A_2, A_3 \in [0,9],\ A_1, A_2, A_3 \in Z)$. Next, four coding values are obtained: $A_1A_2, A_2A_3, A_1A_3, A_1 + A_2 + A_3$. These coding values are used to query their corresponding ArUco codes, which are graphic codes obtained by converting the numeric codes. Finally, the four ArUco codes obtained from the previous step are rotated clockwise by 0°, 90°, 180°, and 270°, respectively, and their position information is added to each coding value through the ArUco code's pose information. The resulting encoding pattern is shown in Figure 4. This encoding method has some redundancy; when decoding, it is only necessary to recognize the position and ID of any two of the four coding values to infer the UUV number.

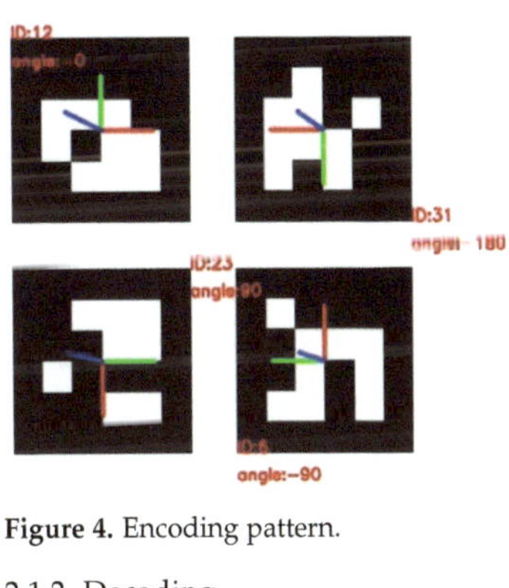

Figure 4. Encoding pattern.

2.1.2. Decoding

Due to the challenging imaging conditions, the ArUco codes in the original images cannot be recognized directly. Therefore, this paper proposes a method for ArUco code detection, which involves image restoration, thresholding, and template-filling techniques to reconstruct the ArUco codes in the respective regions. The specific steps are illustrated in Figure 5.

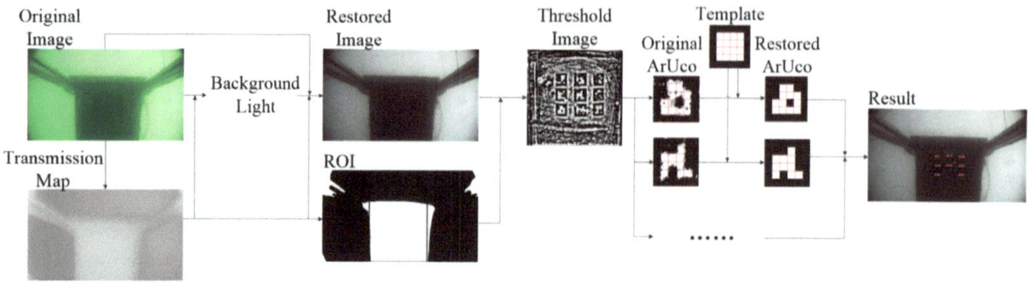

Figure 5. Schematic diagram of ArUco code detection.

In the first step, the original image is subjected to image restoration. The core of the image restoration method involves solving for two unknowns, $t(x,\lambda)$ and $B_\infty(\lambda)$, based on the underwater imaging model as represented by Equation (1). We adopt the method proposed in [19] and use the difference between the red channel and the maximum value of the blue and green channels to estimate the transmission of the red channel. The transmission map of the blue and green channels is then obtained based on statistical analysis [34], as shown in Equation (2). Furthermore, we apply the gray world assumption theory [9] in the field of image restoration to estimate the background light, as shown in Equation (3). By substituting the calculated values of $t(x,\lambda)$ and $B_\infty(\lambda)$ into Equation (1), the restored image can be obtained.

$$I(x,\lambda) = J(x,\lambda)t(x,\lambda) + B_\infty(\lambda)(1 - t(x,\lambda)), \lambda \in \{\mathrm{R,G,B}\} \tag{1}$$

where $I(x,\lambda)$ is the original image, $J(x,\lambda)$ is the restored image, $t(x,\lambda)$ is the transmission map, $B_\infty(\lambda)$ is the background light, λ is the wavelength of light.

$$\begin{gathered} D(x) = \max_{x\in\Omega,\lambda=\mathrm{R}} I(x,\lambda) - \max_{x\in\Omega,\lambda\in\{\mathrm{B,G}\}} I(x,\lambda), \lambda \in \{\mathrm{R,G,B}\} \\ t(x,\mathrm{R}) = D(x) + \left(1 - \max_x D(x)\right) \\ c(\lambda) = -0.00113 \times \lambda + 1.62571, \lambda \in \{\mathrm{R,G,B}\} \\ t(x,\mathrm{G}) = t(x,\mathrm{R})^{\frac{c(\mathrm{G})}{c(\mathrm{R})}} \\ t(x,\mathrm{B}) = t(x,\mathrm{R})^{\frac{c(\mathrm{B})}{c(\mathrm{R})}} \end{gathered} \tag{2}$$

where Ω is a local patch in the image.

$$B_\infty(\lambda) = \frac{\overline{I(x,\lambda)} - M \times \overline{t(x,\lambda)}}{1 - \overline{t(x,\lambda)}}, \lambda \in \{\mathrm{R,G,B}\} \tag{3}$$

where M is a constant value that represents the desired mean gray value of the restored image, which is set to 0.5 in this paper.

In the second step, the restored image is subjected to a thresholding operation. Although the contrast of the restored image has been improved, it still does not meet the recognition criteria for ArUco codes. Traditional methods utilize contrast enhancement and image binarization techniques to assist in ArUco code recognition. However, these operations can amplify the noise in the image and result in the failure of ArUco code recognition. Therefore, this paper proposes an improved approach to local thresholding, as shown in Equation (4). This method creates small window w_1 and large window w_2 around each pixel in the image. It compares the mode of the pixel grayscale values in w_1 with the average grayscale value of the pixels in w_2 and returns the grayscale value of the pixel at the center. Compared to traditional binarization methods, this approach has the advantage of using the mode of the grayscale values in a small window centered around each pixel for comparison, which helps with noise reduction. Additionally, it categorizes all pixels into five classes instead of two (0 and 255), resulting in smoother transitions between grayscale values and enhancing the robustness of the ArUco code recovery process in the third step.

$$\begin{aligned} d_1 &= \text{mode}(w_1) \\ d_2 &= \text{mean}(w_2) \\ I(i,j) &= \begin{cases} 0, & d_2 - d_1 > 0.5 \\ 60, & d_2 - d_1 > 0.1 \\ 180, & d_1 - d_2 > 0.1 \\ 255, & d_1 - d_2 > 0.5 \\ 125, & \text{others} \end{cases} \end{aligned} \tag{4}$$

where w_1 and w_2 are sliding windows with radii of d_{w1} and d_{w2}, as shown in Figure 6. In this paper, $d_{w2} = 10d_{w1}$.

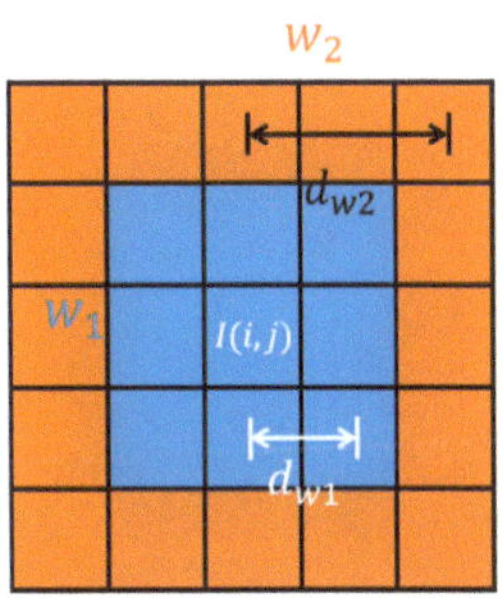

Figure 6. w_1 and w_2 are sliding windows with $I(i,j)$ as the center and d_{w1} and d_{w2} as the radii.

In the last step, the ArUco codes are reconstructed. After obtaining the thresholded image, the regions of interest (ROIs) containing the ArUco codes can be determined. Each ROI is then divided into 36 equally sized rectangles, with the central 16 rectangles containing the encoding information of the ArUco codes. The color of the corresponding blank positions in the template is determined based on the average grayscale value within each rectangle. By filling in the template, the reconstructed ArUco codes are obtained. Finally, the ID and angle of the reconstructed ArUco codes are identified.

2.2. Determination of Target Position

Due to the complex underwater environment, fixing all UUVs to a position that allows the charging stake to be smoothly inserted into their charging port and recording the two-dimensional coordinates of the current target keypoints would require a lot of manpower and resources. Therefore, this paper proposes a method to determine the two-dimensional coordinates of the target keypoints underwater based on the target spraying positions on the UUV. This method first determines the above-water coordinates of the target position based on the target spraying position and then uses the law of refraction to determine the underwater coordinates.

2.2.1. Above-Water Coordinates of the Target Position

The schematic diagram of the two-dimensional coordinates of the water surface target position is shown in Figure 7. The two-dimensional coordinates of the above-water target position refer to the coordinates of the keypoint of the target T in the o-uv coordinate system. During the UUV target spraying, the relative position between the keypoint of the target and the UUV charging port can be obtained, that is, the coordinates (X, Y, Z) of the keypoints of the target in the coordinate system O-XYZ. Since the target position of the underwater charging platform is the position where the charging stake can be inserted into the charging port of the UUV, the coordinates (X_W, Y_W, Z_W) of the keypoint of the target in the coordinate system $O_w - X_W Y_W Z_W$ are $(X, Y, Z + L)$. The value of L is determined by the type of UUV and can be obtained during the decoding process since the clamping device of the charging platform in this method will fix and move the UUV to a specific position. The process of converting the coordinates in the coordinate system $O_w - X_W Y_W Z_W$ to the coordinates in the coordinate system o-uv can be regarded as the camera calibration process. The conversion process is shown in Equation (1):

$$\begin{bmatrix} X_c \\ Y_c \\ Z_c \\ 1 \end{bmatrix} = \begin{bmatrix} R & T \\ 0 & 1 \end{bmatrix} \begin{bmatrix} X_w \\ Y_w \\ Z_w \\ 1 \end{bmatrix}$$
$$Z_c \begin{bmatrix} u \\ v \\ 1 \end{bmatrix} = \begin{bmatrix} \alpha_x f & 0 & u_0 \\ 0 & \alpha_y f & v_0 \\ 0 & 0 & 1 \end{bmatrix} \begin{bmatrix} X_c \\ Y_c \\ Z_c \end{bmatrix} \quad (5)$$

where $\begin{bmatrix} R & T \\ 0 & 1 \end{bmatrix}$ represents the camera's extrinsic matrix, which represents the position relationship between the world coordinate system $O_w - X_W Y_W Z_W$ and the camera coordinate system $O_c - X_c Y_c Z_c$. $\begin{bmatrix} \alpha_x f & 0 & u_0 \\ 0 & \alpha_y f & v_0 \\ 0 & 0 & 1 \end{bmatrix}$ represents the camera's intrinsic matrix, which represents the transformation relationship between the camera coordinate system $O_c - X_c Y_c Z_c$ and the image coordinate system o-uv. The parameters α_x and α_y represent the scaling factors in the x and y directions, respectively. The parameter f represents the camera's focal length, and (u_0, v_0) represents the principal point of the camera, which is the coordinate of the camera's optical center in the image.

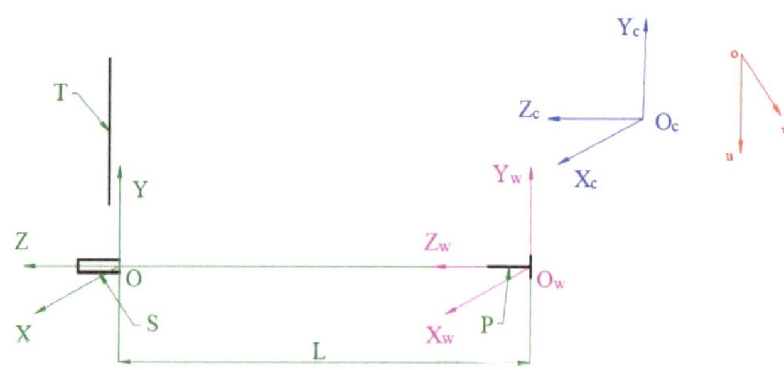

Figure 7. The schematic diagram of the two-dimensional coordinates of the above-water target position is shown above. In the diagram, T represents the spray position of the target on the UUV, S represents the position of the UUV charging port, and P represents the charging pile of the underwater charging platform. The O-XYZ coordinate system has the center of the UUV charging port as the origin, the $O_w - X_W Y_W Z_W$ coordinate system has the center of the charging pile on the underwater charging platform as the origin, the $O_c - X_c Y_c Z_c$ coordinate system has the camera center as the origin, and the o-uv two-dimensional coordinate system has the upper left corner of the image as the origin. L represents the distance between the UUV and the charging platform, which is determined by the UUV model and can be obtained during the decoding process. In this method, the UUV is fixed and moved to a specific position by the clamping device of the charging platform.

The camera's intrinsic parameters can be obtained and image distortion can be corrected using Zhang's calibration method [35]. However, due to the installation errors of the camera, it is necessary to accurately determine the camera's extrinsic matrix through further hand–eye calibration [36].

Inspired by neural network concepts, this paper transforms the camera calibration problem into estimating the function f in the Equation (6). To achieve this, the paper controls the movement of the slider and captures 25 photos, as shown in Figure 8. The coordinates of the four concentric circle centers in the world coordinate system $\left(X_W^i, Y_W^i, Z_W^i\right)$ and their corresponding coordinates in the image coordinate system $\left(u^i, v^i\right)$ are recorded for each photo. This resulted in 100 samples, where $\left(X_W^i, Y_W^i, Z_W^i, 1\right)$ serve as data and $\left(u^i, v^i\right)$ as labels. Fifty percent of the samples were used for training, and the remaining fifty percent for testing. A single-hidden-layer neural network without activation functions was constructed, as shown in Figure 9. The mean squared error (MSE) loss function was employed, and the external parameter matrix estimate was used to initialize the first layer of the network, while the calibrated internal parameters were used to initialize the second layer. The initialized network showed fast convergence, small oscillations, and ultimately converged to a smaller loss value.

$$[u, v]^T = f\left([X_W, Y_W, Z_W]^T\right) \tag{6}$$

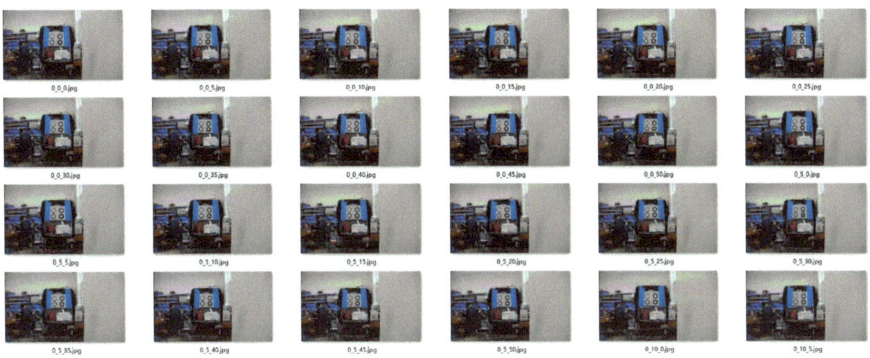

Figure 8. The dataset.

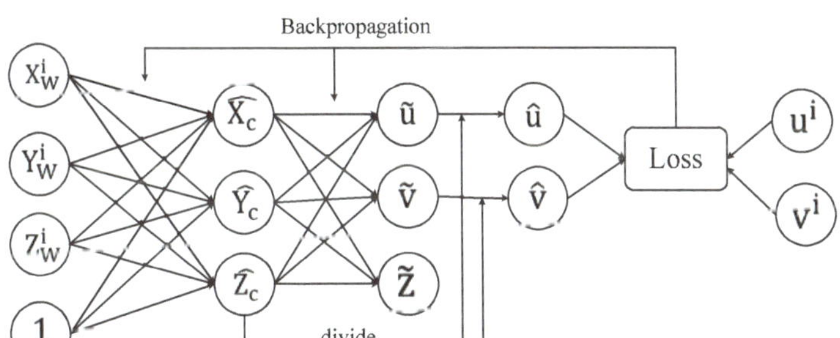

Figure 9. Structure of the neural network.

After training, the network weights consist of two matrices, $\hat{E}$ and $\hat{I}$, with dimensions of 4×3 and 3×3, respectively. Given a coordinate (X_W, Y_W, Z_W) in the world coordinate system, its corresponding coordinate in the image coordinate system can be calculated using Equation (7). It is worth noting that $\hat{E}$ and $\hat{I}$ obtained from training do not have physical meaning, and the intermediate variables $\hat{X}_c$, $\hat{Y}_c$, and $\hat{Z}_c$ in Equation (7) are not the coordinates of the point in the camera coordinate system.

$$\begin{gathered} \left[\hat{X}_c, \hat{Y}_c, \hat{Z}_c\right]^T = \hat{E}[X_W, Y_W, Z_W, 1]^T \\ [\hat{u}, \hat{v}, 1]^T = \frac{\hat{I}\left[\hat{X}_c, \hat{Y}_c, \hat{Z}_c\right]^T}{\hat{Z}_c} \end{gathered} \tag{7}$$

2.2.2. Underwater Coordinates of the Target Location

The imaging principle of the underwater camera is shown in Figure 10. In air, the light reflected by the target propagates in a straight line, and the size of an object with a size of h projected onto the camera imaging plane is a. However, underwater, due to refraction of light between different media, the size of an object with a size of h projected onto the imaging plane is b, and according to Snell's Law [37], as shown in Equation (8), since the refractive index of water is 1.333, $\alpha > \theta$ and $b > a$.

$$n = \frac{\sin\alpha}{\sin\theta} \tag{8}$$

where θ represents the angle of incidence, α represents the angle of refraction, and n represents the refractive index of the medium.

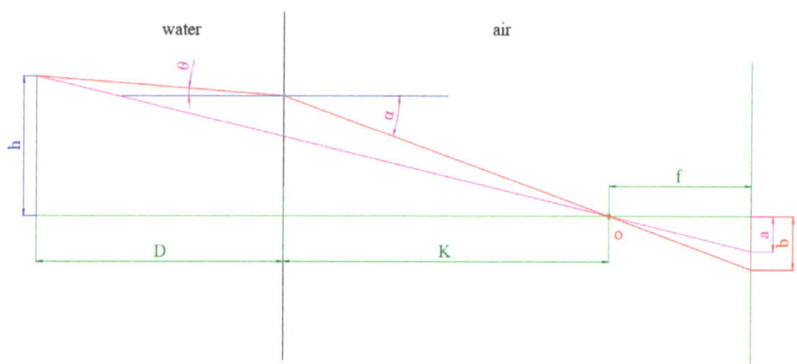

Figure 10. The imaging principle of the underwater camera. In the diagram, o represents the camera's optical center, f represents the camera's focal length, K represents the distance from the refraction surface (the outer glass of the camera) to the camera's optical center, and D represents the distance from the target to the refraction surface. Similar to L mentioned earlier, D is a known constant value determined by the positioning of the clamp and the structure of the underwater charging platform.

It can be inferred that the projection of underwater objects on the imaging plane can be obtained by magnifying the projection of objects above water with the camera center as the projection center by a certain factor. The magnification factor $\frac{b}{a}$ can be obtained from Equation (9).

$$\frac{b}{a} = \frac{(D+K)}{\left(D \times \frac{tan\theta}{tan\alpha} + K\right)} \approx \frac{(D+K)}{\left(D \times \frac{1}{n} + K\right)} \tag{9}$$

2.3. Target Recognition and Instruction Provision

This method utilizes a target designed by Tweddle et al. [38], as shown in Figure 11. The target consists of four concentric circles with area ratios of 1.44, 1.78, 2.25, and 2.94, and the keypoints of the target are the centers of the four concentric circles. During the alignment process, the camera image is preprocessed first, followed by contour detection to identify circles that may originate from the target according to their area and roundness. Then, based on the area ratios, the concentric circles are matched, and the coordinate values of each concentric circle are obtained. Finally, based on the relative position between the visible concentric circle center of the current position and the corresponding target concentric circle center, the charging station is guided to move until the coordinate difference between the current and target positions is less than the maximum allowable error, and the movement of the charging station is controlled to stop. During the alignment process, the x-coordinate of the concentric circle center is compared first to guide the charging station to rotate tangentially for azimuthal alignment, followed by the y-coordinate of the concentric circle center to guide the charging station to move axially for axial alignment.

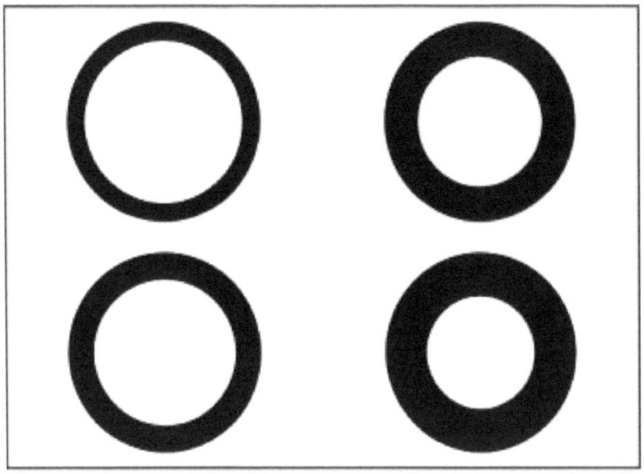

Figure 11. The employed target.

To address the problem of low contrast in underwater images, this paper employs the Niblack binary thresholding method [39]. It calculates the pixel threshold by sliding a rectangular window over the grayscale image [40], as shown in Equation (10). If the pixel value surpasses the threshold, it is set as foreground; otherwise, it is set as background. Because the Niblack binary thresholding algorithm can adaptively adjust the threshold based on local blocks, it can preserve more texture details in the image. However, due to its need for more computational resources and time to process images, the Niblack binary thresholding method is computationally expensive.

$$T = m - k\sqrt{\frac{1}{n}\sum_{i\epsilon\Omega}(p_i - m)^2} \quad (10)$$

where T represents the pixel threshold, Ω represents the sliding window, n represents the number of pixels in the window, p_i represents the grayscale value of each point in the window, m represents the pixel mean value of all points in the window, and k represents the correction parameter.

To accelerate the processing speed of Niblack thresholding, this paper is inspired by the ResNet network [41] and adopts a structure similar to the bottleneck architecture that first reduces and then restores the size of the image. As the processing time required for Niblack thresholding is directly proportional to the size of the image, the size of the image is reduced to one quarter of its original size, and then Niblack thresholding is performed. Finally, the thresholded image is enlarged four times to restore its original size. Although the size of this image is the same as that of the original image, its information content is only one quarter of that of the original image, and its contour details are relatively blurred. Directly using this image for contour detection would lower the detection accuracy. Therefore, inspired by the coarse-to-fine idea in the LoFTR algorithm [42], this paper uses the image to determine the region of interest (ROI) and performs Niblack thresholding on the ROI of the original image. This significantly reduces the number of pixels processed by Niblack thresholding, ensuring the real-time performance of the algorithm.

To address the problem of partial occlusion caused by bubbles and suspended particles in underwater images, this paper enhances the robustness of the algorithm by utilizing the redundancy of target information. During the alignment process, only one of the four concentric circles in the target needs to be identified, and then the two-dimensional coordinates of its center are compared with the center of the concentric circle corresponding to the target position area ratio, thus providing the motion instructions in both the circumferential and axial directions. If multiple concentric circle centers are detected, the center point of multiple circle centers is used for comparison.

3. Results

3.1. Result of the Decoding Experiment

In this paper, decoding experiments were conducted underwater. During the encoding process, assuming the UUV is numbered 123, its corresponding four-digit code would be 12, 23, 31, and 6. These four digits were then converted into ArUco codes and rotated at 0°, 90°, 180°, and 270°, respectively, resulting in the final encoded patterns. The ArUco code recognition result is shown in Figure 12.

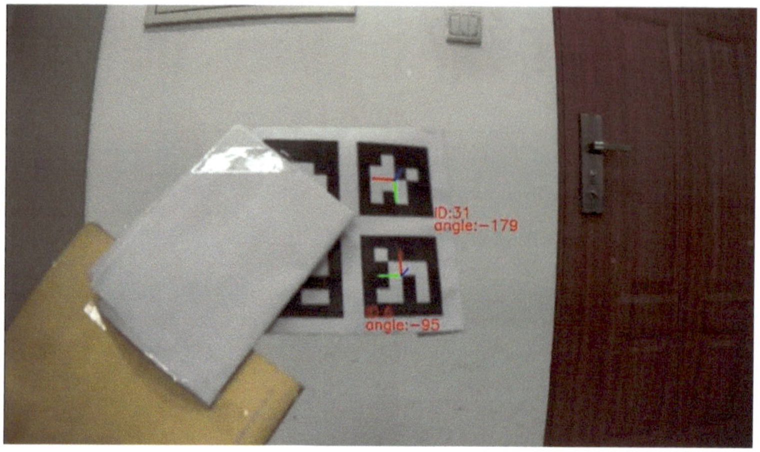

Figure 12. Result of ArUco code recognition.

The proposed image restoration method is compared with image enhancement algorithms based on the gray world assumption theory [43], the UDCP algorithm [44], and the Shallow-UWnet method [28] based on end-to-end convolutional neural networks. The results are shown in Figure 13. Furthermore, the proposed thresholding method is compared with Niblack and Bernsen methods [45]. Table 1 presents the detection results of ArUco codes after being processed by different thresholding methods.

(a) (b) (c) (d) (e)

Figure 13. Result of image processing. (**a**) Original image; (**b**) image enhancement result of the algorithms based on the gray world assumption theory; (**c**) image restoration result of the UDCP algorithm; (**d**) image restoration result of the Shallow-UWnet method; (**e**) image restoration result of the proposed method.

Table 1. ArUco code detection results of different image processing methods.

Thresholding \ Restoration	Gray World Assumption Theory	UDCP	Shallow-UWnet	Proposed
Niblack				
Bernsen				
Proposed	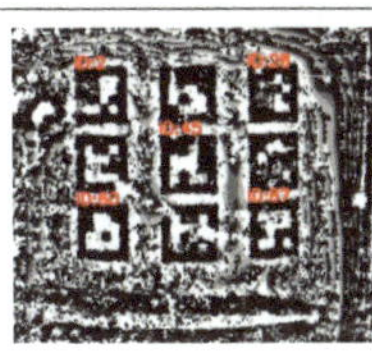		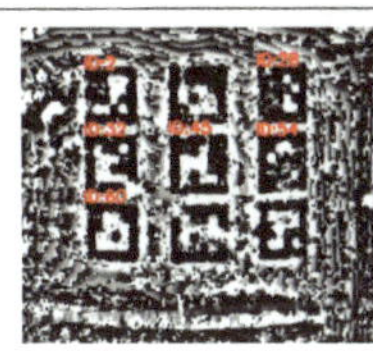	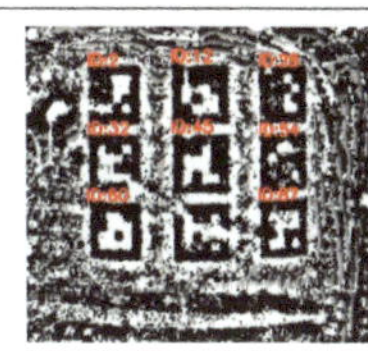

3.2. *Result of the Experiment on Determining Target Position*

3.2.1. Result of the Experiment on Above-Water Target Position Calibration

Figure 14 shows the loss variation of the proposed network architecture and the network architecture proposed by Cao et al. [46] during the training and testing processes.

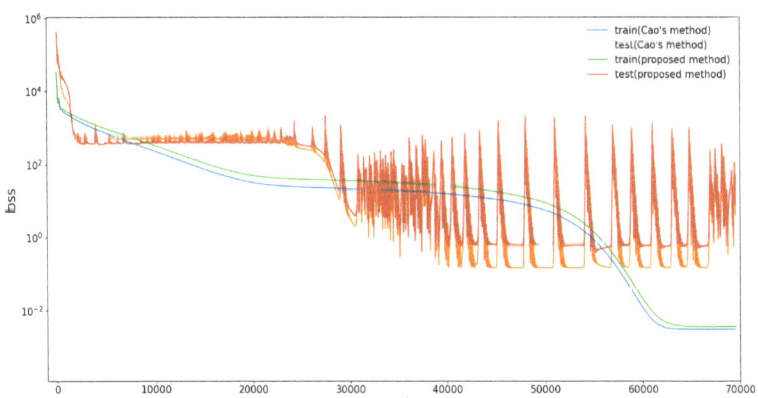

Figure 14. Result of ArUco code recognition.

The training configuration is as follows: there are a total of 100 samples, with labels of (u^i, v^i) and data of $(X_W^i, Y_W^i, Z_W^i, 1)$; half of them are training sets and the other half are test sets. The initial learning rate is set to 0.1, which decreases by half every 1000 epochs to optimize the training process. The MSE loss function is employed as the loss function during the training phase. Training ends when the test loss does not decrease for 500 consecutive epochs.

3.2.2. Result of the Experiment on Underwater Target Position Calibration

Based on Figure 15, the two-dimensional coordinates of the underwater target position can be obtained by enlarging the two-dimensional coordinates of the on-water target position with the projection of the camera center on the imaging plane as the center. The enlargement factor can be obtained from Equation (9). To determine the distance between the camera center and the refraction surface, as well as the projection of the camera center on the imaging plane, the projection of the same target on the camera imaging plane was recorded both underwater and above water. The targets were located 200 mm from the refraction surface, as shown in Figure 11. The corresponding keypoints A_2A_1, B_2B_1, C_2C_1, D_2D_1 were connected and extended, and their intersection point P was taken as the projection of the camera's optical center on the imaging plane. The average values of $\frac{PA_2}{PA_1}, \frac{PB_2}{PB_1}, \frac{PC_2}{PC_1}, \frac{PD_2}{PD_1}$, which were 1.308, were used as $\frac{b}{a}$ in Equation (9). By substituting D = 200 mm into Equation (9), the distance between the camera center and the refraction surface, K was obtained as 10.618 mm.

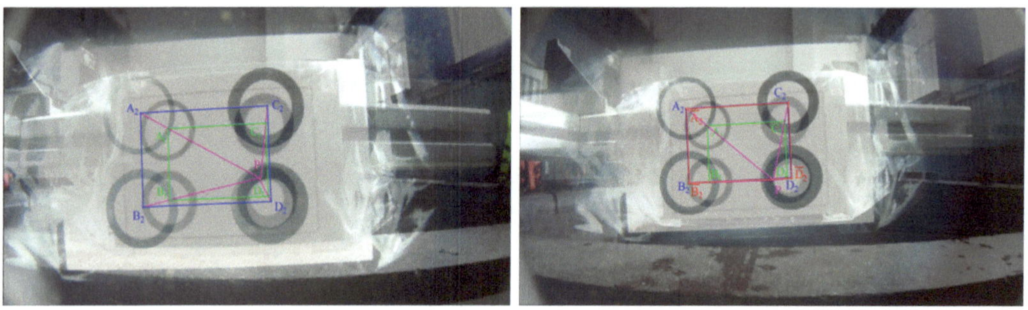

(**a**) (**b**)

Figure 15. Projection of the same target on the imaging plane underwater and above water. (**a**) When D = 200 mm, record the same target's projection on the imaging plane underwater and above water and mark its keypoints. In this figure, A_1, B_1, C_1, D_1 represents keypoints of the above-water image, A_2, B_2, C_2, D_2 represents keypoints of the underwater image. (**b**) When D = 260 mm, the location of the underwater keypoints is predicted from the calculated value of K and compared with the actual value. In this figure, $\overline{A_2}$, $\overline{B_2}$, $\overline{C_2}$, $\overline{D_2}$ represents predicted keypoints of the underwater image.

Once the value of K is calculated, the camera's field of view (FOV) can be determined. The camera model used in this study is LI-IMX185MIPI-CS, with a sensor type of 1/1.9″, which corresponds to a diagonal length of 1/1.9 inches. The aspect ratio of the image is 16:9. Therefore, the sensor size is 7.3396 mm × 4.1285 mm.

In air, using Equation (11), the camera's field of view is calculated to be 173.9073 mm × 309.1704 mm.

$$\begin{aligned} H_a &= \frac{h\times(K+D)}{f} = \frac{4.1285\times(10.618+200)}{5} = 173.9073 \text{ mm} \\ W_a &= \frac{w\times(K+D)}{f} = \frac{7.3396\times(10.618+200)}{5} = 309.1704 \text{ mm} \end{aligned} \quad (11)$$

where W_a and H_a represent the width and height of the FOV, and w and h are the width and height of the sensor, respectively. f denotes the camera's focal length. The definitions of K and D are illustrated in Figure 10.

Underwater, due to the refraction of light between different media, the camera's FOV will be reduced. Using Equation (12), the camera's FOV is calculated to be 132.6537 mm × 235.8289 mm.

$$\begin{aligned} \tan\alpha_h &= \tfrac{h}{f} \\ H_w &= D \times \tan\theta_h + K \times \tan\alpha_h = 132.6537 \text{ mm} \\ \tan\alpha_w &= \tfrac{w}{f} \\ W_w &= D \times \tan\theta_w + K \times \tan\alpha_w = 235.8289 \text{ mm} \end{aligned} \tag{12}$$

where α_h and θ_h represent the refraction angle and incident angle in the height direction of the sensor, and α_w and θ_w represent the refraction angle and incident angle in the width direction of the sensor. These parameters are illustrated in Figure 10.

The accuracy of determining the underwater target position was verified below. The projection of the same target on the camera imaging plane was again recorded both underwater and on-water, with the target located 260 mm from the refraction surface. With the value of K calculated above, $\frac{b}{a}$ was determined as 1.313. Using P and b/a, the predicted two-dimensional coordinates of the underwater target position, $\overline{A_2}$, $\overline{B_2}$, $\overline{C_2}$, and $\overline{D_2}$, were obtained, as shown in Figure 11. The error between the predicted and true values is presented in Table 2.

Table 2. Deviations between the predicted and true values.

Δx (Pixel)	2.35	0.74	2.49	1.79
Δy (Pixel)	0.7	0.17	5.21	1.21

3.3. Result of Image Processing

We downsized the images by different scales, applied Niblack binarization, and then resized them back to their original size. The average processing time for each frame is shown in Table 3. The fluctuations of the keypoint detection results for the target at different scaling ratios over 90 consecutive frames are illustrated in Figure 16.

Table 3. Average processing time per frame.

Scale = 1	Scale = 2	Scale = 4	Ours
0.24592 s	0.08923 s	0.03511 s	0.092 s

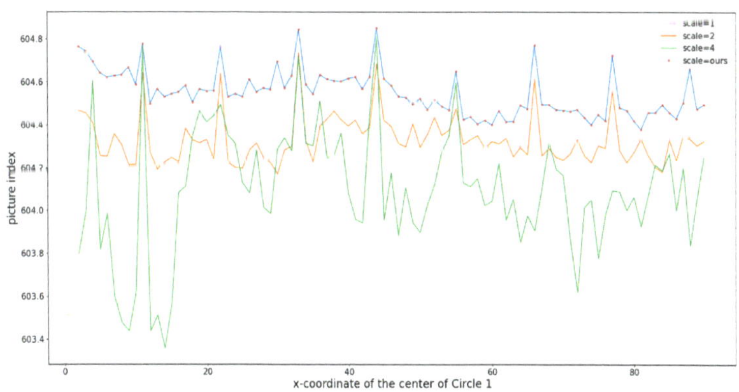

Figure 16. Fluctuations of the keypoints.

3.4. Result of the Actual Alignment Experiment

During the experiment, the first stage is the circumferential alignment process, where the circumferential hydraulic cylinder moves until the difference between the x coordinates of the current position and the target position is within the allowable error range. The second stage is the axial alignment process, where the axial hydraulic cylinder moves until the difference between the y coordinates of the current position and the target position is within the allowable error range. The third stage is the charging process, where the hydraulic cylinder controls the charging stack to rise and insert into the UUV charging port. The distance between the camera and the alignment platform was 200 mm. The displacement changes of each hydraulic cylinder during the alignment process are shown in Figures 17–19, and the distance changes between the current position of the target and the target position are shown in Figures 20 and 21.

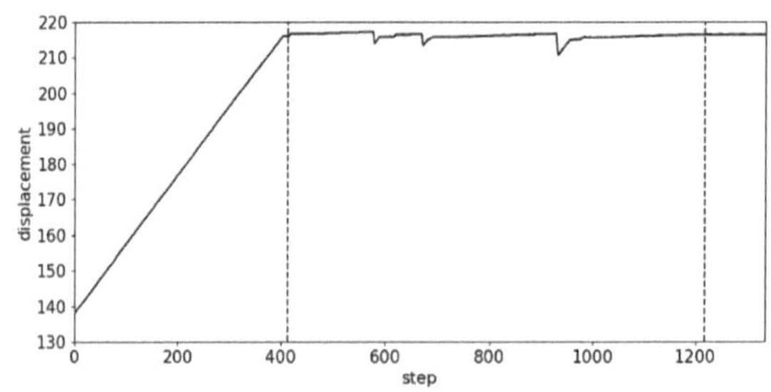

Figure 17. Displacement of the circumferential hydraulic cylinder. In the figure, the first dash line represents the completion of circumferential alignment, transitioning to axial alignment. The second dash line represents the completion of axial alignment, transitioning to the charging process. The following are the same.

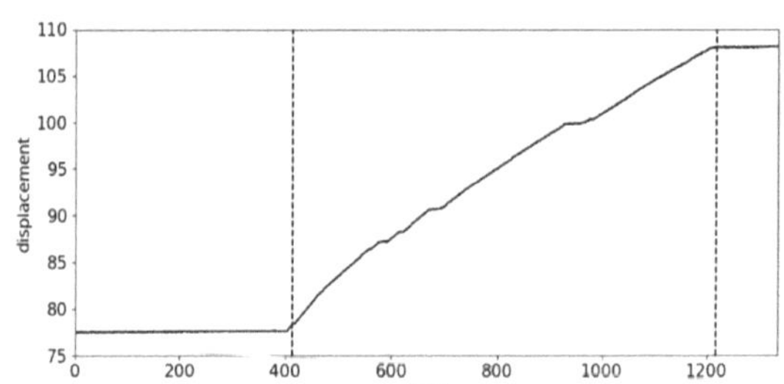

Figure 18. Displacement of the axial hydraulic cylinder.

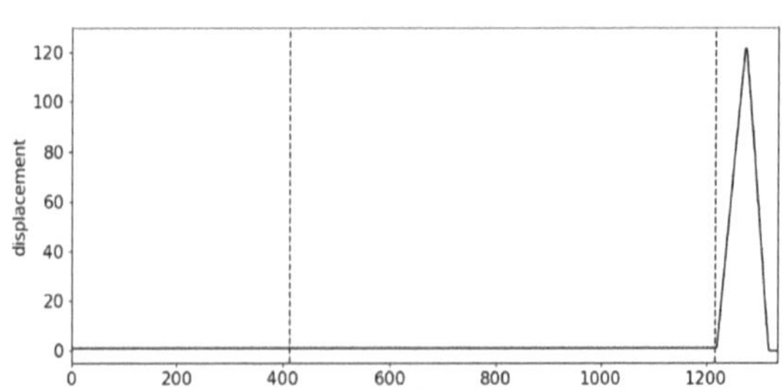

Figure 19. Displacement of the charging hydraulic cylinder.

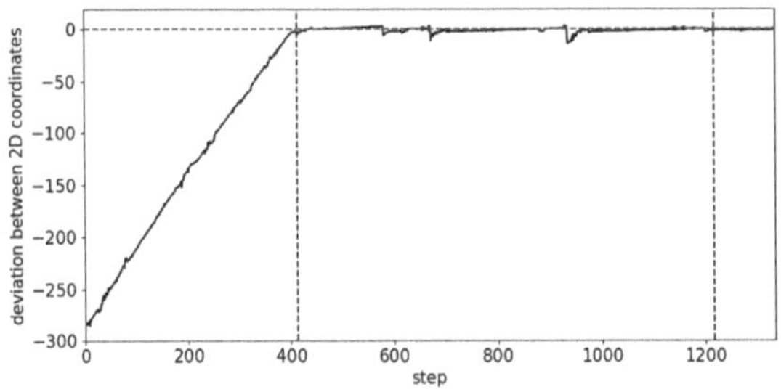

Figure 20. Deviation of the x-coordinate between the current position of the target and the target position.

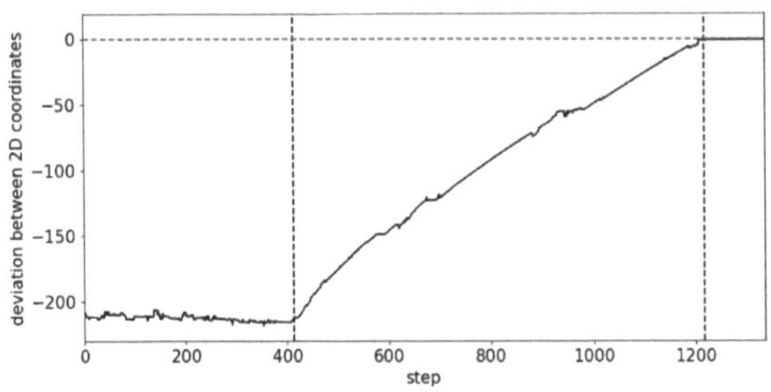

Figure 21. Deviation of the y-coordinate between the current position of the target and the target position.

4. Discussion

The experimental results of underwater encoding are presented in Figure 12. Although some ArUco codes were not recognized due to occlusion, two ArUco codes were identified with IDs 31 and 6 and their yaw angles were 179° and −95°, respectively. Based on machine pose estimation, it can be inferred that ID 31 represents A_3A_1; therefore, $A_1 = 1$ and $A_3 = 3$. ID 6 represents $A_1 + A_2 + A_3$, thus $A_2 = 2$. Decoding yields the UUV number 123.

As shown in Figure 13, the original image suffers from low contrast, blurry contours, and a color shift towards blue and green due to light absorption and scattering in the underwater environment. The method based on the gray world assumption theory [43] corrects the color shift but fails to enhance the image contrast. The UDCP method [44] intensifies the color shift towards green. The Shallow-UWnet method [28] enhances the image contrast but introduces an additional color shift towards yellow. In contrast, the proposed method in this paper corrects the color shift while enhancing image details. As shown in Table 1, by applying the proposed image restoration and thresholding methods, eight out of nine ArUco codes can be detected. When only the proposed image restoration method is applied, a maximum of four codes can be detected, while applying only the proposed image thresholding method can detect up to six codes. The best result of the remaining methods is achieved by combining the Niblack and the gray world assumption theory methods, which detects five codes. Therefore, both the proposed image restoration method and thresholding method effectively improve the detection rate of ArUco codes.

As shown in Figure 14, Cao's method [46] exhibits a fast convergence rate during the initial stages of training, but it also shows significant oscillations during the convergence process. In contrast, the proposed method in this paper has a slower convergence rate but demonstrates a stable convergence process, with the final loss value consistently reaching a smaller value. The minimum testing error achieved by Cao's method [46] is 0.3828 pixels, while the proposed method in this paper achieves a minimum training error of 0.0036 pixels.

Figure 15b presents the difference between the predicted underwater target position and the actual underwater target position. The predicted underwater keypoints and the actual underwater keypoints overlap almost perfectly. The errors are shown in Table 2, with a maximum error of 5.21 pixels, which meets the accuracy requirements for alignment.

Table 3 shows the average processing time per frame under different scaling ratios. Figure 16 shows the data fluctuation under different scaling ratios. The algorithm proposed in this paper has the same processing results as no scaling detection, the data fluctuation is minimal, but the processing time is only 37.41% of the original time. It can meet the real-time requirements in the alignment control process.

The displacement changes of each hydraulic cylinder during the alignment process are shown in Figures 17–19. Finally, the hydraulic cylinder controls the piston rod to move up by 119 mm, indicating that the piston rod has successfully inserted into the hole on the UUV. As shown in Figures 20 and 21, the error between the final alignment position and the target position in the x direction is −0.69661 pixels, and in the y direction it is −0.58738 pixels. Therefore, considering the calibration errors in Figure 14, the underwater coordinate transformation errors in Table 2, and the motion errors in Figures 20 and 21, the maximum alignment error is 4.517 pixels. Based on the calculation of FOV using Equation (12), the maximum alignment error is 0.5548 mm, which meets the accuracy requirements.

Although the method proposed in this paper has successfully achieved the automatic alignment of the UUV underwater charging platform, there is still room for improvement. Firstly, for partially occluded targets, this paper uses redundant information processing methods in both the decoding and target recognition processes. However, accurately completing the occluded parts in real time would further enhance the robustness of the method. Secondly, in determining the camera's intrinsic and extrinsic parameters, this paper still needs to use Zhang's calibration method to determine the distortion coefficients and initialize the network parameters using the calibrated intrinsic parameters. This method is not concise enough. Finally, this paper did not explicitly express the camera's intrinsic and extrinsic parameters. These three points will be our future research directions.

5. Conclusions

This paper presents an automatic alignment method for UUV underwater charging platforms using monocular vision recognition. This method accurately identifies the UUV's identity information and guides the charging stake to smoothly insert into the charging port of the UUV through target recognition. To ensure the accuracy and robustness of decoding, this study introduces an encoding method based on redundant information and proposes an ArUco code reconstruction method specifically designed for underwater imaging environments for decoding purposes. Additionally, a method for determining the target position is proposed to overcome the difficulty of directly determining the underwater target position. The proposed method accurately determines the underwater two-dimensional coordinates of the target keypoints based on the location of the UUV target spray using deep learning and the law of refraction. The experimental results demonstrate that the proposed ArUco code reconstruction method can improve the detection rate of ArUco codes by at least 22.2%. The proposed target detection algorithm has an average processing time of 0.092 s per frame, meeting the requirements for real-time control. The maximum alignment error is 0.5548 mm, meeting the accuracy requirements for alignment.

Author Contributions: Conceptualization, A.Y. and Y.W.; methodology, A.Y.; software, A.Y.; validation, A.Y., Y.W. and H.L.; formal analysis, A.Y. and B.Q.; investigation, Y.W.; resources, H.L. and B.Q.; data curation, H.L.; writing—original draft preparation, A.Y.; writing—review and editing, A.Y.; visualization, A.Y.; supervision, Y.W.; project administration, Y.W.; funding acquisition, Y.W. All authors have read and agreed to the published version of the manuscript.

Funding: This research received no external funding.

Institutional Review Board Statement: Not applicable.

Informed Consent Statement: Not applicable.

Data Availability Statement: Not applicable.

Conflicts of Interest: The authors declare no conflict of interest.

References

1. Zuo, M.; Wang, G.; Xiao, Y.; Xiang, G. A unified approach for underwater homing and docking of over-actuated AUV. *J. Mar. Sci. Eng.* **2021**, *9*, 884. [CrossRef]
2. Wang, T.; Zhao, Q.; Yang, C. Visual navigation and docking for a planar type AUV docking and charging system. *Ocean Eng.* **2021**, *224*, 108744. [CrossRef]
3. Bharti, V.; Wang, S. Autonomous Pipeline Tracking Using Bernoulli Filter for Unmanned Underwater Surveys. In Proceedings of the IEEE/RSJ International Conference on Intelligent Robots and Systems (IROS), Kyoto, Japan, 23–27 October 2022; pp. 7129–7136.
4. Zhang, Y.-X.; Zhang, Q.-F.; Zhang, A.-Q.; Chen, J.; Li, X.-G.; He, Z. Acoustics-Based Autonomous Docking for A Deep-Sea Resident ROV. *China Ocean Eng.* **2022**, *36*, 100–111. [CrossRef]
5. Chen, Y.; Duan, Z.; Zheng, F.; Guo, Y.; Xia, Q. Underwater optical guiding and communication solution for the AUV and seafloor node. *Appl. Opt.* **2022**, *61*, 7059–7070. [CrossRef]
6. Vandavasi, B.N.J.; Gidugu, A.R.; Venkataraman, H. Deep Learning Aided Magnetostatic Fields Based Real-Time Pose Estimation of AUV for Homing Applications. *Ieee Sens. Lett.* **2023**, *7*, 22814400. [CrossRef]
7. Lin, R.; Zhao, Y.; Li, D.; Lin, M.; Yang, C. Underwater Electromagnetic Guidance Based on the Magnetic Dipole Model Applied in AUV Terminal Docking. *J. Mar. Sci. Eng.* **2022**, *10*, 995. [CrossRef]
8. Chhatpar, S.R.; Branicky, M.S. Search strategies for peg-in-hole assemblies with position uncertainty. In Proceedings of the 2001 IEEE/RSJ International Conference on Intelligent Robots and Systems. Expanding the Societal Role of Robotics in the the Next Millennium (Cat. No. 01CH37180), Maui, HI, USA, 29 October–3 November 2001; pp. 1465–1470.
9. Wang, X.; Zhao, X.; Liu, Z. Underwater Optical Image Enhancement Based on Color Constancy and Multiscale Wavelet. *Laser Optoelectron. Prog.* **2022**, *59*, 1601002. [CrossRef]
10. Kong, S.; Zhou, K.; Huang, X. Online measurement method for assembly pose of gear structure based on monocular vision. *Meas. Sci. Technol.* **2023**, *34*, 065110. [CrossRef]
11. Fan, J.; Jing, F.; Yang, L.; Long, T.; Tan, M. An initial point alignment method of narrow weld using laser vision sensor. *Int. J. Adv. Manuf. Technol.* **2019**, *102*, 201–212. [CrossRef]
12. Chen, B.; Liu, Y. Antenna alignment by using computer vision technology. *Microw. Opt. Technol. Lett.* **2020**, *62*, 1267–1269. [CrossRef]
13. Kim, J. New Wafer Alignment Process Using Multiple Vision Method for Industrial Manufacturing. *Electronics* **2018**, *7*, 39. [CrossRef]
14. Ke, K.; Zhang, C.; Wang, Y.; Zhang, Y.; Yao, B. Single underwater image restoration based on color correction and optimized transmission map estimation. *Meas. Sci. Technol.* **2023**, *34*, 55408. [CrossRef]
15. Jaffe, J.S. Computer modeling and the design of optimal underwater imaging systems. *IEEE J. Ocean. Eng.* **1990**, *15*, 101–111. [CrossRef]
16. McGlamery, B. A computer model for underwater camera systems. In *Ocean Optics VI*; SPIE: Washington, DC, USA, 1980; pp. 221–231.
17. He, K.; Sun, J.; Tang, X. Single Image Haze Removal Using Dark Channel Prior. *IEEE Trans. Pattern Anal. Mach. Intell.* **2011**, *33*, 2341–2353. [CrossRef] [PubMed]
18. He, K.; Sun, J.; Tang, X. Guided image filtering. In Proceedings of the European Conference on Computer Vision, Crete, Greece, 5–11 September 2010; pp. 1–14.
19. Carlevaris-Bianco, N.; Mohan, A.; Eustice, R.M. Initial results in underwater single image dehazing. In Proceedings of the Oceans 2010 Mts/IEEE Seattle, Seattle, WA, USA, 20–23 September 2010; pp. 1–8.
20. Li, C.; Quo, J.; Pang, Y.; Chen, S.; Wang, J. Single underwater image restoration by blue-green channels dehazing and red channel correction. In Proceedings of the 2016 IEEE International Conference on Acoustics, Speech and Signal Processing (ICASSP), Shanghai, China, 20–25 March 2016; pp. 1731–1735.
21. Galdran, A.; Pardo, D.; Picon, A.; Alvarez-Gila, A. Automatic Red-Channel underwater image restoration. *J. Vis. Commun. Image Represent.* **2015**, *26*, 132–145. [CrossRef]
22. Hou, G.; Pan, Z.; Wang, G.; Yang, H.; Duan, J. An efficient nonlocal variational method with application to underwater image restoration. *Neurocomputing* **2019**, *369*, 106–121. [CrossRef]
23. Hu, J.; Shen, L.; Sun, G. Squeeze-and-excitation networks. In Proceedings of the IEEE Conference on Computer Vision and Pattern Recognition, Salt Lake City, UT, USA, 18–22 June 2018; pp. 7132–7141.
24. Bochkovskiy, A.; Wang, C.-Y.; Liao, H.-Y.M. Yolov4: Optimal speed and accuracy of object detection. *arXiv* **2020**, arXiv:2004.10934.
25. Kirillov, A.; Wu, Y.; He, K.; Girshick, R. Pointrend: Image segmentation as rendering. In Proceedings of the IEEE/CVF Conference on Computer Vision and Pattern Recognition, Seattle, WA, USA, 13–19 June 2020; pp. 9799–9808.
26. Wang, K.; Shen, L.; Lin, Y.; Li, M.; Zhao, Q. Joint Iterative Color Correction and Dehazing for Underwater Image Enhancement. *Ieee Robot. Autom. Lett.* **2021**, *6*, 5121–5128. [CrossRef]
27. Wang, Y.; Yu, X.; An, D.; Wei, Y. Underwater image enhancement and marine snow removal for fishery based on integrated dual-channel neural network. *Comput. Electron. Agric.* **2021**, *186*, 106182. [CrossRef]
28. Naik, A.; Swarnakar, A.; Mittal, K.; Assoc Advancement Artificial, I. Shallow-UWnet: Compressed Model for Underwater Image Enhancement (Student Abstract). *Proc. AAAI Conf. Artif. Intell.* **2021**, *35*, 15853–15854. [CrossRef]

29. Song, W.; Wang, Y.; Huang, D.; Tjondronegoro, D. A rapid scene depth estimation model based on underwater light attenuation prior for underwater image restoration. In Proceedings of the Pacific Rim Conference on Multimedia, Hefei, China, 21–22 September 2018; pp. 678–688.
30. Meng, C.; Li, Z.; Sun, H.; Yuan, D.; Bai, X.; Zhou, F. Satellite Pose Estimation via Single Perspective Circle and Line. *IEEE Trans. Aerosp. Electron. Syst.* **2018**, *54*, 3084–3095. [CrossRef]
31. Luckett, J.A. *Comparison of Three Machine Vision Pose Estimation Systems Based on Corner, Line, and Ellipse Extraction for Satellite Grasping*; West Virginia University: Morgantown, WV, USA, 2012.
32. Zhang, H.; Meng, C.; Bai, X.; Li, Z. Rock-ring detection accuracy improvement in infrared satellite image with sub-pixel edge detection. *IET Image Process.* **2019**, *13*, 729–735. [CrossRef]
33. Huang, B.; Sun, Y.; Zeng, Q. General fusion frame of circles and points in vision pose estimation. *Optik* **2018**, *154*, 47–57. [CrossRef]
34. Gould, R.W.; Arnone, R.A.; Martinolich, P.M. Spectral dependence of the scattering coefficient in case 1 and case 2 waters. *Appl. Opt.* **1999**, *38*, 2377–2383. [CrossRef]
35. Zhang, Z. A flexible new technique for camera calibration. *IEEE Trans. Pattern Anal. Mach. Intell.* **2000**, *22*, 1330–1334. [CrossRef]
36. An, Y.; Wang, X.; Zhu, X.; Jiang, S.; Ma, X.; Cui, J.; Qu, Z. Application of combinatorial optimization algorithm in industrial robot hand eye calibration. *Measurement* **2022**, *202*, 111815. [CrossRef]
37. Ma, Y.; Zhou, Y.; Wang, C.; Wu, Y.; Zou, Y.; Zhang, S. Calibration of an underwater binocular vision system based on the refraction model. *Appl. Opt.* **2022**, *61*, 1675–1686. [CrossRef]
38. Tweddle, B.E. *Computer Vision Based Navigation for Spacecraft Proximity Operations*; Massachusetts Institute of Technology: Cambridge, MA, USA, 2010.
39. Niblack, W. *An Introduction to Digital Image Processing*; Strandberg Publishing Company: Hovedstaden, Denmark, 1985.
40. Khurshid, K.; Siddiqi, I.; Faure, C.; Vincent, N. Comparison of Niblack inspired binarization methods for ancient documents. In *Document Recognition and Retrieval XVI*; SPIE: Washington, DC, USA, 2009; pp. 267–275.
41. He, K.; Zhang, X.; Ren, S.; Sun, J. Deep residual learning. *Image Recognit.* **2015**, *7*, 770–778.
42. Sun, J.; Shen, Z.; Wang, Y.; Bao, H.; Zhou, X.; Ieee Comp, S.O.C. LoFTR: Detector-Free Local Feature Matching with Transformers. In Proceedings of the IEEE/CVF Conference on Computer Vision and Pattern Recognition (CVPR), Nashville, TN, USA, 20–25 June 2021; pp. 8918–8927.
43. Lam, E.Y. Combining gray world and Retinex theory for automatic white balance in digital photography. In Proceedings of the 9th International Symposium on Consumer Electronics (ISCE 2005), Taipa, Macao, 14–16 June 2005; pp. 134–139.
44. Drews, P., Jr.; do Nascimento, E.; Moraes, F.; Botelho, S.; Campos, M. Transmission Estimation in Underwater Single Images. In Proceedings of the 2013 IEEE International Conference on Computer Vision Workshops, Sydney, Australia, 2–8 December 2013; pp. 825–830.
45. Hashem, A.-R.A.; Idris, M.Y.I.; Ahmad, A.E.-B.A. Comparative study of different binarization methods through their effects in characters localization in scene images. *Data Knowl. Eng.* **2018**, *117*, 216–224. [CrossRef]
46. Cao, Y.; Wang, H.; Zhao, H.; Yang, X. Neural-Network-Based Model-Free Calibration Method for Stereo Fisheye Camera. *Front. Bioeng. Biotechnol.* **2022**, *10*, 955233. [CrossRef]

Journal of
Marine Science and Engineering

MDPI

Article

Collision Avoidance Strategy for Unmanned Surface Vessel Considering Actuator Faults Using Kinodynamic Rapidly Exploring Random Tree-Smart and Radial Basis Function Neural Network-Based Model Predictive Control

Yunxuan Song, Yimin Chen *, Jian Gao, Yazhou Wang and Guang Pan

School of Marine Science and Technology, Northwestern Polytechnical University, Xi'an 710072, China
* Correspondence: yimin.xjtu@outlook.com

Abstract: Path planning and tracking are essential technologies for unmanned surface vessels (USVs). The kinodynamic constraints and actuator faults, however, bring difficulties in finding feasible paths and control efforts. This paper proposes a collision avoidance strategy for USV by developing the kinodynamic rapidly exploring random tree-smart (kinodynamic RRT*-smart) algorithm and the fault-tolerant control method. By utilizing the triangular inequality and the intelligent biased sampling strategy, the kinodynamic RRT*-smart shows its advantages in terms of path length, cost and running time. With consideration of kinodynamic constraints, a feasible and collision-free trajectory can be provided. Then, a radial basis function neural network-based model predictive control (RBF-MPC) method was designed that compensates for the model's uncertainties by developing the radial basis function neural network (RBF-NN) approximator and by constructing a feedback-state training dataset in real time. Furthermore, two types of fault situation were analyzed considering the thruster failure. We established the faults' mathematical models and investigated the fault-tolerant strategies for different fault types. The simulation studies were conducted to validate the effectiveness of the proposed strategy. The results show that the proposed planning and control methods can avoid obstacles in faulty conditions.

Keywords: path planning; unmanned surface vessel; learning-based MPC; fault-tolerance strategy

Citation: Song, Y.; Chen, Y.; Gao, J.; Wang, Y.; Pan, G. Collision Avoidance Strategy for Unmanned Surface Vessel Considering Actuator Faults Using Kinodynamic Rapidly Exploring Random Tree-Smart and Radial Basis Function Neural Network-Based Model Predictive Control. *J. Mar. Sci. Eng.* **2023**, *11*, 1107. https://doi.org/10.3390/jmse11061107

Academic Editor: Fabio Bruno

Received: 28 April 2023
Revised: 18 May 2023
Accepted: 21 May 2023
Published: 23 May 2023

1. Introduction

Unmanned surface vessels (USVs), as an intelligent water work platform, have been widely used in oceanic exploration, personnel search and rescue, water surface cleaning and other aspects [1–5]. Compared with man-operated ships, USVs are not only more economical and safe but are also more suitable for harsh environments, which attracts enormous attention from both the academic and industrial fields. Planning a path that aligns harmoniously with its own motion properties is an indispensable prerequisite for an unmanned surface vehicle (USV) to fulfill its objectives. Once a reference path has been assigned, the paramount challenge lies in tracing the reference trajectory amidst the intricate interplay of maritime elements. However, these considerations alone prove insufficient for that the USV's propulsion systems are often susceptible to compromise. Therefore, how to achieve stable control of USVs in the face of modeling error and thruster failure is also crucial for accomplishing the designated missions.

By obtaining the environmental information, the path planning component is able to provide a collision-free path that guides the USV from the start position to the destination. Intensive research efforts have been devoted to the path planning of USVs over the past decades. The A* algorithm is proposed to generate a fast node path from its heuristics by searching the road graph [6]. Ref. [7] proposes a dynamic A* method that generates the reference trajectories considering the moving obstacles in the dynamic environment.

Furthermore, sampling-based path planning algorithms, such as rapidly exploring randomized trees (RRTs) and the probabilistic roadmap method (PRM) [8,9], have been developed to solve planning problems. The asymptotic optimal variant of RRT and PRM, namely RRT* and PRM* [10], almost surely converges. However, these algorithms place more emphasis on geometric optimality, without considering the motion characteristics of the planning object. In response to this issue, kinodynamic RRT* has been proposed for linear dynamics, which have extended the RRT* to kinodynamic systems [11]. These kinodynamic planning algorithms require solving a large number of two-point boundary value problems (TPBVPs), resulting in high computational costs. To avoid this, Li et al. [12] designed SST and SST*, which provide asymptotic near-optimal and optimal paths for kinodyanmic planning without the TPBVP solver. Li et al. [13] developed a near-optimal RRT (NoD-RRT) to avoid solving the TPBVP by utilizing the neural networks as an approximator considering nonlinear kinodynamic constraints. Zheng et al. [14] proposed the Kino-RRT*, which shows faster convergence compared with kinodynamic RRT*, by reducing the sampling dimension. When it comes to the USVs' path planning, Mao et al. [15] proposed a state prediction RRT for USVs by taking into account complete dynamic constraints. In [16], a novel planning algorithm based on the artificial vector field method and RRT* was proposed for USV low-cost path planning. Zhang et al. [17] improved the feasibility and efficiency of the planned path by utilizing the dual sampling space strategy and the Dubins curve. Inspired by DWA, Han et al. [18] presented an extended dynamic window approach (EDWA) for the automatic docking of USVs. However, the above planning algorithms are limited by the high computational costs and kinodynamic constraints of the USVs.

The accurate path-tracking control of USVs is a guarantee for completing tasks in the complex and ever-changing oceanic environment. Model predictive control (MPC) has become a rising control method in recent years. It can systematically handle multi-input, multi-output problems with constraints in the controller design processes. At present, the research on MPC has become one of the mainstream research directions in the field of motion control of USVs [19]. USVs need to face the problems of modeling errors and external disturbances such as waves, wind, and ocean currents [20,21]. To deal with these uncertainties, Kabzan et al. [22] presented an online learning model predictive control algorithm. This method utilized the data points for Gaussian process predictions to implement online learning. In [23], the dynamic sliding mode control (DSMC) theory was adopted to improve the system robustness under the effects of the ocean current and model uncertainties. Ning et al. [24] proposed a novel online-learning-based risk averse stochastic MPC framework, which utilized external disturbance data in real time to update the training data set. In [25], the Lyapunov filtered probabilistic model predictive control (LFP-MPC) was developed based on reinforcement learning. Shen et al. [26] utilized the adaptive unscented Kalman filter (AUKF) to estimate the full states of USV in real time for parameter identification of the MPC algorithm. In [27], the authors studied the path-following control problem of USVs. A novel control method based on the wavelet neural network and heading-surge (HS) guidance scheme was proposed.

The thrusters during USV path tracking in fault conditions may be entangled by plastic bags, branches, aquatic plants and fishing nets in the water [28–30]. There are many studies in the field of USV fault-tolerant control. In [31], an adaptive fuzzy sliding mode tracking control strategy was developed for Takagi–Sugeno fuzzy model-based nonlinear systems. Guo et al. [32] presented a fault estimator via the extended Kalman filter. Then, the MPC scheme utilized the estimation to realize the function of fault-tolerant control. Ding et al. [33] designed a novel fault-tolerant control strategy combining the MPC and the fault-tolerant reconstruction algorithm. This method uses the weighted pseudo inverse and quantum particle swarm optimization (QPSO) to achieve hybrid fault-tolerant control for different degrees of thruster fault. In [34], Zhang et al. presented an adaptive neural fault-tolerant control algorithm for the path-following activity of a USV using the novel output-based triggering approach. Some other unmanned platforms also have thruster failures. A fault-tolerant controller based on nonlinear model predictive control (NMPC) was developed for a quadrotor subjected to the complete failure of a single rotor [35]. In [36], Chen et al. proposed a fault-tolerant control (FTC) method and a fault-prevention control (FPC) method for vehicle motion control considering motor thermal protection. However, the actuator faults and model errors were not handled simultaneously by the control methods mentioned above.

The depicted methodology for obstacle avoidance is visually presented in Figure 1. Initially, a novel approach referred to as "Kindoyanic RRT*-smart" was devised, leveraging the optimization techniques and a biased sampling strategy. The method generates a reference path that adheres to the kinodynamic constraints. Subsequently, a model predictive controller, employing radial basis function neural networks, was introduced. This controller capitalizes on real-time feedback data to facilitate online training of the neural networks, thus rectifying the nominal models within the MPC framework. Lastly, drawing from the inherent constraint characteristics of MPC, fault-tolerant strategies are designed to cater to various fault types.

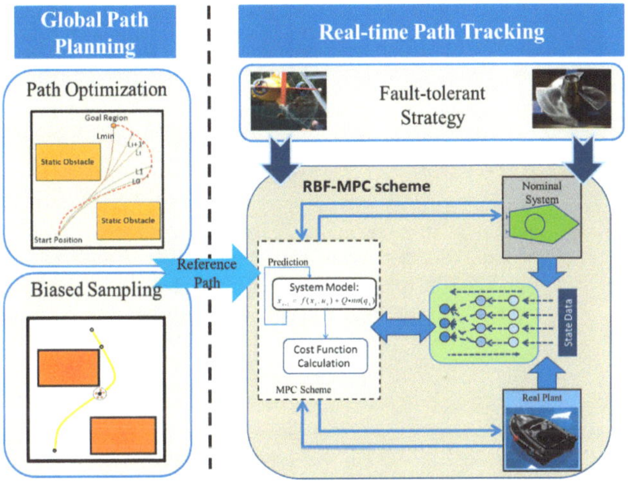

Figure 1. The proposed obstacle avoidance strategy.

The contributions of this paper lie in the following aspects: (1) A kinodyanmic RRT*-smart algorithm is proposed by synthesizing the RRT*-smart and kinodyanmic RRT* methods. This algorithm utilizes the triangular inequality concept and the biased sampling strategy to optimize the paths and to reduce the computational cost. (2) An online-learning MPC scheme based on RBF-NNs is utilized to address the modeling errors. The real-time state-feedback data are selected for neural network online training. (3) Two types of USV thruster faults are modeled and considered in the controller design. The corresponding fault-tolerant control strategies are designed through the proposed RBF-MPC method for different fault types.

The remaining parts of this paper are structured as follows. The main procedure of the kinodynamic RRT*-smart algorithm is explained in Section 2. The RBF-MPC with fault-tolerant strategy is illustrated in Section 3. Simulation results are shown in Section 4 and the conclusion is given in Section 5.

2. Collision Avoidance Path Generation

A practical and feasible reference path is generated for the USV to avoid obstacles. The path not only evades obstacles but also considers the motion characteristics of the USV. The subsequent section illustrates the processes of planning the collision-free paths.

2.1. Preliminaries

In this paper, we address the problem of kinodynamic motion planning for a USV on the basis of a sampling-based algorithm. The state space is defined as $\chi \in \mathbb{R}^n$, and the obstacle space is set as χ_{obs}. Consequently, we obtain the obstacle-free space as:

$$\chi_{free} = \chi \Big/ \chi_{obs} \tag{1}$$

The initial state of the USV is defined as x_{init}, and the goal state is defined as x_{goal}. Let $B_r(p) = \{x \in \chi | d(x,p) < r\}$ represent the ball centered at p with a radius r. Then, we obtain the definition of goal region around the target goal point as

$$\mathbb{Z}(x_{goal}, t) = B_r(x_{goal}) \cap \chi_{free} \tag{2}$$

The goal of this part is to find a path:

$$\pi : [0, T] \mapsto \chi_{free} \tag{3}$$

where the initial state $\pi(0) = x_{init}$, the terminal state $\pi(T) \in \mathbb{Z}(x_{goal}, T)$, and all states $\pi(t) \in \chi_{free}(t), \forall t \in [0, T]$. Let $U \in \mathbb{R}^m$ and ε be the control space and the control parameters of the USV. The path corresponds to a series of control inputs $u : [0, T] \mapsto U$; therefore, $\forall t \in [0, T]$

$$\begin{cases} \pi(t) = f(\pi(t-1), u(t)) \\ u(t) \in U(u(t-1), \varepsilon) \end{cases} \tag{4}$$

where $f(\cdot)$ represents the dynamic function of the USV, and $U(\cdot)$ denotes the set of the possible control outputs from the last node on the path.

Let Λ be the set of all nontrivial trajectories. The optimal motion planning problem is then formally defined as searching the feasible path π^* that minimizes a given cost function c: $\Lambda \mapsto \mathbb{R} \geq 0$. The path connects x_{init} to x_{goal} through X_{free} with corresponding control inputs u under the control constraints of the USV:

$$\begin{aligned} &\pi^* = \underset{\sigma \in \Sigma}{\text{argmin}}\ c(\pi) \\ s.t.\ \ &\pi(0) = x_{init} \\ &\pi(T) \in \mathbb{Z}(x_{goal}, T) \\ &\pi(t) \in X_{free}(t), \forall t \in [0, T] \end{aligned} \tag{5}$$

2.2. State Equations and Control Inputs

By analyzing the kinodynamic property of the USV, the dynamic model is defined by the following equations:

$$\dot{X} = \begin{bmatrix} \dot{x} \\ \dot{y} \\ \dot{\theta} \\ \dot{v} \\ \dot{w} \end{bmatrix} = \begin{bmatrix} v\cos(\theta) \\ v\sin(\theta) \\ w \\ 0 \\ 0 \end{bmatrix} + \begin{bmatrix} 0 \\ 0 \\ 0 \\ 1 \\ 0 \end{bmatrix} u_1 + \begin{bmatrix} 0 \\ 0 \\ 0 \\ 0 \\ 1 \end{bmatrix} u_2 \tag{6}$$

$$a_{\min} \leq u_1 \leq a_{\max} \tag{7}$$

$$\alpha_{\min} \leq u_2 \leq \alpha_{\max} \tag{8}$$

Equation (6) describes the state transition equation for the two-thruster surface vessel. The state vector $X = (x, y, \theta, v, w)^T$ consists of the Cartesian coordinates (x, y), yaw angle θ, forward speed v and angular speed w. The control inputs are bounded by dynamic constraints. The USV in this research cannot move backward. Thus, the velocities are bounded as $v \in (0, v_{\max})$.

2.3. Kindynamic RRT*-Smart Algorithm

In this section, the kinodynamic RRT*-smart is designed based on RRT*-smart [37]. The procedure of our algorithm is similar to the RRT* algorithm except for two distinct points. The first point is that the proposed algorithm aims to find a path that satisfies the USVs' kinodynamic constraints. The second point is the intelligent biased sampling strategy employed in the algorithm.

Algorithm 1 shows the main procedures of the proposed algorithm. First, the original state should be initialized as the root of the exploring tree. The main loop (lines 6 to 14) is similar to the RRT* algorithm. The termination condition for the algorithm is based on two criteria: either reaching the maximum number of iterations (K) or reaching the goal region. Once either of these conditions is met, the algorithm terminates its execution. Some essential steps in the progress are illustrated as follows.

Nearest Neighbor Given the vertex set V and considering metric function, the function *Nearest_Neighbor* returns the vertex in V that is "closest" to x_{rand}. In this paper, the Euclidean distance is used, and hence

$$Nearest(T = (V, E), x) = \text{argmin}_{v \in V} \|x - v\| \tag{9}$$

Parent Given the set of vertices V, the function returns the best "parent" vertex x_{parent} in V that minimizes the value of $distance(x_{parent}, x_{new})$.

Algorithm 1 Kinodynamic RRT*-Smart

Input: Goal, Map, K
Output: The feasible path

```
1   Trajectory ← ∅
2   V ← {x_init}, E ← {∅}, Tree1 = (V, E)
3   V_2 ← {x_init}, E_2 ← {∅}, Tree2 = (V_2, E_2)
4   x_goal ← Goal, Map ← Load()
5   for k = 1 to K do
6      x_rand ← Random Sampling()
7      x_near ← Nearest Neighbor(x_rand, Tree1)
8      if Obstacle free(x_nearest, x_new) then
9      x_parent ← Parent(V_nearest, x_new),
10     V ← V ∪ {x_new}, E ← E ∪ {V_nearest, x_new}
11     Tree1 = (V, E)
12  if x_new ∈ B_r(x_goal) then
13     Break
14  end
15  if x_goal is already reached then
16     Tree2(V_2, E_2) = Path_Optimization(Tree1)
17  Return Trajactory = Tree2(V_2, E_2)
```

Path_Optimization Given the feasible path *Tree*1 in a geometric sense, the function returns a path *Tree*2 satisfying the kinodynamic properties of the USV. More details about this part are described in Algorithms 2 and 3.

Algorithm 2 Path_Optimization

```
    Input : Tree1(V, E)
    Output : Tree2(V_2, E_2)
1   s_begin = x_init
2   for i = 1 to i = N do
3       s_end = V(i)
4       if Obstacle Free(s_begin, s_end) then
5           continue
6       else
7           s_end = V(i − 1)
8           V_2 = V_2 ∪ {s_end}
9           E_2 = E_2 ∪ {s_begin, s_end}
10          s_begin = s_end
11  end
12  Tree2 = Rewiring Edges(V_2, E_2)
13  Return Tree2
```

Algorithm 2 outlines the optimization process based on the RRT*-Smart algorithm, utilizing the triangular inequality concept to optimize the path. In essence, if the path node x_1 is visible to x_i, they can be directly connected.

From line 1 to line 11, an optimized path is already acquired. However, this path is only reachable to the goal region in the perspective of planar geometry. In other words, when applied to the real plant, it may not be practical to execute the path. To address this issue, the edges of *Tree* are rewired, as detailed in algorithm 3. This rewiring process ensures that the optimized path is practically executable by the USVs, considering its kinodynamic constraints.

Algorithm 3 Rewiring Edges

```
Input: E ← ∅, V
Output: The feasible path
s_begin = V(1)
for i = 1 to i = N do
    s_end = V(i)
    if Trajectory Feasible(s_begin, s_end) then
        for j = 1 to j = k do
            S_new = IntelligentBiasedSampling(S_end)
            s_end = CostJudgement(s_begin, s_end, s_new)
            s_begin = s_end
            E = E ∪ FeasibleTrajctory(s_begin, s_end)
        end
    else
        s_end = FindFeasiblePoint(s_end)
        for j = 1 to j = k do
            s_new = IntelligentBiasedSampling(s_end)
            s_end = CostJudgement(s_begin, s_end, s_new)
            s_begin = s_end
            E = E ∪ FeasibleTrajctory(s_begin, s_end)
        end
end
Return Tree(V, E)
```

The main loop of the *Rewiring Edges* algorithm is shown above. First, the vertices of *Tree* will be checked if they can reach each other in a kinodynamic sense. If not, the *Find Feasible Point* will be utilized to find a feasible point by biased sampling. After that, the edges will be calculated by solving the two-point boundary problem (TPBVP). Then, we designed *IntelligentBiasedSampling* to find a more appropriate point with a lower path cost. Some key steps in the algorithm are given as follows:

Path Feasible Given the initial state $s_{initial}$ and final state s_{goal}, the path is calculated by utilizing Equations (4)–(6).

IntelligentBiasedSampling The main idea of this step is to find an optimized path by selecting nodes as close as possible to the vertices generated by the triangular inequality concept aforementioned. The biased ratio is defined to set the biasing radius around the selected node of the sampling strategy. Furthermore, the generated node will be checked for if it is *Path Feasible* to the other nodes. The Ratio of this strategy is given by the Equation (10).

$$BiasRatio = \left(\frac{n}{X_{free}}\right) * B \tag{10}$$

where B is the designed constant inflecting the convergence speed of the algorithm, and n represents the iteration times.

CostJudgement The main purpose of this step is to determine whether the new node can reduce the path cost [38], as shown in Formulas (11) and (12).

$$x_{check} = \underset{x_i}{\text{argmin}} Cost(x_i, x_{goal}) \tag{11}$$

$$Cost(x_1, x_2) = \omega_1 \|x_2 - x_1\| + \omega_2 \text{arccos} \frac{\vec{v}_1 \cdot x_1 x_2}{\left|\vec{v}_1\right| |x_1 x_2|} + \omega_3 t(x_1, x_2) \tag{12}$$

FindFeasiblePoint This step is designed for finding a reachable point around the s_{error} that is in the set of vertices but is checked unreachable by *Path Feasible*.

3. RBF-MPC for USV with Actuator Faults

The RBF-MPC approach is developed to track the predetermined collision-free path considering the modeling error and actuator faults. The actuator faults are modeled and introduced in the controller design for motion control in the presence of actuator faults.

3.1. Assumptions

In order to simplify the problem, this paper focuses solely on the horizontal plane motion of the USV [39]. The planar motion of the USV is illustrated in Figure 2. To facilitate computer simulations of the USV motion, certain simplifications were made, which are outlined as follows:

a. Neglecting roll, pitch, and heave motion: The motion of the USV in the roll, pitch, and heave directions is disregarded for simplicity.
b. Neutral buoyancy and body-fixed coordinate: The USV is assumed to have neutral buoyancy, and the origin of the body-fixed coordinate system is positioned at the center of mass of the USV.
c. Three planes of symmetry: The USV possesses three planes of symmetry, which aid in simplifying the analysis and modeling process.

By adopting these simplifications, the focus is placed on the planar motion of the USV, allowing for more manageable simulations and analyses.

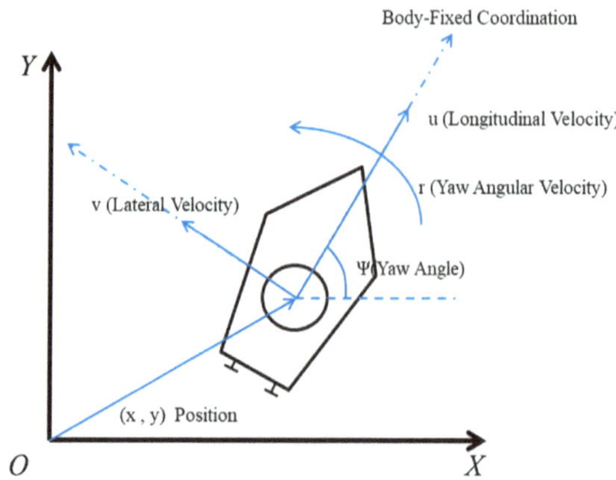

Figure 2. Vessel motion system.

3.2. Vehicle Kinematic Model and Dynamic Model

The vessel's kinematic model is utilized to describe its planar motion, as follows:

$$\begin{cases} \dot{X} = u\cos\psi - v\sin\psi \\ \dot{Y} = u\sin\psi + v\cos\psi \\ \dot{\psi} = r \end{cases} \tag{13}$$

where u and v represent the longitudinal and lateral velocities. X is the longitudinal position, and Y is the lateral position. ψ is the yaw angle, and r is the yaw rate.

The kinematic model illustrates the correlation between the vehicle's potion and velocity. Furthermore, the dynamic model of the USV can be expressed as

$$M\dot{v} = \tau_T + \tau_{Tw} - C(v)v - D(v)v \tag{14}$$

where $M = diag\{m_{11}, m_{22}, m_{33}\}$ is the inertial matrix, and $D = diag\{d_{11}, d_{22}, d_{33}\}$ is the damping matrix. The centripetal force matrix of the kinematic model is

$$C = \begin{bmatrix} 0 & 0 & -m_{22}v \\ 0 & 0 & m_{11}u \\ m_{22}v & -m_{11}u & 0 \end{bmatrix} \tag{15}$$

$\tau_T = \begin{bmatrix} \tau_u & 0 & \tau_r \end{bmatrix}$ contains forward thrust τ_u and the yawing moment τ_r. $\tau_{Tw} = \begin{bmatrix} \tau_{Twu}, \tau_{Twv}, \tau_{Twr} \end{bmatrix}$ represents the external disturbance's effect on $[u, v, r]$.

From the aforementioned information, the dynamic model can be written as:

$$\begin{cases} \dot{u} = (\tau_u - d_{11}u + m_{22}vr)/m_{11} \\ \dot{v} = (-d_{22}v - m_{11}ur)/m_{22} \\ \dot{r} = (\tau_r - d_{33}r - m_{22}uv + m_{11}uv)/m_{33} \end{cases} \tag{16}$$

where $d_{11} = -(X_u + X_{u|u|}|u|)$, $d_{22} = -(Y_v + Y_{v|v|}|v|)$, $d_{33} = -(N_r + N_{r|r|}|r|)$.$X_u$, $X_{u|u|}$, Y_v, $Y_{v|v|}$, N_r, $N_{r|r|}$ are the first-order and second-order fluid coefficients, respectively.

3.3. RBF Neural Network-Based Model Predictive Control

The above part follows the classical kinematic model and dynamic model of USV. However, due to the different characteristics of each USV in practical applications, the vessel's motion cannot be accurately described by the universal model alone.

To enhance the accuracy of the vehicle model, we introduce an online learning neural network component to take the residual model uncertainty into account. This method is improved based on the measurement tools and the neural network. The main idea is shown in Figure 3.

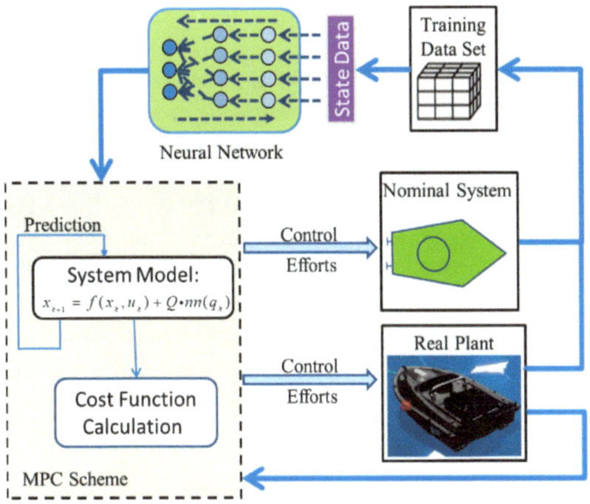

Figure 3. Architecture of NN-MPC.

Here, we identify an unknown function $d_{true}: \mathbb{R}^{n_z} \rightarrow \mathbb{R}^{n_d}$ from a collection of inputs $z_k \in \mathbb{R}^{n_z}$ and outputs $y_k \in \mathbb{R}^{n_d}$. Then, the considered model used for control can be expressed as:

$$x_{k+1} = f(x_k, u_k) + Q \cdot nn(q_k) \tag{17}$$

where $f(x_k, u_k)$ describes the nominal vehicle dynamics, and $nn(\cdot)$ is the additional neural network part, estimating the modeling error of the nominal model mentioned above. $u_k = [T_l \quad T_r]$ is the control efforts. The learned part $nn(\cdot)$ of the dynamics is assumed to only affect the subspace spanned by Q, corresponding to the velocity states of the vehicle. The learning part depends on a set of features q_k that are relevant to the neural network.

Based on physical assumptions, we presume that the vehicle model error is mainly caused by the dynamic part of the system equations. In the following research, a learning vector was drawn into the USV model independently, as:

$$q = [u; v; r; T_l; T_r] \tag{18}$$

The difference between the nominal model predictions and the actual measurement state will be used as a database for online learning of the neural network, which is

$$y_k - Q^{\dagger} \cdot (x_{k+1} - f(x_k, u_k)) = nn(q_k) \tag{19}$$

where $Q^{\dagger}$ is the Moore–Penrose pseudo-inverse, and the true modeling error function is $nn(q_k)$.

Therefore, the USV path tracking control problem can be described as the following constrained dynamic optimization problem:

$$\min J(k) = \min \sum_{i=0}^{N-1} \|e(k+i|k)\|_R^2 + \|u(k+i\,|\,k)\|_Q^2 \tag{20}$$

$$s.t. \boldsymbol{x}_{k+1} = f_{nom}(\boldsymbol{x}_k, \boldsymbol{u}_k) + Q \cdot nn(q_k) \tag{21}$$

$$\boldsymbol{e}(\boldsymbol{k}+1) = \boldsymbol{x}_{k+1} - \boldsymbol{x}_{ref} \tag{22}$$

$$u_{min} \leq u_k \leq u_{max}, k = 1, \ldots, N_c \tag{23}$$

$$x_0 = x_k \tag{24}$$

where R is the weight matrix of the path tracking state deviation, and Q is the control input weight matrix. x_{k+1} represents the model state after the compensation of the neural network. Let x_{ref} be the reference state. Let $u_k = [T_l \quad T_r]$ be the control input, x_0 be the initial states of prediction and x_k be the current state feedback of the real plant.

The radial basis function neural network (RBF-NNs) approximator was chosen to tackle the structural and parameter uncertainties. The structure of RBF-NNs is depicted in Figure 4, consisting of three layers: the input layer, the hidden layer and the output layer. The first layer of structure, the input layer, is expressed as $X = [x_i]^T$. This layer assigns n inputs to m nodes of the second hidden layer. The output of the second hidden layer is $H = [h_j]^T$, where x_i is the ith input of the input layer, and h_j is jth output of the hidden layer. In this paper, the Gaussian function is chosen as the activation function of the neural network, which is

$$h_j(x) = \exp\left(-\frac{\| x - c_j \|^2}{2\sigma_j^2}\right) \tag{25}$$

where $\|x - c_j\|$ represents the Euclidean distance between x and c_j. Let c_j and σ_j be the center and width of the Gauss basis function.

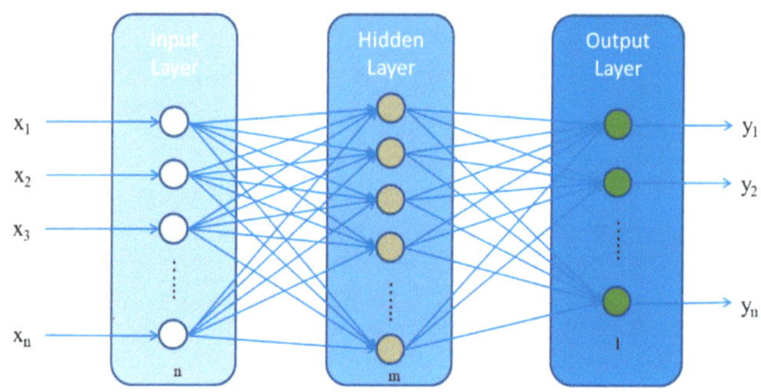

Figure 4. RBF-NN structure.

The third output layer represents the overall output of the neural network. The prediction output of the network is calculated through a linear combination of hidden layers' output. Thus, we obtain:

$$y_q(t) = \sum_{j=1}^{m} h \cdot w_{jq}{}^T = h_1 w_{1q} + h_2 w_{2q} + \cdots + h_m w_{mq} \tag{26}$$

where w_{jq} represents the weight vector from the jth hidden layer node to the qth output layer node.

By combining Equation (26), the uncertainty of USV can be expressed as:

$$nn(q_k) = W^T H(x_k) + \varepsilon \tag{27}$$

where W is the weight matrix from the hidden layer to the output layer. H is the vector calculated by Equation (25), and ε is the deviation term. The dimensions of the weight matrix W and H are $m \times l$ and $n \times m$. Let n, m and l be the number of nodes in the input layer, hidden layer and output layer.

3.4. Fault-Tolerance Strategy

The studied USV is equipped with two propellers that generate the forward force. However, in the fault condition, the propellers are unable to provide the desired output torque. Thus, the USV is incapable of tracking the reference path, potentially resulting in collisions. The actuator fault is taken into account in the controller design process.

Regarding the proportional fault, the propeller torque under the faulty conditions is proportional to that of the health state. Let λ represent the fault parameter, and the torque of the propeller under the faulty condition is written as

$$T_{if} = \lambda T_i \tag{28}$$

where T_{if} is the output torque of the i th thruster and $i = 1, r$. For the reduced output torque under the fault condition, the fault parameter λ is confined as

$$0 \leq \lambda \leq 1 \tag{29}$$

In the fault condition, the control efforts need to be modified. Therefore, the final output of the controller is changed under consideration of the proportion case mentioned above in order to guarantee the USV's path tracking. This can be specifically expressed as

$$u_f = [T_h \quad \lambda T_f] \tag{30}$$

Then, we substitute the above equation into Equation (21) and use the updated model, shown as Equation (31), to adaptively adjust the calculation output of MPC.

$$x_{k+1} = f_{nom}(x_k, u_k) + Q \cdot nn(q_k),\ u_{kf} = [T_h \quad \lambda T_f] \tag{31}$$

Furthermore, the limiting fault is considered and introduced in the following equations. The output torque is limited to a fixed value that is reduced when the fault occurs. This can be expressed as

$$T_{i\max} = T_{fault\max} < T_{h\max} \tag{32}$$

where $T_{fault\max}$ is the maximum of the faulty thrusters' torque, and $T_{h\max}$ is the maximum of the healthy thrusters' torque.

In this case, we set the maximum thrust of the faulty thruster as $T_{if\max}$, as shown in the following equation:

$$0 < T_{if} \leq T_{if\max} < T_{h\max} \tag{33}$$

Then, we obtain Equation (34) by substituting the new thrust constraint into Equation (23), as follows:

$$u_{fmin} \leq u_{kf} \leq u_{fmax}, k = 1, \ldots, N_c \tag{34}$$

4. Simulation Results

For the simulations, a computer with MATLAB 2019a on an Intel i7-127000 CPU with 32 GB of RAM was utilized. In order to evaluate the performance of the proposed method, we conducted a series of simulations and compared the results with those obtained from the other methods. First, the collision avoidance performance of the Kino-RRT* algorithm is shown as a comparison of the proposed algorithm. Then, we show the advantage of RBF-MPC with the fault-tolerance strategies by comparing it with the traditional MPC under faulty conditions. Furthermore, we demonstrate the well-tracking performance of the designed fault-tolerant controller for the planned path.

4.1. Collision Avoidance Path Performance Comparison in Simulation Studies

In the simulation environment, the global map is represented as a rectangular area with pixels of 500 × 500 representing an ocean area of 500 × 500 m. The obstacles are randomly set to represent floats or observation stations to be avoided. The results of the

Kino-RRT* algorithm and the kinodynamic RRT*-smart algorithm are given in Figure 5. In Figure 5, we can see that the proposed algorithm found a better path under the same sampling conditions.

Furthermore, due to the unique randomness of the RRT algorithm, we conducted ten sets of simulation experiments and took their average values. The data are shown in Table 1. Upon analyzing the data, it becomes apparent that increasing the number of sampling times in the RRT algorithm leads to longer planning times. However, it also results in improved planning effectiveness. From Table 1, we can draw a conclusion that the path length is 20.48% less than Kino-RRT*'s result when the sampling times are 5000. This is because the triangular inequality concept optimizes the path cost directly. The running time is reduced by 31.34% for less TPBVP problem solving. At the same time, our biased sampling strategy also reduces the path costs via the kinodynamic constraints, and the cost is 29.59% lower than Kino-RRT*. This also resulted in a 2.98% reduction in our path length compared to RRT*-smart.

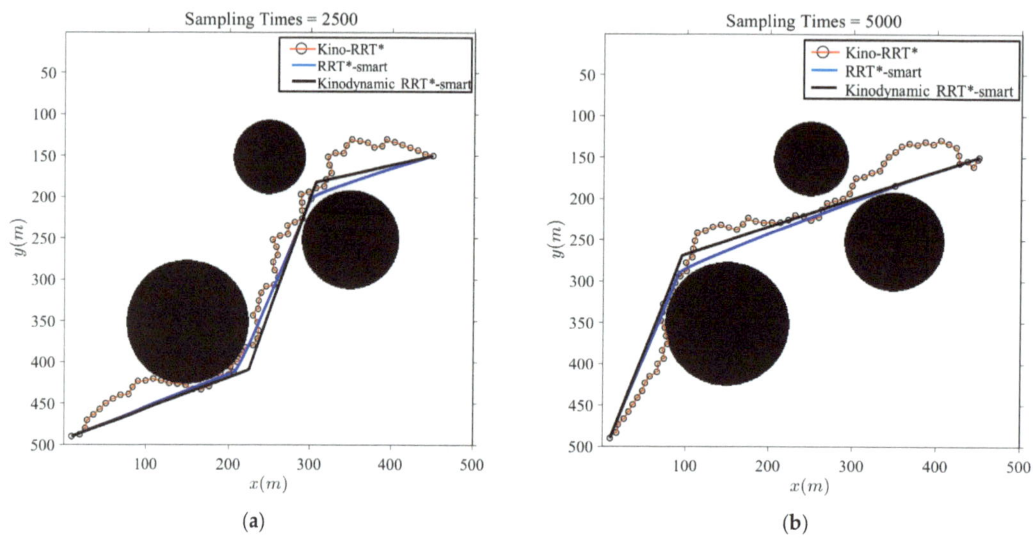

Figure 5. Planning simulation results: (**a**) sampling times = 2500; (**b**) sampling times = 5000.

Table 1. Summary of comparative analysis.

Sampling Times	Algorithm	Running Time	Path Cost	Path Length
2500	RRT*-smart	0.45	×	608.73
	Kino-RRT*	0.67	748.65	752.49
	proposed method	0.56	534.57	612.31
5000	RRT*-smart	0.63	×	572.46
	Kino-RRT*	1.34	647.55	698.65
	proposed method	0.92	455.93	555.58

To be more practical, we chose a part of the Dalian Port image as the planning map from the Electronic Chart Display and Information System (ECDIS), as shown in Figure 6.

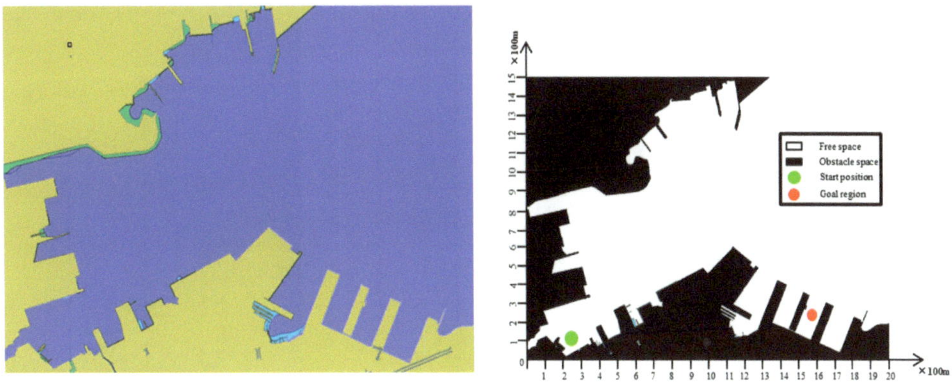

(**a**) (**b**)

Figure 6. Conversion from the ECDIS image to the binarization map: (**a**) Dalian Port partial (location: 38°57′ N, 121°38′ E, width: 2000 m, height: 1500 m); (**b**) binarization map.

We set the initial position of the USV as [250, 100] m and the goal region as [1560, 260] m. The planning result in the real environment is shown in Figure 7.

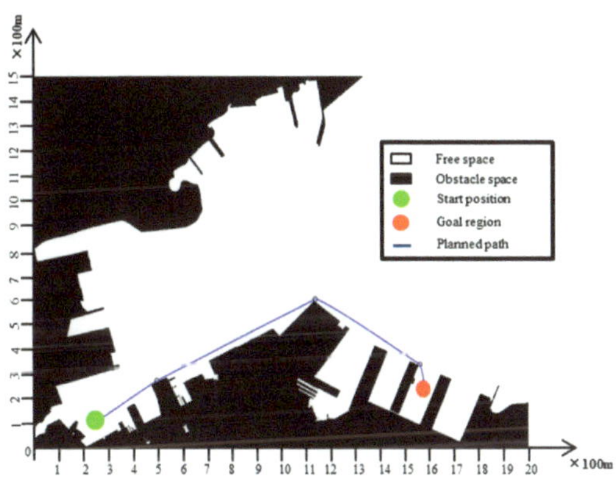

Figure 7. The planning path in a real environment.

From Figure 7, we see that the planning path is free from collisions, indicating that the proposed algorithm is capable of navigating in the actual ocean environment. As depicted in the aforementioned illustrations, we can deduce that the planning path is viable for the USV. This achievement is attributed to our incorporation of motion characteristics and kinodynamic constraints into the planning process. Consequently, the proposed algorithm is applicable for practical USV path planning.

4.2. RBF-MPC with Fault-Tolerance Strategy Performance Comparison in Simulation Studies

In order to evaluate the effectiveness of the proposed algorithm in terms of model correction and fault tolerance, comparative simulations were conducted between the proposed method and the traditional MPC. The specific kinetic parameters of the USV used in the simulations are presented in Table 2. The nominal model parameters listed in the table intentionally differ from the actual model parameters. This mismatch is performed to simulate the modeling errors that commonly exist in real-world scenarios. By introducing such modeling errors, the simulations can realistically assess the capability of the proposed algorithm to handle model inconsistencies and to provide accurate control. The external disturbance is set as

$$\begin{cases} \tau_{Twu} = -3\cos(0.5t)\cos(t) + 1.5\cos(0.5t)\sin(0.5t) \\ \tau_{Twv} = 0.6\sin(0.1t) \\ \tau_{Twr} = 0.9\sin(1.1t)cos(0.3t) \end{cases}$$

Table 2. The parameters of the nominal model and real plant.

Parameter	m_{11}	m_{22}	m_{33}	X_u	$X_{u\|u\|}$	Y_v	$Y_{v\|v\|}$	N_r	$N_{r\|r\|}$
Nominal Model	141.85	197.75	15.6	−45.6	−67.26	−29.54	−73.85	−10.71	−5.59
Real Plant	153.65	204.35	18.2	−40.3	−67.26	−30.54	−70.25	−10.71	−5.59

In order to ensure the intuitive optimization effect, we chose linear path tracking for simulation. The original position of the USV was set as [0, 5] m and its original speed as [0.5, 0] m/s. In the first fault condition, the fault parameter was set as $\lambda = 0.5$. The fault of the right thruster occurred at $x = 75$ m. The simulation results in the first fault case are as follows.

From Figure 8c and the locally enlarged images in Figure 8a, it can be seen that RBF-MPC can enable the USV to achieve the line tracking faster. This observation demonstrates the effectiveness of online RBF neural network learning in compensating for model errors. From Figure 8b, it can be seen that external interference has an impact on the USV yaw angle. From the local enlarged image, it can be seen that under the same interference conditions, RBF-MPC is less affected by disturbances. Therefore, we can analyze that compared to the traditional MPC, our designed RBF-MPC has stronger anti-interference ability. As shown in Figure 8b,c, after the fault occurs, the yaw angle of RBF-MPC with the fault-tolerant strategy does not change significantly, and it can still stably track the preset trajectory, but the traditional MPC cannot. This verifies the effectiveness of the fault-tolerant strategy.

In Figure 9a, the fault tolerant strategy makes the USV's forward speed change smoother. Figure 9b shows that the RBF-MPC with the fault-tolerant strategy can make the USV's lateral speed converge to 0 m/s faster and changes little when the fault occurs compared with the traditional MPC. From Figure 10d, it is evident that during the initial operation stage of the USV (0–20 m), the torque variation in the computed output of RBF-MPC is smaller. By analyzing the other figures, the traditional MPC exhibits more significant changes in control output compared to RBF-MPC integrated with the fault-tolerant strategy when a fault occurs. Moreover, the traditional MPC is more susceptible to external disturbances. In Figure 11b, we can see that the real force of the right thruster is half of the MPC output, which represents the proportional failure. From Figures 10 and 11, a conclusion can be drawn that the MPC with the fault-tolerant strategy can autonomously adapt to the proportional fault and adjust thrust output to ensure the normal operation of the USV.

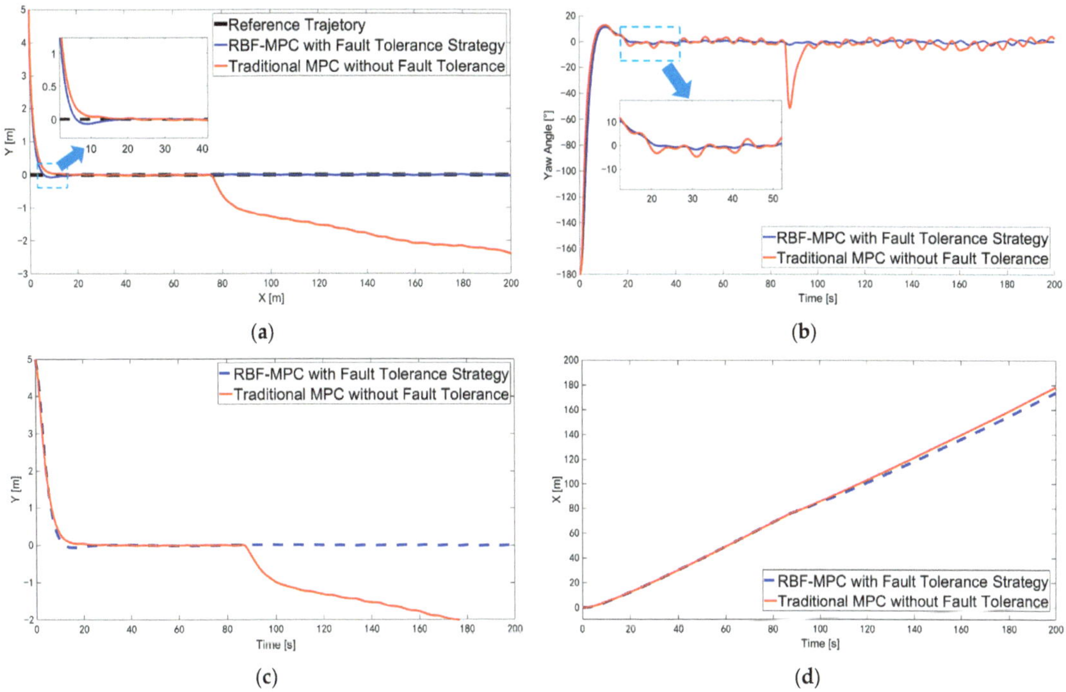

Figure 8. Trajectory tracking result in the first fault case. (**a**) position; (**b**) yaw angle; (**c**) X-direction displacement; (**d**) Y-direction displacement.

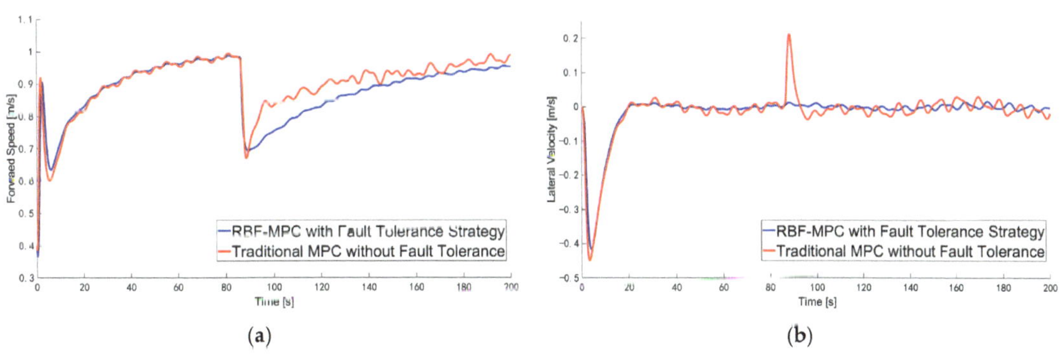

Figure 9. The speed of USV in the first fault case: (**a**) forward speed; (**b**) lateral speed.

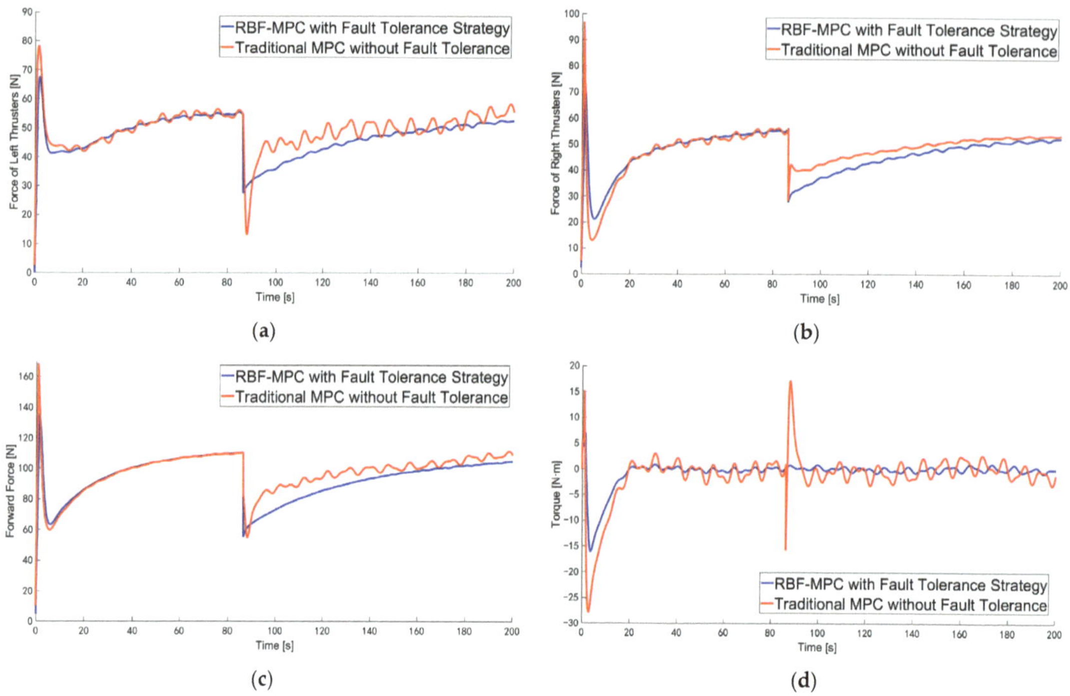

Figure 10. The force of the thrusters in the first fault case: (**a**) left thruster; (**b**) right thruster; (**c**) forward force; (**d**) torque.

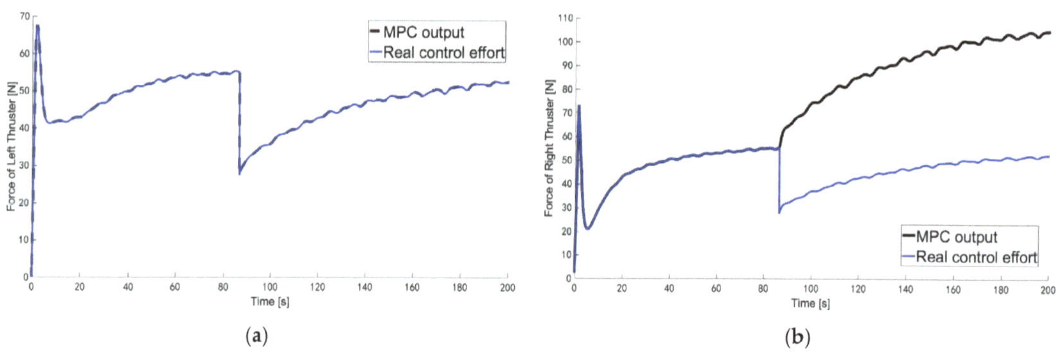

Figure 11. MPC output and real control effort in the first fault case: (**a**) left thruster; (**b**) right thruster.

In the second fault case, the fault parameter is set as $T_{f\max} = 50$ N. The fault of the right thruster occurs at $x = 100$ m. The simulation results of the second type of fault are as follows.

From Figure 12c and the magnified images in Figure 12a, it is evident that RBF-MPC effectively modifies the nominal model used within the MPC framework. The magnified image in Figure 12b demonstrates that the yaw angle of the USV under RBF-MPC control is less affected by external disturbances. Hence, we conclude that RBF-MPC exhibits a certain degree of disturbance attenuation. By analyzing Figure 12b–d collectively, we can deduce that after the fault occurs, the yaw angle of the USV controlled by the traditional MPC changes greatly, and the displacement in the Y direction becomes unstable. In contrast, the RBF-MPC demonstrates the effectiveness of the designed fault-tolerant strategy as it maintains relatively stable yaw angles and stable displacement even in the presence of faults. The simulation proves that when the limiting fault occurs, the MPC controller incorporating a fault-tolerant strategy enables the USV to stably track straight lines, which verifies the effectiveness of the fault-tolerant strategy for limiting faults.

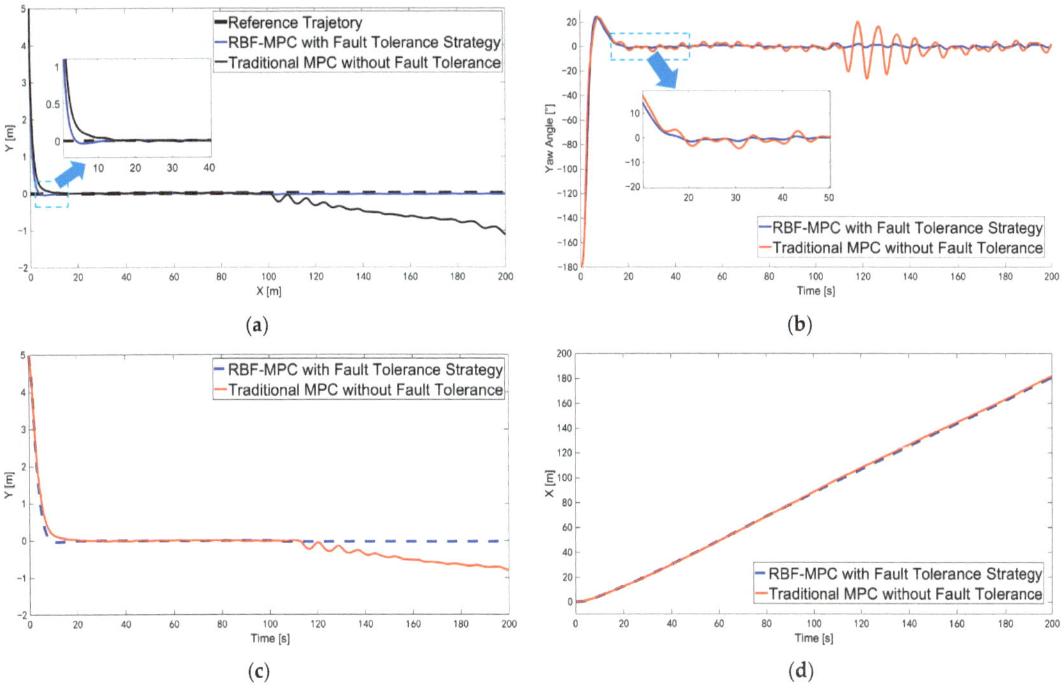

Figure 12. Trajectory tracking result in the second fault case. (**a**) position; (**b**) yaw angle; (**c**) X-direction displacement; (**d**)Y-direction displacement.

From Figure 13, it can be observed that the introduced external disturbances have an impact on both the forward velocity and longitudinal velocity, but RBF-MPC is less affected by these disturbances. Additionally, in the event of a fault, the velocity of the USV under MPC control without the fault-tolerant strategy exhibits increased oscillations. In Figure 14a,b, it can be observed that when a fault occurs in the right thruster, the thrust is constrained to 50 N. However, RBF-MPC with the fault-tolerant strategy adjusts the thrust output of the left thruster to ensure the normal operation of the USV. From Figure 14c,d, it can be deduced that RBF-MPC with the fault-tolerant strategy exhibits a small variation in thruster output after a fault occurs. However, the traditional MPC, due to the combined effect of the fault and external disturbances, fails to achieve stable control of the USV.

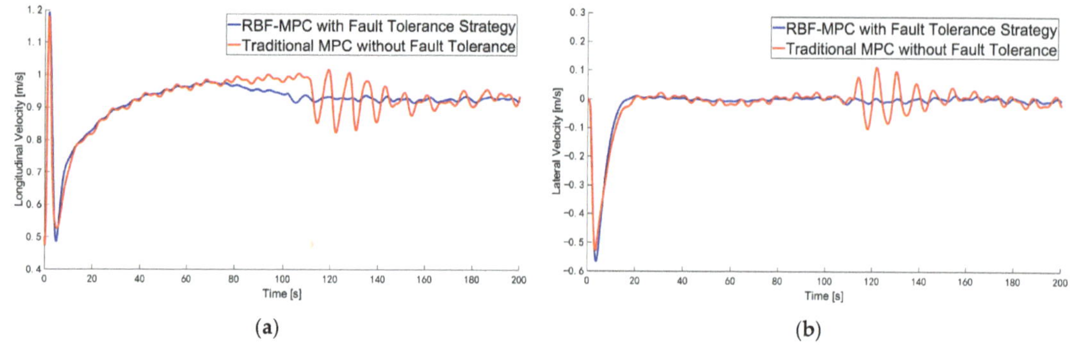

Figure 13. The speed of the USV in the second fault case: (**a**) forward speed; (**b**) lateral speed.

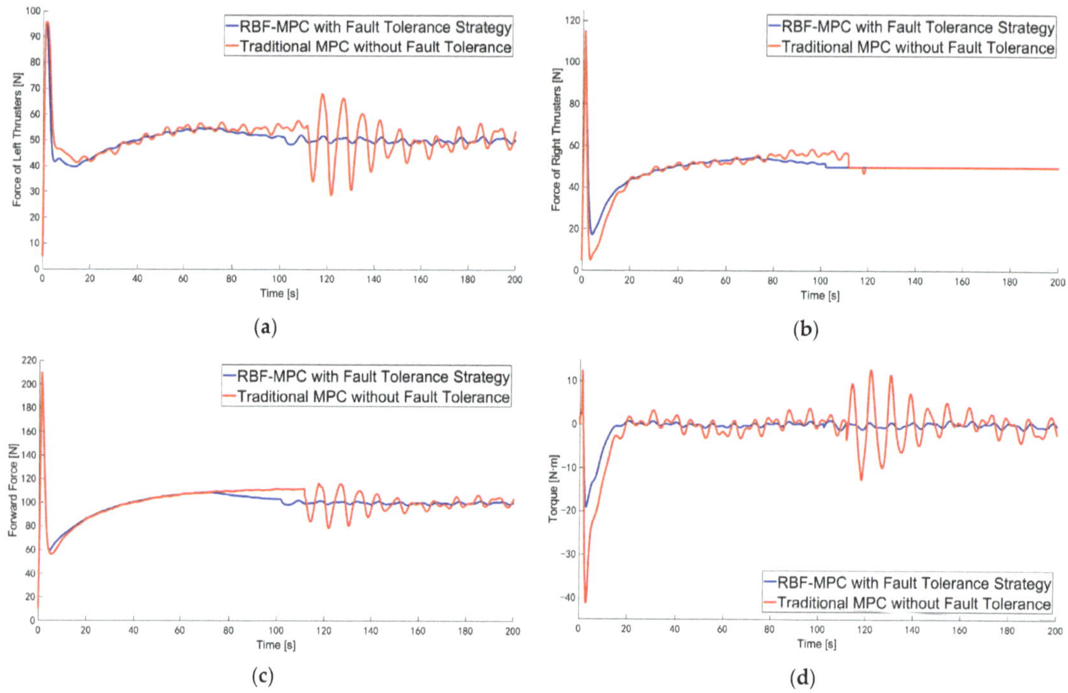

Figure 14. The force of the thrusters in the second fault case: (**a**) left thruster; (**b**) right thruster; (**c**) forward force; (**d**) torque.

In Figure 15, the simulation illustrates that RBF-MPC with the fault-tolerant strategy makes the USV track the planned path even under the thruster-fault condition. From the partial enlarged view, the traditional MPC cannot control the USV well, and the USV collides with an obstacle when the fault occurs.

In summary, the proposed RBF-MPC combined with the fault-tolerant strategy has shown the effective performances in terms of model error correction and fault-tolerant control.

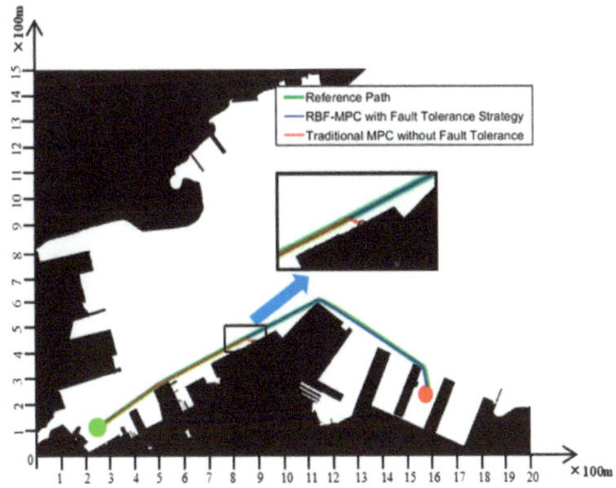

Figure 15. The planned path tracking result.

5. Conclusions

This paper focuses on two key issues in USV technology: path planning and tracking control. We developed the kinodynamic RRT*-smart algorithm and an online-learning-based MPC algorithm for the USV, considering the modeling error and actuator faults. The planning algorithm uses geometric optimization theory to optimize the paths generated by random trees in a geometric sense. Then, the optimized path points are utilized for biased sampling and generating paths that adhere to the kinodynamic constraints. Compared to the traditional methods, the proposed approach exhibits a lower computational cost for solving the optimized path. After the reference path was given, we designed the RBF-MPC scheme, which utilizes the learning ability of neural networks to approximate modeling errors. This method introduces the training results of RBF-NNs into the cost function calculation of MPC, making the nominal model used in MPC approximate the actual model, and achieving accurate control of the USV. Building upon this method, different fault-tolerant strategies were developed, taking into account the potential occurrence of thruster failures. The strategies were integrated into the RBF-MPC framework, and their effectiveness was verified by comparing them with the traditional MPC in the numerical simulation.

In our future research endeavors, we will prioritize the following issues:

(1) Application to a USV physical platform: We aim to implement and validate the effectiveness of the proposed method on an actual USV physical platform through experimentation.
(2) Learning-based MPC: We will develop the learning-based MPC algorithm that obviates the need for explicit fault-tolerant strategy design. By leveraging the power of neural networks, we aim to approximate the changes in the USV dynamic model induced by faults, thereby circumventing the task of fault-tolerant strategy design.
(3) Optimization of sampling strategy: We will delve deeper into the optimization of the sampling strategy employed in our path planning algorithm. Additionally, we will try to provide evidence of the algorithm's probability completeness and asymptotic optimality.

Author Contributions: Conceptualization, Y.S. and Y.C.; methodology, Y.S. and Y.C.; software, Y.S.; validation, J.G. and Y.W.; formal analysis, Y.W.; writing—original draft preparation, Y.S.; writing—review and editing, Y.C., J.G., Y.W. and G.P.; funding acquisition, G.P. All authors have read and agreed to the published version of the manuscript.

Funding: This work is supported by in part by the National Natural Science Foundation of China under Grants of 51979228 and 52102469, in part by the Fundamental Research Funds for the Central Universities under Grant of 3102021HHZY030007, and in part by the National Basic Scientific Research Program under Grant JCKY2019207A019.

Institutional Review Board Statement: Not applicable.

Informed Consent Statement: Not applicable.

Data Availability Statement: Not applicable.

Conflicts of Interest: The authors declare no conflict of interest.

References

1. Zhang, M.; Liu, Z.; Cai, W.; Yan, Q. Design of Low-cost Unmanned Surface Vessel for Water Surface Cleaning. In Proceedings of the 2021 China Automation Congress (CAC), Beijing, China, 22–24 October 2021; pp. 2290–2293.
2. Mun, A.; Long, C.; Shen, K.; An, O.; Nugroho, H.; Rusydi, M.I. Fuzzy-based Control System of Unmanned Surface Vehicle (USV) for Floating Garbage Collection. In Proceedings of the 2022 IEEE International Conference on Artificial Intelligence in Engineering and Technology (IICAIET), Kota Kinabalu, Malaysia, 13–15 September 2022; pp. 1–6.
3. Ntakolia, C.; Lyridis, D.V. Path Planning in the Case of Swarm Unmanned Surface Vehicles for Visiting Multiple Targets. *J. Mar. Sci. Eng.* **2023**, *11*, 719. [CrossRef]
4. Mishra, R.; Koay, T.B.; Chitre, M.; Swarup, S. Multi-USV Adaptive Exploration Using Kernel Information and Residual Variance. *Front. Robot. AI* **2021**, *8*, 572243. [CrossRef] [PubMed]
5. Cho, H.; Jeong, S.-K.; Ji, D.-H.; Tran, N.-H.; Vu, M.T.; Choi, H.-S. Study on Control System of Integrated Unmanned Surface Vehicle and Underwater Vehicle. *Sensors* **2020**, *20*, 2633. [CrossRef] [PubMed]
6. Hart, P.E.; Nilsson, N.J.; Raphael, B. A Formal Basis for the Heuristic Determination of Minimum Cost Paths. *IEEE Trans. Syst. Sci. Cybern.* **1968**, *4*, 100–107. [CrossRef]
7. Stentz, A. Optimal and efficient path planning for partially-known environments. In Proceedings of the 1994 IEEE International Conference on Robotics and Automation, San Diego, CA, USA, 8–13 May 1994; pp. 3310–3317.
8. LaValle, S.M. Randomized Kinodynamic Planning. *Int. J. Robot. Res.* **2001**, *20*, 378–400. [CrossRef]
9. LaValle, S.M. On the Relationship Between Classical Grid Search and Probabilistic Roadmaps. *Int. J. Robot. Res.* **2004**, *23*, 673–692. [CrossRef]
10. Karaman, S.; Frazzoli, E. Sampling-based Algorithms for Optimal Motion Planning. *Int. J. Robot. Res.* **2011**, *30*, 846–894. [CrossRef]
11. Webb, D.J.; Berg, J. Kinodynamic RRT*: Asymptotically optimal motion planning for robots with linear dynamics. In Proceedings of the 2013 IEEE International Conference on Robotics and Automation, Karlsruhe, Germany, 6–10 May 2013; pp. 5054–5061.
12. Li, Y.; Littlefield, Z.; Bekris, K.E. Asymptotically Optimal Sampling-based Kinodynamic Planning. *Int. J. Robot. Res.* **2016**, *35*, 528–564. [CrossRef]
13. Li, Y.; Cui, R.; Li, Z.; Xu, D. Neural Network Approximation Based Near-Optimal Motion Planning with Kinodynamic Constraints Using RRT. *IEEE Trans. Ind. Electron.* **2018**, *65*, 8718–8729. [CrossRef]
14. Zheng, D.; Tsiotras, P. Accelerating Kinodynamic RRT* Through Dimensionality Reduction. In Proceedings of the 2021 IEEE/RSJ International Conference on Intelligent Robots and Systems (IROS), Prague, Czech Republic, 27 September–October 2021; pp. 3674–3680.
15. Mao, S.; Yang, P.; Gao, D.; Bao, C.; Wang, Z. A Motion Planning Method for Unmanned Surface Vehicle Based on Improved RRT Algorithm. *J. Mar. Sci. Eng.* **2023**, *11*, 687. [CrossRef]
16. Wen, N.; Zhang, R.; Liu, G. Online planning low-cost paths for unmanned surface vehicles based on the artificial vector field and environmental heuristics. *Int. J. Adv. Robot. Sys.* **2020**, *17*, 846–894. [CrossRef]
17. Wen, N.; Zhang, R.; Wu, J. Online planning for relative optimal and safe paths for USVs using a dual sampling domain reduction-based RRT* method. *Int. J. Mach. Learn. Cyber.* **2020**, *11*, 2665–2687. [CrossRef]
18. Han, S.; Wang, Y.; Wang, L.; He, H. Automatic berthing for an underactuated unmanned surface vehicle: A real-time motion planning approach. *Ocean Eng.* **2021**, *235*, 109352. [CrossRef]
19. Esfahani, H.N.; Szlapczynski, R. Model Predictive Super-Twisting Sliding Mode Control for an Autonomous Surface Vehicle. *Polish Mar. Res.* **2019**, *26*, 163–171. [CrossRef]
20. Liang, X.; Qu, X.; Hou, Y.; Li, Y.; Zhang, R. Distributed coordinated tracking control of multiple unmanned surface vehicles under complex marine environments. *Ocean Eng.* **2020**, *205*, 107328. [CrossRef]
21. Qu, X.; Liang, X.; Hou, Y. Fuzzy State Observer-Based Cooperative Path-Following Control of Autonomous Underwater Vehicles with Unknown Dynamics and Ocean Disturbances. *Int. J. Fuzzy Syst.* **2021**, *23*, 1849–1859. [CrossRef]
22. Kabzan, J.; Hewing, L.; Liniger, A.; Zeilinger, M.N. Learning-Based Model Predictive Control for Autonomous Racing. *IEEE Robot. Autom. Lett.* **2019**, *4*, 3363–3370. [CrossRef]
23. Vu, M.T.; Le, T.-H.; Thanh, H.L.N.N.; Huynh, T.-T.; Van, M.; Hoang, Q.-D.; Do, T.D. Robust Position Control of an Over-actuated Underwater Vehicle under Model Uncertainties and Ocean Current Effects Using Dynamic Sliding Mode Surface and Optimal Allocation Control. *Sensors* **2021**, *21*, 747. [CrossRef]
24. Ning, C.; You, F. Online learning based risk-averse stochastic MPC of constrained linear uncertain systems. *Automatica* **2021**, *125*, 109402. [CrossRef]
25. Xia, L.; Shao, C.; Li, H.; Cui, Y. Robust Model-based Reinforcement Learning USV System Guided by Lyapunov Neural Networks. In Proceedings of the 2022 IEEE International Conference on Robotics and Biomimetics (ROBIO), Jinghong, China, 5–9 December 2022; pp. 1976–1981.

26. Shen, H.; Lv, Y.; Zhou, J.; Wang, L. An Online Parameter Estimation Method Based on Adaptive Unscented Kalman Filter for Unmanned Surface Vessel. In Proceedings of the 2022 34th Chinese Control and Decision Conference (CCDC), Hefei, China, 15–17 August 2022; pp. 2584–2589.
27. Qu, X.; Liang, X.; Hou, Y. Path-following control of unmanned surface vehicles with unknown dynamics and unmeasured velocities. *J. Mar. Sci. Technol.* **2021**, *26*, 395–407. [CrossRef]
28. Choo, K.-B.; Cho, H.; Park, J.-H.; Huang, J.; Jung, D.; Lee, J.; Jeong, S.-K.; Yoon, J.; Choo, J.; Choi, H.-S. A Research on Fault Diagnosis of a USV Thruster Based on PCA and Entropy. *Appl. Sci.* **2023**, *13*, 3344. [CrossRef]
29. Yu, X.; Hao, L.; Wang, X. Fault Tolerant Control for an Unmanned Surface Vessel Based on Integral Sliding Mode State Feedback Control. *Int. J. Control Autom. Syst.* **2022**, *20*, 2514–2522. [CrossRef]
30. Tanakitkorn, K.; Phoemsapthawee, S. Impacts of thruster configurations on the task performance of an unmanned surface vehicle. *Ocean Eng.* **2022**, *256*, 111465. [CrossRef]
31. Wang, Y.; Jiang, B.; Wu, Z.-G.; Xie, S.; Peng, Y. Adaptive Sliding Mode Fault-Tolerant Fuzzy Tracking Control with Application to Unmanned Marine Vehicles. *IEEE Trans. Syst. Man Cybern.* **2021**, *51*, 6691–6700. [CrossRef]
32. Guo, W.; Lin, X.; Wang, Y. Predictive Fault-tolerant Control for Trajectory Tracking of Unmanned Surface Vehicle. In Proceedings of the 2021 CAA Symposium on Fault Detection, Supervision, and Safety for Technical Processes (SAFEPROCESS), Chengdu, China, 17–18 December 2021; pp. 1–6.
33. Ding, X.; Zhu, D.; Yan, M. Research on static fault-tolerant control method of thruster based on MPC. *J. Mar. Sci. Technol.* **2021**, *26*, 861–871. [CrossRef]
34. Zhang, G.; Chu, S.; Zhang, W.; Liu, C. Adaptive Neural Fault-Tolerant Control for USV With the Output-Based Triggering Approach. *IEEE Trans. Veh. Technol.* **2022**, *71*, 6948–6957. [CrossRef]
35. Nan, F.; Sun, S.; Foehn, P.; Scaramuzza, D. Nonlinear MPC for Quadrotor Fault-Tolerant Control. *IEEE Robot. Autom. Lett.* **2022**, *7*, 5047–5054. [CrossRef]
36. Chen, Y.; Wang, J. In-Wheel-Motor-Driven Electric Vehicles Motion Control Methods Considering Motor Thermal Protection. *J. Dyn. Syst. Meas. Control* **2019**, *141*, 011015. [CrossRef]
37. Nasir, J.; Islam, F.; Malik, U. RRT*-Smart: A rapid convergence implementation of RRT*. *Int. J. Adv. Robot. Syst.* **2013**, *10*, 299–310. [CrossRef]
38. Chi, W.; Wang, C.; Wang, J.; Meng, M.Q.-H. Risk-DTRRT-Based Optimal Motion Planning Algorithm for Mobile Robots. *IEEE Trans. Autom. Sci. Eng.* **2019**, *16*, 1271–1288. [CrossRef]
39. Vu, M.T.; Van, M.; Bui, D.H.P.; Do, Q.T.; Huynh, T.-T.; Lee, S.-D.; Choi, H.-S. Study on Dynamic Behavior of Unmanned Surface Vehicle-Linked Unmanned Underwater Vehicle System for Underwater Exploration. *Sensors* **2020**, *20*, 1329. [CrossRef] [PubMed]

Journal of
*Marine Science
and Engineering*

MDPI

Article

An Accurate Detection Model of *Takifugu rubripes* Using an Improved YOLO-V7 Network

Siyi Zhou [1], Kewei Cai [2], Yanhong Feng [1,*], Xiaomeng Tang [1], Hongshuai Pang [1], Jiaqi He [1] and Xiang Shi [1]

[1] College of Information and Engineering, Dalian Ocean University, Dalian 116000, China
[2] College of Mechanical and Electrical Engineering, Dalian Minzu University, Dalian 116600, China
* Correspondence: fengyh@dlou.edu.cn

Abstract: In aquaculture, the accurate recognition of fish underwater has outstanding academic value and economic benefits for scientifically guiding aquaculture production, which assists in the analysis of aquaculture programs and studies of fish behavior. However, the underwater environment is complex and affected by lighting, water quality, and the mutual obscuration of fish bodies. Therefore, underwater fish images are not very clear, which restricts the recognition accuracy of underwater targets. This paper proposes an improved YOLO-V7 model for the identification of *Takifugu rubripes*. Its specific implementation methods are as follows: (1) The feature extraction capability of the original network is improved by adding a sizeable convolutional kernel model into the backbone network. (2) Through ameliorating the original detection head, the information flow forms a cascade effect to effectively solve the multi-scale problems and inadequate information extraction of small targets. (3) Finally, this paper appropriately prunes the network to reduce the total computation of the model; meanwhile, it ensures the precision of the detection. The experimental results show that the detection accuracy of the improved YOLO-V7 model is better than that of the original. The average precision improved from 87.79% to 92.86% (when the intersection over union was 0.5), with an increase of 5.07%. Additionally, the amount of computation was reduced by approximately 35%. This shows that the detection precision of the proposed network model was higher than that for the original model, which can provide a reference for the intelligent aquaculture of fishes.

Keywords: *Takifugu rubripes*; accurate identification; improved YOLO-V7 network; large convolution kernel

Citation: Zhou, S.; Cai, K.; Feng, Y.; Tang, X.; Pang, H.; He, J.; Shi, X. An Accurate Detection Model of *Takifugu rubripes* Using an Improved YOLO-V7 Network. *J. Mar. Sci. Eng.* **2023**, *11*, 1051. https://doi.org/10.3390/jmse11051051

Academic Editor: Ernesto Weil

Received: 13 April 2023
Revised: 9 May 2023
Accepted: 10 May 2023
Published: 15 May 2023

1. Introduction

Triggerfish, commonly known as "puffer fish" [1], are a kind of bony fish with a high economic value and are an important aquaculture group in northeast Asia. The usual species are *Takifugu rubripes* (*T. rubripes*), *Takifugu obscurus*, and *Takifugu pseudommus*, etc. The mariculture species are mainly *T. rubripes*, and the production mode of combining sea cage and land-based industrialization is adopted. *Takifugu rubripes* is delicious and nutritious and, as a high-quality food, it is in great demand for exports. With the development of digitalization and informatization, the traditional aquaculture fishery management model relying on human resources and experience now encounters limitations [2]. Problems of manual observation and purely empirical methods such as misdetection, missing detection, and untimely feedback happen occasionally. At present, the fish movement posture in water is variable. The underwater environment, lights, and the mutual occlusion between fish bodies reduce the accuracy of static fish identification in breeding ponds, which leads to the problem of a low accuracy of fish identification [3–5]. Traditional methods can no longer meet the needs of precision and intelligence for modern aquaculture [6]. The accurate identification of fish requires more and more attention and the automatic detection and identification of fish underwater are significant for fishery resource assessment and ecological environment monitoring [7–9]. Therefore, this paper focuses on the accurate identification algorithms for fish in underwater images to solve these above problems.

Traditional target detection algorithms typically use a sliding window model, in which a window sequentially slides on the detected image [10]. Feature extraction is carried out on each sliding window, respectively, and machine learning algorithms are used for the extracted features to determine whether the window contains the object. This method means that the feature extraction and matching have certain defects. Meanwhile, the adaptability, accuracy, and detection speed of traditional algorithms are relatively poor [11]. With the development of deep learning [12–14], Hinton's group won the ImageNet image recognition competition in 2012 by building the AlexNet [15] convolutional neural network to crush the performance of the second-place SVM classifier. As a result, many scholars have shifted their attention from traditional image processing to deep learning target recognition [16,17]. With the advantages of a simple structure, higher efficiency, and higher accuracy, the target detection algorithm based on deep learning has quickly caught up with the traditional target detection algorithms [18,19]. It has become the most mainstream target detection algorithm.

There are currently two main types of mainstream object detection algorithms. The first are known as two-stage algorithm models, such as R-CNN [20], Fast R-CNN [21], and Faster R-CNN [22]. This model first uses heuristics or RPN structures to generate a series of candidate boxes. Subsequently, it uses convolutional neural networks to regress and classify the samples. Using this process, it gains a higher precision, but a lower inference speed. The other algorithm is the single-stage algorithmic model, which uses regression ideas to input images into the convolutional neural network and output the result directly after the detection, such as YOLO [23], SSD [24], and Retina-Net [25]. This algorithm lacks the screening and optimization process of the prediction frameworks, which reduces the accuracy of the detection. Despite this, its detection speed is faster and higher than those of the two-stage methods.

With the fast development of deep learning, it has quickly been applied in fish detection, but there are still many challenges in the model for the accurate identification of fish. Liu et al. [26] detailed a fish recognition detection algorithm based on the FML-Centernet, which introduces a feature fusion module in a Centernet algorithm network structure to fuse low-level feature information and high-level feature information. On this basis, they put forward a more complete feature map, but the detection accuracy was not ideal. Cai et al. [27] constructed a CNN model for fish identification, using the ReLU function as the activation function through dropout and regularization, but increased the detection time. Dong et al. [28] detailed a network that mixed the spatial domain attention mechanism and hierarchical streamlined bilinear features together. Its feature extraction network was initialized with the parameters trained on the ImageNet dataset and further fine-tuned using the fish dataset, while the amount of computation was increased.

YOLO (You Only Look Once) [29–31], a classical single-stage detection algorithm, has achieved a good balance between accuracy and speed and is widely used in various target detection tasks. For example, Wu et al. [32] used a modified YOLO model to detect how pine nematode disease affected trees at different stages of infection. Wang et al. [33] used improved YOLO-V4 and binocular positioning for real-time vehicle identification and tracking during an agricultural operation. Qiu et al. [34] used a YOLO-based method to detect sidewalk cracks in real-time drone images. The above results show good accuracy when detecting targets that are more dispersed from the background. However, there will be still many problems if this model is to be directly applied to the accurate recognition of *T. rubripes*:

(1) Compared to common scenarios, underwater images are affected by lighting, water flow, and water quality, etc., and the fish bodies in the images form a relatively complex background due to overlapping and occlusion, which increases the difficulty of the detection and causes inaccurate detection results.
(2) In the feature extraction and fusion, the feature map output from each node is not fully utilized and the feature extraction ability can be further strengthened during training.

(3) Due to the high density of cultured *T. rubripes* and the different target sizes in the images, the detection head of the YOLO-V7 needs to be improved.

In response to the above issues, based on the YOLO-V7 [35] algorithm framework, this paper proposes an accurate identification and detection algorithm for *T. rubripes* to solve the problem of the low accuracy of fish recognition in images.

2. Materials and Methods

2.1. Dataset

2.1.1. Data Acquisition and Image Features

This paper collected experimental data from the breeding ponds of the Dalian Tianzheng Breeding Factory, which raises different sizes of *T. rubripes* in ponds, and finally collected videos of the *T. rubripes*. The light in the breeding ponds was relatively fixed and soft. To avoid the influence of vertical light, the camera shot the water surface from bottom to top at a 30-degree angle. Considering the changes in light and turbidity, etc., the camera used a zoom lens and was kept in auto mode. The captured video size was 1920 × 1080 and one frame was extracted from the video every six seconds, eventually selecting 3870 images of *T. rubripes*.

2.1.2. Image Annotation and Dataset Production

The open-source script LabelImg (https://github.com/tzutalin/labelImg, accessed on 11 June 2021) on GitHub was used to annotate the dataset. After running the LabelImg script, the target samples in each image were marked. With these produced datasets, an XML file containing the target type and coordinate information was generated and trained. An example of an annotated image is shown in Figure 1. In the production of the dataset, 3870 images of *T. rubripes* were used. These comprised 3096 pictures used for the training dataset and 774 images used for the validation dataset.

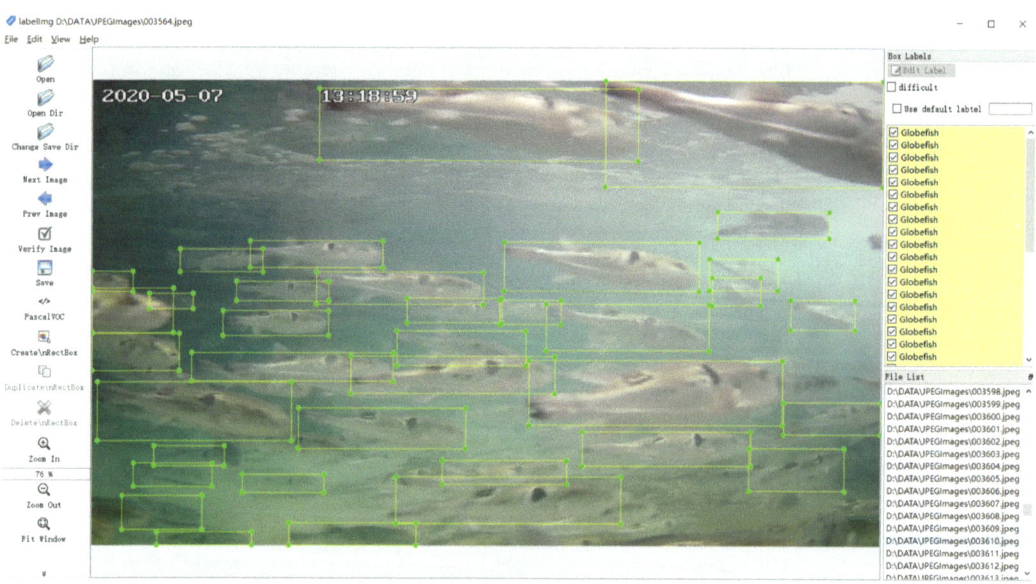

Figure 1. Example of annotated image. (The green frame represents the labeled *Takifugu rubripes*).

2.2. Related Works

2.2.1. YOLO-V7

As the most typical representation of a single-stage object detection algorithm, the YOLO algorithm is based on deep neural networks for object recognition and localization. It uses a single CNN model to achieve end-to-end target detection, takes the whole image as the input into the network structure, and directly regresses the location of the bounding box and the category to which it belongs in the output layer. The YOLO-V7 network represents a continuous improvement over the previous the YOLO series, which provides a good balance between the accuracy and operating speed. The YOLO-V7 network consists of four main modules: input, backbone, head, and prediction. It adopts strategies such as extended efficient layer aggregation networks (E-ELAN), model scaling for concatenation-based models [36], re-parameterized convolution [37], and other techniques. The algorithm structure of YOLO-V7 is shown in Figure 2.

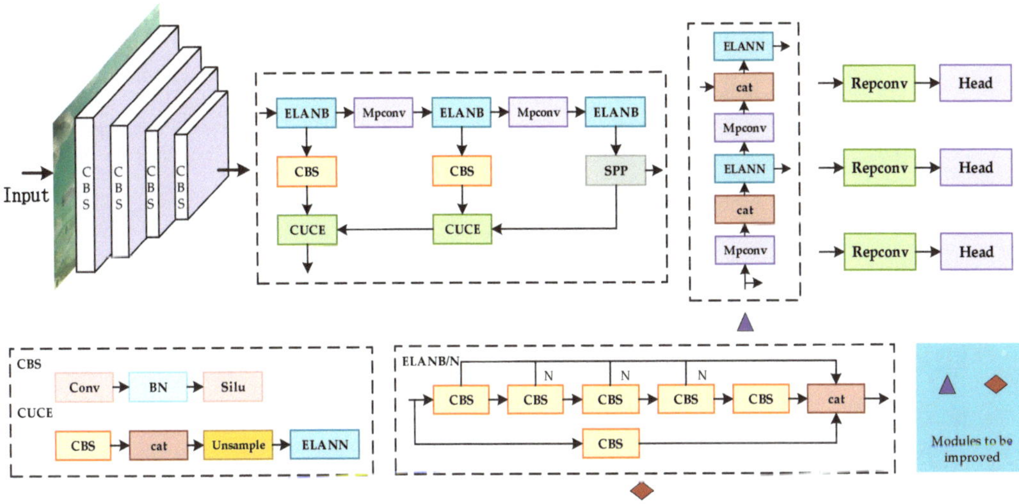

Figure 2. The overview architecture of YOLO-V7.

E-ELAN is a computational block in the YOLO-V7 backbone network that can guide different groups of computation blocks to learn more diverse characteristics. In large-scale ELANs, the network always reaches an equilibrium state, regardless of the gradient direction, path length, and total number of blocks. However, such equilibrium states may be destroyed and the usage of the main parameters will be reduced if the blocks are stacked endlessly. The E-ELAN algorithm uses expansion, random scrambling, and merging cardinality to continuously enhance the ability of the network learning without destroying the original gradient path, and to also guide the different computational block groups to learn more diverse features. The primary purpose of the model scaling is to adjust the specific properties of the model and generate models of different sizes to meet the needs of varying inference speeds. When scaling the model for a cascade-based model, only the depth in the computational block needs to be scaled, and the rest of the transport layer is scaled with the corresponding width. When scaling the depth factor of a calculated block, the change in the output channel of the block is calculated and makes the same changes to the transition layer. RepConv without constant connection is used to redesign the architecture of the reparametrized convolution and proposed to generate coarse to acceptable hierarchical labels with guidance from the prediction results of the guidance head, which are used to assist the learning of the guidance head. This paper presents an improved algorithm based on YOLO-V7, and the research content flowchart is shown in Figure 3.

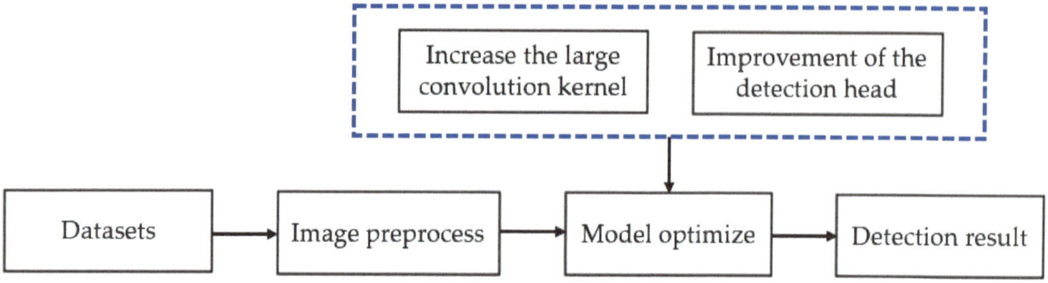

Figure 3. Research content flowchart.

2.2.2. Evaluation Metrics

An experiment needs performance indexes to evaluate an algorithm model. According to the evaluation indexes of the neural network model [38], this paper uses accuracy, recall rate, F_1 score, and average precision as its evaluation indicators. The calculations of the accuracy, recall, F_1 score, and average precision are shown in Equations (1)–(4).

$$\text{Precision} = \frac{\text{TP}}{\text{TP} + \text{FP}} \quad (1)$$

$$\text{Recall} = \frac{\text{TP}}{\text{TP} + \text{FN}} \quad (2)$$

$$F_1 = \frac{2 \times \text{Precision} \times \text{Recall}}{\text{Precision} + \text{Recall}} \quad (3)$$

$$\text{AP} = \int_0^1 \text{Precision} \times \text{Recall dr} \quad (4)$$

$$\text{AP}_{@0.5:0.95} = \frac{1}{10}\left(\text{AP}_{@0.50} + \text{AP}_{@0.55} + \ldots + \text{AP}_{@0.90} + \text{AP}_{@0.95}\right) \quad (5)$$

TP is the real example, that is, the sample correctly identified as *T. rubripes*; FP is the false positive example, which is the incorrectly identified sample of the *T. rubripes*; FN is the false counter example, that is, the sample wrongly identified as the background; TN is the true counter example, that is, the sample correctly identified as the background; 'r' represents the integral variable, which is used to determine the integration of precision $*$ recall and is between 0 and 1; AP is the size of the area under the curve drawn by the accuracy–recall ratio (P-R); $\text{AP}_{@0.5}$ is the average of the accuracy at different recall values when the Intersection Over Union (IOU) is 0.5; and $\text{AP}_{@0.95}$ is the average of the accuracy at different recall values when the IOU is 0.95. $\text{AP}_{@0.5:0.95}$ is the average of the ten values, $\text{AP}_{@0.50}$, $\text{AP}_{@0.55}$, … , $\text{AP}_{@0.90}$, and $\text{AP}_{@0.95}$, and the calculation formula is shown in Equation (5).

2.3. The Proposed Algorithm

This paper proposes an algorithm that is an improvement on YOLO-V7, based on data set characteristics. The sample code and pseudocode of the proposed algorithm are provided in Appendix A.

To solve the problem of the receptive field not being significantly improved after adding depth, it is necessary to increase the convolutional kernel. Compared to a large number of small convolutional kernels, a small number of large convolutional kernels can improve the receptive field and optimize the network backbone model. By these means, the proposed algorithm will capture more effective information and enhance the ability of the feature extraction (Figure 4).

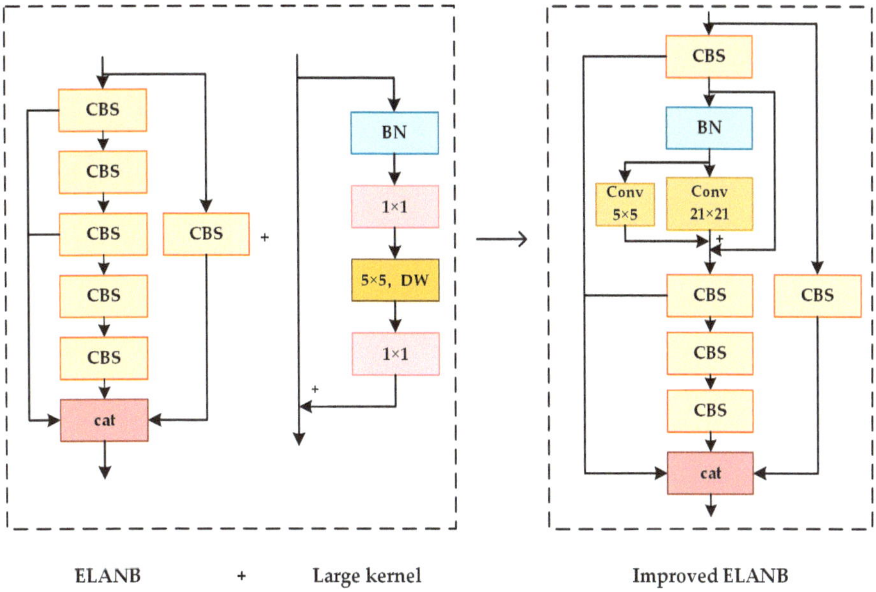

Figure 4. Improvements in feature extraction capability.

Based on the difficulty of the feature loss caused by the excessive occlusion between the targets, this paper upgrades the original detection head. With the progressive information flow, it clears up issues such as the multi-scale and insufficient extraction of small targets, thereby improving the accuracy of the target detection task (Figure 5).

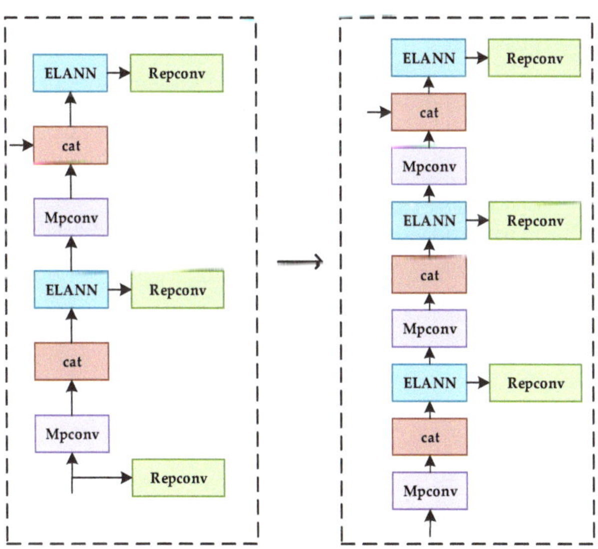

Figure 5. Improvements to the detection head.

2.3.1. Improvements in Feature Extraction Capabilities

During the experiment, there were missed and incorrect results when the original algorithm was used to detect the *T. rubripes*. After the evaluation, we found that most of the false detection or missed detection occurred in the area of dense fish. In order to capture the features clearly, this study hoped to expand the effective receptive field of the model. According to the effective receptive field theory, the size of the receptive field is proportional to the size of the convolution kernel, and the square root of the convolution kernel layers has a positive effect as well. Therefore, we considered that increasing the receptive field by adding depth was more effective for improving the receptive field than directly expanding the convolution kernel. For example, while ResNet [34] looks like it can go very deep, or go up to hundreds or thousands of layers, its effective depth is not very deep and a lot of its signals are from the shortcut layer, which does not increase its effective depth. However, if the convolution kernel size is increased, its effective receptive field tends to go up very significantly.

This paper added a large convolution kernel into an ELANB module to expand the effective receptive field of YOLO-V7, so that we could train a larger area. The structure of the improved algorithm is shown in Figure 4. We increased the convolution kernel of the second layer from 3×3 to 21×21. This upgraded the extraction ability and achieved a more precise recognition result.

2.3.2. Improvement of the Detection Head

The head of the original YOLO-V7 algorithm has three sizes: large, medium, and small. Based on the actual distribution and sample situation of *T. rubripes*, the experiment found that obscured targets increased the error rate during the feature extraction. Furthermore, short distances caused interference and reduced the precision of the target detection, making the target extraction insufficient. Thus, this study proposed adding an object detection layer, abandoning the initially extracted feature map, deepening the layer based on the original, and further informing the information flow. The improved feature map found it easier to inform the feature information of the target and achieve the purpose of improving the detection accuracy, as shown in Figure 5.

3. Results

The software environment of our experiment is shown in Table 1. Considering the GPU memory limitation after adding large convolution kernels during the training, the batch size was set to 16. In order to analyze the training process perfectly, our study selected 300 iterations in the experiment. During the test, a batch of images was chosen with the same resolution in the training phase to verify the algorithm.

Table 1. Experimental environment.

Configuration	Parameter
CPU	Intel Xeon(R) Gold 5128R
GPU	Nvidia RTX 3090 Ti
Operating system	Ubuntu 20.04
Development environment	Pycharm 2022.2
Accelerated environment	CUDA11.1

3.1. Analysis of Training Results

Our study statistically analyzed the test results of the verification set. The sample number of *T. rubripes* in the verification set was 2280, the conf threshold (target confidence threshold) was set to 0.25, and the IOU was set to 0.45 for verification. The results are displayed in Table 2. In the improved model, the accuracy, recall rate, and F1 score were all improved by varying degrees. Among them, the accuracy increased by five percent, the recall by eight percent, and the F1 score by seven percent. The main reasons for this increase in the recall rate were the significant increase in the TP and the significant decrease in the FN, which meant that more and more *T. rubripes* were correctly identified. This demonstrates the effectiveness of the improved algorithm.

Table 2. Comparison of evaluation indicators before and after improvement.

Conf-Thresh = 0.25 IOU = 0.5	Precision	Recall	F1-Score	TP	FP	FN
YOLO-V7	0.91	0.82	0.86	1859	174	421
Improved YOLO-V7	0.96	0.94	0.95	2154	79	126

3.2. *Algorithm Performance Evaluation*

3.2.1. Pre-Training

The pre-training dataset was composed of 3870 *T. rubripes* images, for which a weight file was obtained and the parameters during the pre-training set were as follows: firstly, this study set the initial learning rate to 0.01, made the weight attenuation 0.0005, set the batch size to 16, and performed 300 iterations to generate the pre-training weight files. The dataset was trained using transfer learning. Figure 6 shows the difference between using these transfer learning strategies and not using them. It points out that transfer learning improved the AP by 1.02%, which indicates that this strategy is effective.

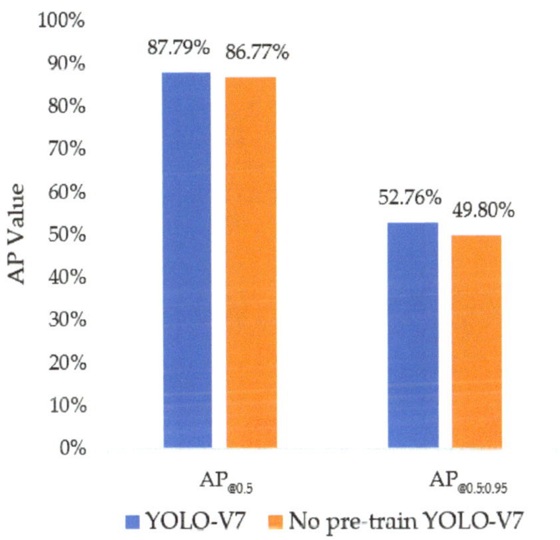

Figure 6. Pre-training comparison chart.

3.2.2. Performance Comparison to Improve Feature Extraction Capabilities

The experiment changed the effective receptive field of YOLO-V7 by changing the network structure of the ELANB. Although adding convolution kernels improved the precision, it also required more running memory. We selected convolution kernels with the sizes of 13 × 3, 17 × 17, 21 × 21, and 27 × 27 for comparison. We comprehensively compared the number of parameters and running speeds and finally decided on a size of 21 × 21 for the convolution kernel. The following experiments were conducted to add the number of improved modules, replacing one, two, and four enhanced modules. The results of this experiment are shown in Figure 7. Considering the running speed and accuracy, we finally replaced two ELANB blocks. After strengthening the feature map, the trained AP increased from 87.7% to 91.3%. These results show that the enhanced ELANB block improved the correction of the training. At the same time, the large convolution kernel strengthened the ability of the feature extraction and positively reduced the occurrence of error detection.

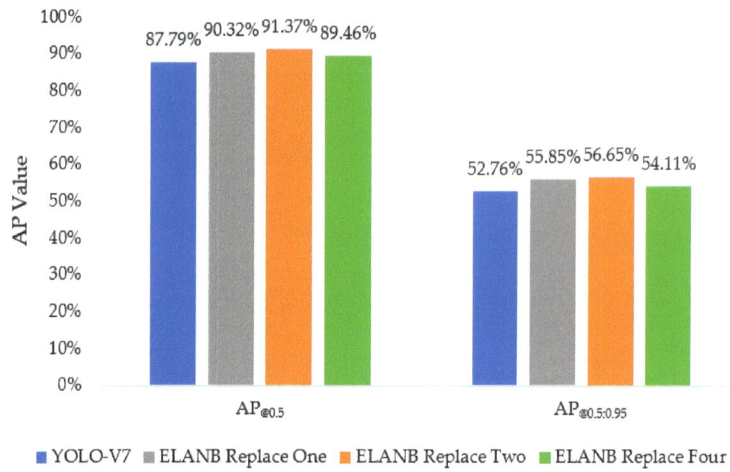

Figure 7. Improved feature extraction capability performance comparison.

3.2.3. Performance Comparison to Improve Head

This paper improved the detection head of the original YOLO-V7, abandoned the feature map of the original, and deepened it based upon the last feature extraction. Through this, the progressive information flow, multi-scale problem [39,40], and inadequate extraction of the small target information were effectively solved during the detection. The results of the enhanced comparative experiment are shown in Figure 8. The AP of the improved training increased from 87.79% to 89.81%. This shows that the enhanced head improved the detection accuracy.

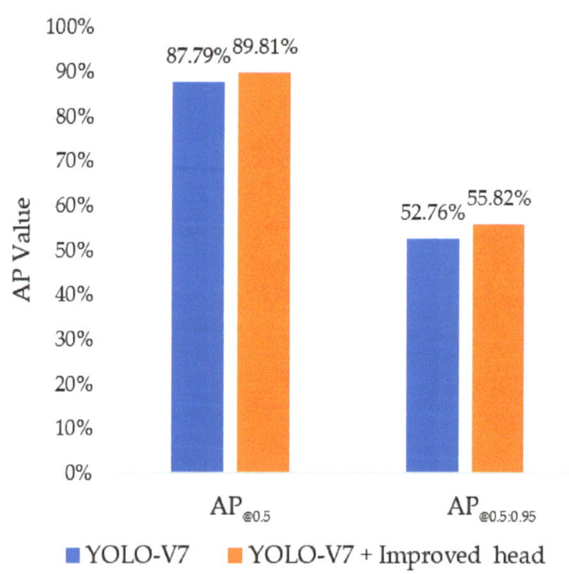

Figure 8. Comparison of the results of the improved detection head.

3.2.4. Comparison of Network Pruning Performance

Takifugu rubripes detection is a challenging detection task. Still, the network structure of YOLO-V7 is large and its accuracy has been improved by increasing the convolution kernel. This task selected the depth of the convolutional layer. Under the premise of ensuring the detection accuracy, we pruned the network appropriately. Table 3 shows the total amount of computation for each algorithm in a billion floating-point operations (GFlops). The results show that, after pruning the network, the whole calculation amount of the model was reduced by about 35%.

Table 3. Comparison of the number of network pruning parameters.

Model	YOLO-V7	Feature Extraction	Improved Head	Network Pruning	GFLOPs
1	√				104.8
2	√	√			109.9
3	√		√		119.4
4 (Ours)	√	√	√	√	68.2

3.3. *Performance Comparison of the Overall Algorithm*

Table 4 compares the results of the different improvement strategies. The results show that the AP50 increased from 87.79% to 92.86%, with an increase of 5.07%. Through a series of improved operations, we successfully improved the detection accuracy for *T. rubripes*.

Table 4. Ablation experiment comparison results.

Model	YOLO-V7	Feature Extraction	Improved Head	Network Pruning	$AP_{@0.5}$ (%)	$AP_{@0.5:0.95}$ (%)
1	√				87.79%	52.76%
2	√	√			91.37%	55.82%
3	√		√		89.81%	56.65%
4 (Ours)	√	√	√	√	92.86%	57.94%

In order to further analyze the performance of the proposed method, we compared it with YOLO-V5, Faster R-CNN, and SSD. We used the same training, verification, and test set to compare the five networks, the results of which are shown in Table 5. The improved YOLO-V7 has a higher accuracy than the other models.

Table 5. Comparison with current mainstream detection algorithms.

Model	$AP_{@0.5}$ (%)	$AP_{@0.5:0.95}$ (%)
YOLO-V5	87.11%	51.80%
Faster R-CNN	88.71%	53.55%
SSD	82.26%	46.43%
YOLO-V7	87.79%	52.76%
Ours	92.86%	57.94%

Figure 9 compares the PR curves of the *T. rubripes* before and after the improvement of YOLO-V7. The closed area of the PR curve before the improvement was much smaller.

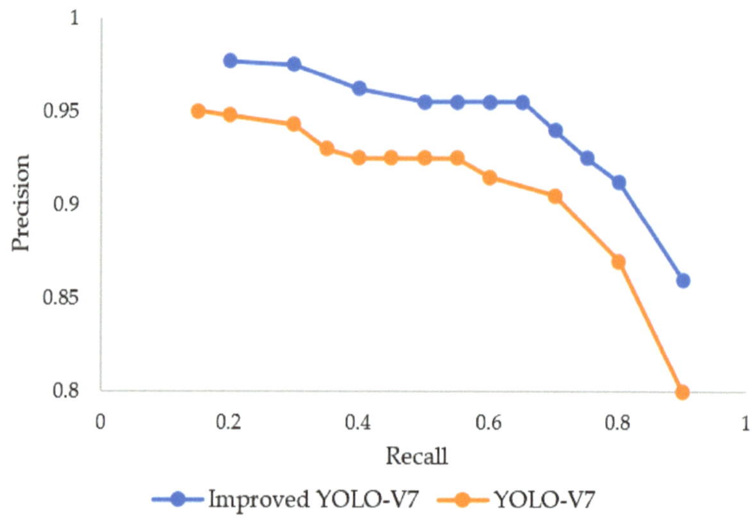

Figure 9. Comparison of PR curves of the model.

Figure 10 shows an example image of the before and after comparison results of some of the test picture improvements. The red box represents the detection result of the original YOLO-V7 and the purple box represents the result of the improved YOLO-V7. The yellow box indicates that the *T. rubripes* could be correctly detected after improving the algorithm. The modified YOLO-V7 network could effectively improve the detection accuracy for the *T. rubripes.* The reason for this was that the enhanced network increased the ability to process feature maps. In this way, *T. rubripes* could be detected correctly when they occurred at higher densities, with an overlapping environment between the fish.

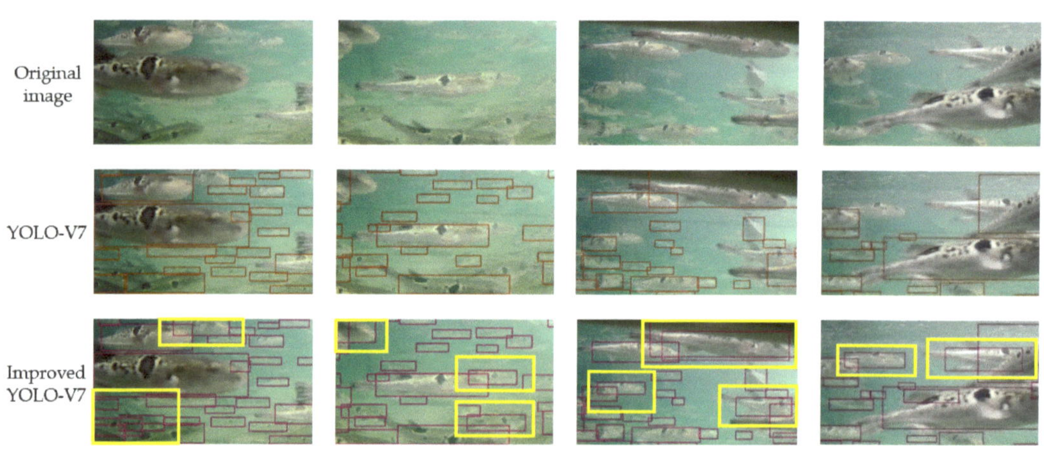

Figure 10. Comparison of test picture results.

The results of the whole screen, before and after the improvement, are shown in Figure 11. Under the interference of high density, the grid units in the feature map extraction network did not perform accurately enough to predict the target, resulting in omissions and error detection in the original YOLO-V7. After the improvement, the detection effect was improved, 47 targets were obtained from the original analysis, 56 targets were obtained after the improvement in picture (a), 40 targets were obtained from the original analysis, and 47 targets were obtained after the improvement in picture (b). This shows that the improved network was better than the unimproved one.

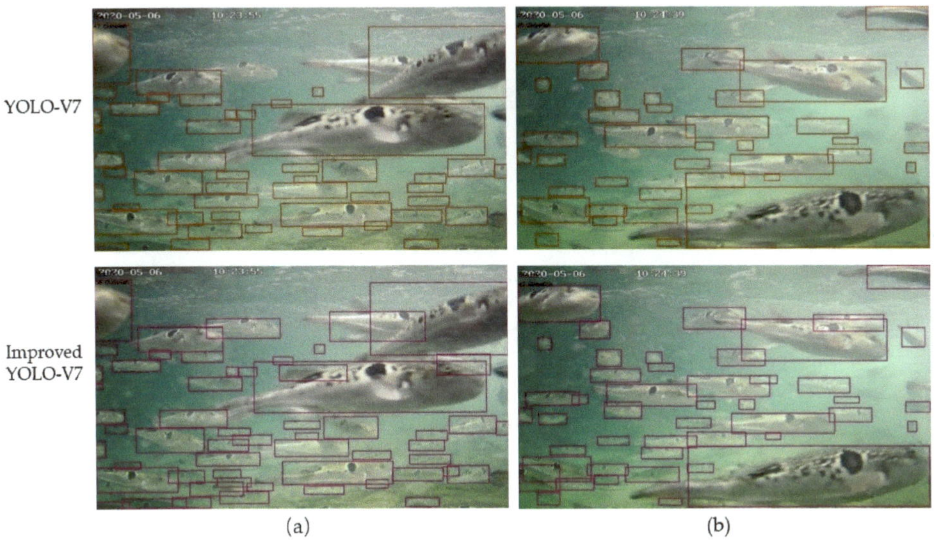

Figure 11. Comparison of test results. (**a**) 47 targets were obtained from the original analysis, 56 targets were obtained after the improvement and (**b**) 40 targets were obtained from the original analysis, and 47 targets were obtained after the improvement.(The first line shows the detection result of the original YOLO-V7 and the second line shows the result of the improved YOLO-V7).

4. Conclusions and Discussion

In this paper, we proposed an improved YOLO-V7 network for accurately detecting *T. rubripes*. The feature extraction ability and detection head of YOLO-V7 were all modified in our new model. The new model ameliorated the situation of low-quality underwater images and more overlapping and dense identification targets, leading to a higher precision. The experimental results show that the AP increased from 87.79% to 92.86%, with a total increase of 5.07%, which means that the improved YOLO-V7 network was better the original version.

This paper effectively solved the problem of fish identification needing to be improved in the cultivation of *T. rubripes* when the background is relatively simple. The recognition of fish images with a complex background is not as good as that with a single background in a practical situation. The intelligence of aquaculture still needs more research and exploration, and our next step will be studying fish identification in complex backgrounds.

Author Contributions: Conceptualization, S.Z.; methodology, S.Z. and K.C.; software, S.Z. and K.C.; validation, S.Z. and X.T.; formal analysis, S.Z. and K.C.; writing—original draft preparation, S.Z.; writing—review and editing, S.Z.; visualization, S.Z., J.H. and X.S.; supervision, Y.F. and H.P. All authors have read and agreed to the published version of the manuscript.

Funding: This research was funded by the province scientific research project Education Department of Liaoning [JL201917].

Institutional Review Board Statement: Not applicable.

Informed Consent Statement: Not applicable.

Data Availability Statement: The data presented in this study are available upon request from the corresponding author.

Acknowledgments: The authors are grateful for the support from the Education Department of Liaoning Province, China.

Conflicts of Interest: The authors declare no conflict of interest.

Appendix A

Algorithm A1: The pytorch style code of the improved LgConv & LKDeXt module

```
class LgConv(nn.Module):
    def __init__(self, in_channels, dw_channels, block_lk_size, small_kernel, drop_path, small_kernel_merged=False):
        super().__init__()
        self.pw1 = conv_bn_relu(in_channels, dw_channels, 1, 1, 0)
        self.pw2 = conv_bn(dw_channels, in_channels, 1, 1, 0)
        self.large_kernel = ReparamLargeKernelConv(dw_channels, dw_channels, block_lk_size, 1, dw_channels, small_kernel, small_kernel_merged)
        self.lk_nonlinear = nn.ReLU()
        self.prelkb_bn = get_bn(in_channels)
        self.drop_path = DropPath(drop_path) if drop_path > 0. else nn.Identity()

    def forward(self, x):
        out = self.prelkb_bn(x)
        out = self.pw1(out)
        out = self.large_kernel(out)
        out = self.lk_nonlinear(out)
        out = self.pw2(out)
        return x + self.drop_path(out)

class LKDeXt(nn.Module):
    def __init__(self, c1, c2, n=1, True, g=1, e=0.5):
        super().__init__()
        c_ = int(c2 * e)
        self.cv1 = Conv(c1, c_, 1, 1)
        self.cv2 = Conv(c1, c_, 1, 1)
        self.cv3 = Conv(2 * c_, c2, 1)
        self.m = nn.Sequential(*(LgConv(c_, c_, 21, 5, 0.0, False) for _ in range(n)))
    def forward(self, x):

        return self.cv3(torch.cat((self.m(self.cv1(x)), self.cv2(x)), dim=1))
```

Algorithm A2: The pseudocode of the improved ConvBlock & LKDeXt

```
def ConvBlock(x):
        x = Conv(x)
        x = Batch_norm(x)
        x = ReLU(x)
        return x
def LgConv(x):
        y = Batch_norm(x)
        y = ConvBlock(y)
        y = ReparamLargeKernelConv (y)
        y = ReLU(y)
        y = ConvBlock(y)
        return x + dropout(y)

def LKDeXt(x):
        y = Conv(concat(LgConv(Conv(x)),Conv(x),dim=1))
        return y
```

References

1. Guo, R.; Zhang, X.; Su, H.; Liu, H. The research status of nutrition value and by-products ultilization of puffer fish. *J. Food Sci. Technol.* **2018**, *3*, 113–116.
2. Yang, D.; Zhang, S.; Tang, X. Research and development of fish species identification based on machine vision technology. *Fish. Inf. Strategy* **2019**, *31*, 112–120.
3. Sun, L.; Wu, Y.; Wu, Y. Multi-objective fish object detection algorithm is proposed to study. *J. Agric. Mach.* **2019**, *50*, 260–267.
4. Tu, B.; Wang, J.; Wang, S.; Zhou, X.; Dai, P. Research on identification of freshwater fish species based on fish back contour correlation coefficient. *Comput. Eng. Appl.* **2016**, *52*, 162–166.
5. Wan, P.; Zhao, J.; Zhu, M.; Tan, H.; Deng, Z.; Huang, S.; Wu, W.; Ding, A. Freshwater fish species identification method based on improved ResNet50 model. *J. Agric. Eng.* **2021**, *12*, 159–168.
6. Liu, S.; Li, G.; Tu, X.; Meng, F.; Chen, J. Research on the development of aquaculture production information technology. *Fish. Mod.* **2021**, *48*, 64–72.
7. Zhao, Z.; Liu, Y.; Sun, X.; Liu, J.; Yang, X.; Zhou, C. Composited FishNet: Fish detection and species recognition from low-quality underwater videos. *IEEE Trans. Image Process.* **2021**, *30*, 4719–4734. [CrossRef]
8. Li, S.; Yang, L.; Yu, H.; Chen, Y. Underwater fish species identification model and real-time recognition system. *J. Intell. Agric.* **2022**, *4*, 130–139.
9. Wang, W.; Jiang, H.; Qiao, Q.; Zhu, H.; Zheng, H. Research on fish recognition and detection algorithm based on deep Learning. *J. Inf. Technol. Netw. Secur.* **2020**, *33*, 6157–6166.
10. Sun, S.; Zhao, J. Pattern Recognition and Machine Learning. *J. Sci. Technol. Publ.* **2021**, *322*, 154.
11. Li, J.; Xu, L. Research hot trend prediction model based on machine learning algorithm comparison and analysis, the BP neural network, support vector machine (SVM) and LSTM model. *Mod. Intell.* **2019**, *33*, 23–33.
12. Amanullah, M.; Selvakumar, V.; Jyot, A.; Purohit, N.; Fahlevi, M. CNN based prediction analysis for web phishing prevention. In Proceedings of the International Conference on Edge Computing and Applications (ICECAA), Tamilnadu, India, 1–3 December 2022; pp. 1–7.
13. Althubiti, S.A.; Alenezi, F.; Shitharth, S.; Reddy, C.V.S. Circuit manufacturing defect detection using VGG16 convolutional neural networks. *Wirel. Commun. Mob. Comput.* **2022**, *2022*, 1070405. [CrossRef]
14. Alyoubi, K.H.; Shitharth, S.; Manoharan, H.; Khadidos, A.O.; Khadidos, A.O. Connotation of fuzzy logic system in underwater communication systems for navy applications with data indulgence route. *Sustain. Comput. Inform. Syst.* **2023**, *38*, 100862. [CrossRef]
15. Krizhevsky, A.; Sutskever, I.; Hinton Geoffrey, E. ImageNet classification with deep convolutional neural networks. *Commun. ACM* **2017**, *60*, 84–90. [CrossRef]
16. Naseer, A.; Baro, E.N.; Khan, S.D.; Vila, Y. A novel detection refinement technique for accurate dentification of nephrops norvegicus burrows in underwater imagery. *Sensors* **2022**, *12*, 4441. [CrossRef] [PubMed]
17. Shitharth, S.; Prasad, K.M.; Sangeetha, K.; Kshirsagar, P.R.; Babu, T.S.; Alhelou, H.H. An enriched RPCO-BCNN mechanisms for attack detection and classification in SCADA systems. *IEEE Access* **2021**, *9*, 156297–156312. [CrossRef]
18. Sun, H.; Li, Y.; Lin, Y. Significant target detection based on deep learning review. *J. Data Acquis. Process.* **2023**, *38*, 21–50.
19. Qian, C. Target detection algorithm based on depth of learning research progress. *J. Wirel. Commun. Technol.* **2022**, *31*, 24–29.
20. Girshick, R.; Donahue, J.; Darrell, T.; Malik, J. Rich feature hierarchies for accurate object detection and semantic segmentation. In Proceedings of the IEEE Conference on Computer Vision and Pattern Recognition, CVPR, Washington, DC, USA, 23–28 June 2014; IEEE: Pitscatway, NJ, USA, 2014.
21. Girshick, R. Fast R-CNN. In Proceedings of the 2015 IEEE International Conference on Computer Vision (ICCV), Santiago, Chile, 7–13 December 2015; pp. 1440–1448.
22. Ren, S.; He, K.; Girshick, R.; Sun, J. Faster R-CNN: Towards real—Time object detection with region proposal networks. *IEEE Trans. Pattern Anal. Mach. Intell.* **2017**, *39*, 1137–1149. [CrossRef] [PubMed]
23. Redmon, J.; Farhadi, A. YOLOv3: An incremental improvement. *arXiv* **2018**, arXiv:1804.02767.
24. Liu, W.; Anguelov, D.; Erhan, D.; Szegedy, C., Reed, S.; Fu, C.Y.; Berg, A.C. SSD: Single shot multibox detector. In Proceedings of the 14th European Conference on Computer Vision, Amsterdam, The Netherlands, 11–14 October 2016; Springer: Cham, Switzerland, 2016; pp. 21–37.
25. Liu, S.; Huang, D.; Wang, Y. Learning spatial fusion for single-shot object detection. *arXiv* **2019**, arXiv:1911.09516.
26. Liu, Y.; Wang, Y.; Hunag, L. Fish recognition and detection based on FML-Centernet algorithm. *Laser Optoelectron. Prog.* **2022**, *59*, 317–324.
27. Cai, W.; Pang, H.; Zhang, Y.; Zhao, J.; Ye, Z. Recognition model of farmed fish species based on convolutional neural network. *J. Fish. China* **2022**, *46*, 1369–1376.
28. Dong, S.; Liu, W.; Cai, W.; Rao, Z. Fish recognition based on hierarchical compact bilinear attention network. *Comput. Eng. Appl.* **2022**, *5*, 186–192.
29. Bochkovskiy, A.; Wang, C.Y.; Liao, H.Y.M. YOLOv4: Optimal speed and accuracy of object detection. *arXiv* **2020**, arXiv:2004.10934.
30. Zhu, X.; Lyu, S.; Wang, X.; Zhao, Q. TPH-YOLOv5: Improved YOLOv5 based on transformer prediction head for object detection on drone-captured scenarios. In Proceedings of the 2021 IEEE/CVF International Conference on Computer Vision Workshops (ICCVW), Montreal, BC, Canada, 11–17 October 2021; pp. 2778–2788.

31. Ge, Z.; Liu, S.; Wang, F.; Li, Z.; Sun, J. YOLOX: Exceeding YOLO series in 2021. *arXiv* **2021**, arXiv:2107.08430.
32. Wu, K.; Zhang, J.; Yin, X.; Wen, S.; Lan, Y. An improved YOLO model for detecting trees suffering from pine wilt disease at different stages of infection. *Remote Sens. Lett.* **2023**, *14*, 114–123. [CrossRef]
33. Wang, L.; Li, L.; Wang, H.; Zhu, S.; Zhai, Z.; Zhu, Z. Real-time vehicle identification and tracking during agricultural master-slave follow-up operation using improved YOLO v4 and binocular positioning. *Proc. Inst. Mech. Eng.* **2023**, *237*, 1393–1404. [CrossRef]
34. Qiu, Q.; Lau, D. Real-time detection of cracks in tiled sidewalks using YOLO-based method applied to unmanned aerial vehicle (UAV) images. *Autom. Constr.* **2023**, *147*, 104745. [CrossRef]
35. Wang, C.Y.; Bochkovskiy, A.; Liao, H.Y.M. YOLOv7: Trainable bag-of-freebies sets new state-of-the-art for real-time object detectors. *arXiv* **2022**, arXiv:2207.02696.
36. Wang, C.Y.; Bochkovskiy, A.; Liao, H.Y.M. Scaled-YOLOv4: Scaling cross stage partial network. In Proceedings of the 2021 IEEE/CVF Conference on Computer Vision and Pattern Recognition (CVPR), Nashville, TN, USA, 20–25 June 2021; pp. 13024–13033.
37. Ding, X.; Zhang, X.; Ma, N.; Han, J.; Ding, G.; Sun, J. RepVGG: Making VGG-style convnets great again. In Proceedings of the 2021 IEEE/CVF Conference on Computer Vision and Pattern Recognition (CVPR), Nashville, TN, USA, 20–25 June 2021; pp. 13728–13737.
38. Goutte, C.; Gaussier, E. A probabilistic interpretation of precision, recall and F score, with implication for evaluation. In Proceedings of the European Conference on Information Retrieval, Santiago de Compostela, Spain, 21–23 March 2005; pp. 345–359.
39. Khan, S.D.; Basalamah, S. Multi-Scale person localization with multi-stage deep sequential framework. *Int. J. Comput. Intell. Syst.* **2021**, *14*, 1217–1228. [CrossRef]
40. Khan, S.D.; Alarabi, L.; Basalamah, S. A unified deep learning framework of multi-scale detectors for Geo-spatial object detection in high-resolution satellite images. *Arab. J. Sci. Eng.* **2022**, *47*, 9489–9504. [CrossRef]

Journal of Marine Science and Engineering

Article

An Improved S-Plane Controller for High-Speed Multi-Purpose AUVs with Situational Static Loads

Chunmeng Jiang [1], Jinhua Lv [1], Lei Wan [2,*], Jianguo Wang [3], Bin He [4] and Gongxing Wu [5]

[1] Wuhan Institute of Shipbuilding Technology, Wuhan 430050, China; 2020017@mail.wspc.edu.cn (C.J.); 1997926@mail.wspc.edu.cn (J.L.)
[2] School of Naval Engineering, Harbin Engineering University, Harbin 150001, China
[3] China Ship Development and Design Center, Wuhan 430064, China
[4] Wuhan Second Ship Design and Research Institute, Wuhan 430205, China
[5] College of Ocean Science and Engineering, Shanghai Maritime University, Shanghai 201306, China
* Correspondence: wanlei@hrbeu.edu.cn

Abstract: The classic S-plane control method combines PD structure with fuzzy control theory, with the advantages of a simple control structure and fewer parameters to be adjusted. It has been proved as a practical method in an autonomous underwater vehicle (AUV) motion control at low and medium speeds, but it takes no account of the situational static load and varying hydrodynamic forces which influence the control quality and even result in a "dolphin effect" at the time of high-speed movement. For this reason, an improved S-plane controller is designed based on the sliding mode variable structure, sliding mode surface, and control items in order to respond to the situational static load and high-speed movement. The improved S-plane controller is verified by Lyapunov stability analysis. The thrust allocation strategies are also discussed with constraints introduced in accordance with task requirements. In order to verify the practicability and effectiveness of the improved S-plane controller, both simulation experiments and field trials of AUV motion control, long-range cruise, and path point following were carried out. The results have demonstrated the superiority of the improved S-plane controller over the classic S-plane controller.

Keywords: autonomous underwater vehicle; S-plane control; high speed; situational static load; sliding mode variable structure

Citation: Jiang, C.; Lv, J.; Wan, L.; Wang, J.; He, B.; Wu, G. An Improved S-Plane Controller for High-Speed Multi-Purpose AUVs with Situational Static Loads. *J. Mar. Sci. Eng.* **2023**, *11*, 646. https://doi.org/10.3390/jmse11030646

Academic Editor: Tieshan Li

Received: 27 February 2023
Revised: 13 March 2023
Accepted: 17 March 2023
Published: 19 March 2023

1. Introduction

With the advancements in automation technology, artificial intelligence, deep learning, and system identification, notable breakthroughs have been made in intelligent underwater vehicles [1–4]. Autonomous underwater vehicles (AUVs) have played a decisive part in workplaces where human divers find it difficult or impossible to access. So far, AUVs have been widely used in underwater pipeline detection, undersea cable maintenance, dam detection, deep-sea resource exploration, etc. [5–7]. AUVs have presented a bright prospect in civilian and military fields as a practical tool for special or demanding underwater operations [8].

AUVs are subject to significant coupling and high nonlinear characteristics [9]. The different modes of movement, complex environments, and uncertain factors jointly make it difficult to establish an accurate AUV motion model. For this reason, the design of the AUV motion control system must take the control quality into consideration for the purpose of the admirable completion of operational tasks. A number of control methods have been put forward over the past years to meet different requirements for AUV motion control.

Shi et al. [10] proposed a fuzzy PID method by combining fuzzy logic with PID control in the design of a motion control system for a novel underwater vehicle. The 3D simulation experiments of motion control were conducted to prove the superiority of the proposed method over the conventional PID method in robustness and dynamic response. Hasan

Mustafa Wassef et al. [11] developed an adaptive fuzzy nonlinear PID controller for underwater vehicles to eliminate the effect of the disturbances caused by ocean currents. Its anti-interference advantage was verified in contrast simulation experiments with the fuzzy PID controller and the conventional PID controller. The simulation results showed that the proposed controller improved the vehicle's confrontation against deep-water and near-surface wave disturbances. Keymasi Khalaji Ali et al. [12] proposed a finite time sliding mode controller to reduce external disturbances. The performance and stability of the method were compared with the sliding mode controller. The numerical comparison results showed that the proposed approach was effective and applicable in practice. A fuzzy control system was designed by Zhilenkov Anton [13]. The comparison with the PD controller showed that the designed fuzzy controller provides a higher quality of control of the plant under uncertainties. Cruz Ulloa Christyan et al. [14] combined closed-loop fuzzy theory with visual control and proposed a fuzzy visual control method. The simulation results showed that the robot movements were very close to the morphological behavior of a real jellyfish regarding the curves of displacements, speeds, and accelerations. Keymasi Khalaji Ali et al. [15] put forward a tracking control approach based on passive self-adaption. The adaptive rules were used to estimate and reduce the influence of unknown interference. The stability of the proposed method was analyzed using passivity properties and the Lyapunov theory. The comparative results showed the advantages of the proposed method. Xue et al. [16] developed a model-independent adaptive controller for an underwater vehicle manipulator system. The proposed controller showed good trajectory tracking performance without a precise dynamic model of the manipulator system, which was of great importance to applications in engineering. A robust nonlinear predictive method was proposed by Nikou Alexandros et al. [17]. The proposed controller fell within the tube-based non-linear model predictive control methodology and could handle the rich expressivity in both safety and reachability specifications. Oliveira Everton L. et al. [18] came up with a model predictive method based on interference observation whose robustness and performance were verified in simulation experiments of underwater vehicle manipulator systems. Shojaei Khoshnam [19] worked on an adaptive neural network controller. The combination of multi-layer neural networks (NNs) and adaptive robust control techniques were designed to handle the compensation of model uncertainties including unknown parameters, time-varying environmental disturbances induced by waves and ocean currents, and NN approximation errors. Simulation results showed the effective performance of the method for practical applications. Tony Salloom et al. [20] proposed an adaptive neural network for parameter regulation via the genetic algorithm. The contrast simulation experiments showed the advantages of the proposed network in regulation efficiency and control accuracy over manual parameter regulation. Deep reinforcement learning was applied to adaptive path planning and intelligent control of underwater vehicles [21]. The study proposed an adaptive motion planning and obstacle avoidance technique based on deep reinforcement learning for an AUV. The research adopted a twin-delayed deep deterministic policy algorithm. The simulation results showed that the motion planning system could precisely guide an AUV with six degrees of freedom dynamics toward the target. Elhaki Omid et al. [22] combined a saturated adaptive robust neural network with reinforcement learning, and the stability of the proposed closed-loop system was investigated by Lyapunov's direct methodology, and simulations along with a comparative study certified the contributions. The radial basis function was combined with sliding mode control [23] to improve the performance of target tracking. Lyapunov stability analysis and homogeneity theory proved that tracking errors can converge on a small region that contained the origin with prescribed performance in finite time. A comparative study with sliding mode control verified the superiority of the proposed method in steady-state error and convergence time. Liang et al. [24] completed the design of a PID neural network sliding mode controller based on nonlinear high-order observation, together with a comparative analysis with the neural network sliding mode controller. The strength of the proposed controller in tracking accuracy and anti-interference was verified in simulation experiments. Karpenko Mark et al. [25] designed a bias feedback attitude control system based on point following, and the simulation experiments showed that the proposed system was able to reduce the overall resources together with quality control performance. Cortes Perez Noel et al. [26] designed a mirror active vision system for underwater vehicles and applied it to target tracking experiments. The active vision system is tested by an experiment. The experiment results showed that the target could be detected and tracked under low light conditions. Shi et al. [27] proposed an underwater vehicle dynamic target tracking control method based on deep reinforcement learning, whose performance was verified in simulation experiments. Simulation experiment results demonstrated good control performance regarding both stability and computational complexity, indicating the effectiveness of the proposed algorithm in target tracking tasks.

Simulation experiments have been carried out for all the above-mentioned control methods, some of which have been practically applied in engineering. The existing methods primarily deal with motion control of underwater vehicles at low and medium speed ($\leq$1.0 m/s). However, there are few studies on high-speed underwater vehicle motion control and operations demanding high speeds. In addition, some of the existing control methods are complicated in structure or involve many parameters that require a long adjustment process, which makes them ineffective in engineering practices.

The classic S-plane method has been recognized as a practical approach in engineering fields [28]. In practical operations, however, the static loads such as buoyancy and gravity of an AUV may vary due to the sensors or facilities it carries as required by the operational tasks. The classic S-plane method fails to consider the variable static loads and the changing hydrodynamic force at the time of high-speed movement, causing difficulty in high-speed control or even failure in convergence. In addition to motion control, there are more demanding operation tasks in different situations that expect better performance and higher control accuracy in which the classic S-plane is currently incompetent. According to the analysis of field trial results, the average overshoot of velocity control based on the classic S-plane method in the horizontal plane is approximately 0.1 m/s and that in heading control is approximately 3.9°. In regard to operations with high requirements on control quality, such as an overshoot of less than 0.05 m/s in velocity control or an overshoot of less than 2.0° in heading control, the classic S-plane method would find it difficult to meet the requirements.

In order to deal with the influence of variable static load and high-speed movement in AUV motion control, an improved S-plane controller is developed. The improved controller introduces the concept of sliding mode surface and sliding mode variable structure and considers the above factors in form of control items to counter the influence of situational static load and high-speed movement. A Lyapunov analysis is conducted to prove the stability of the improved S-plane controller. Both simulation experiments and field trials of AUV motion control, long-range cruise, and path point following were carried out. The results have demonstrated the superiority of the improved S-plane controller over the classic S-plane controller. The improved S-plane controller admirably coped with the impacts of situational static loads and high-speed movement and improved the high-speed control quality in different operation tasks.

This paper is organized as follows. In the second section, the underwater vehicle platform is introduced, including the structure design and propulsion system, as well as software and hardware architectures. In the third section, the motion coordinate system and the dynamical model are established. In the fourth section, an improved S-plane controller is designed. The stability analysis of the controller is carried out based on the Lyapunov function, together with the thrust allocation strategies for the redundant propulsion system. In the fifth and sixth sections, a detailed analysis of the results of the simulation experiments and field trials is carried out to verify the feasibility and effectiveness of the improved method.

2. Underwater Vehicle Platform

2.1. AUV Structure

As shown in Figure 1, the research is carried out on a composite platform composed of a fundamental control system, target detection system, emergency system etc. [29]. With the top priority to control quality in multiple functions including motion control, long-range cruise and path point following [30], the platform is driven with a thruster-based layout.

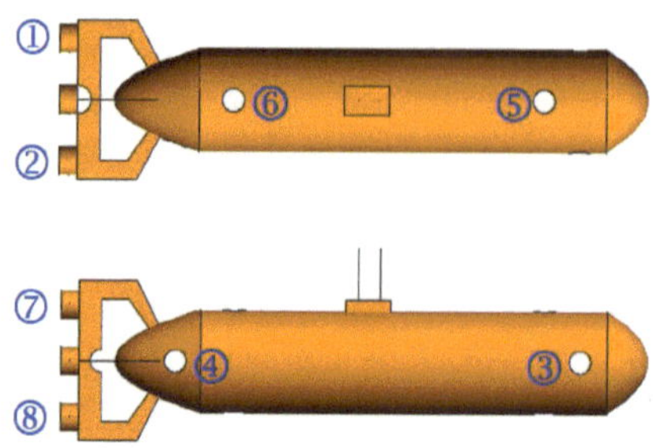

Figure 1. Profile of the AUV.

The platform weighs over 2 t and is approximately 5 m long with a maximum radius of 0.45 m. It is designed into a streamlined droplet with slightly positive buoyancy. The platform is in a double shell structure with the outer layer as the protective shell and the inner layer as the pressure shell. The streamlined protection shell is designed to reduce drag force with composite materials for the bow and the stern. The pressure shell bears high strength with waterproof sealing rings. The batteries and electronic control cabin are placed inside the pressure shell which is made up of a ball head in front, a main housing, and a ball head at the stern.

2.2. Propulsion System

Without a rudder or fin, the propulsion system includes eight thrusters and corresponding accessories, including four conduit thrusters and four slot thrusters. ① and ② are the left thruster and right thruster, respectively. ③ and ④ are the side thruster at the bow and the side thruster at the stern, respectively. ⑤ and ⑥ are the vertical thruster at the bow and the vertical thruster at the stern, respectively. ⑦ and ⑧ are the upward thruster and downward thruster, respectively.

2.3. Hardware Architecture

The hardware architecture shown in Figure 2 contains the intelligent planning sector, the autonomous navigation sector, and the motion control sector. These sectors allow real-time data transmission with each other via a PC/104 ISA bus [31]. The hardware architecture has been tested on several underwater vehicles of the same series in lake trials and sea trials, whose feasibility and effectiveness have been verified. The intelligent planning sector interprets and delivers target instructions. The autonomous navigation sector is connected with a GPS and an inertial navigation system (INS), providing high-precision navigation data [32]. The motion control sector deals with sensor data sampling and decoding, control algorithm calculation, thrust allocation strategy, thrust instruction sending, etc. The target instructions from the intelligent planning sector are transmitted through UDP. The longitude and latitude data from the autonomous navigation sector is transmitted through TCP. The attitude angle from the fiber optical gyrocompass (FOG) and velocity information from the Doppler velocity logger are collected through a serial card. The depth and leakage data are acquired through the A/D card, and the D/A card sends analog voltage instructions to the thrusters.

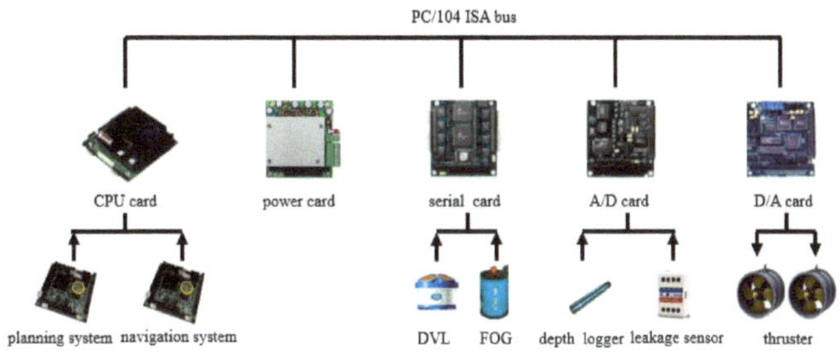

Figure 2. Hardware architecture of the platform.

2.4. Software Architecture

The control program is developed in C programming language [33]. The modular design aims to improve the expansibility and portability of the software [34,35]. The controlling software architecture fulfills functions of sensor data acquisition, data preprocessing and fusion, control algorithm calculation, thrust allocation, thrusting instruction transmission, and data storage.

The test platform follows the information flow as shown in Figure 3 at the time of motion control. The unit of target generating and processing receives the objective instructions and classifies the instructions according to the tasks. The control target is initialized in the unit of control commands, including the selection of the motion control model and the type and parameters of motion control. The aforesaid information is then sent to the unit of control algorithm where the data from different sensors gather after being pre-processed, filtered, and fused. The unit of the control algorithm is the core section that has the control target processed, such as remote control or automatic control. In automatic control, the target is calculated by the control algorithm and then the results are sent to the thrust allocation module. After strategy selection and force calculation, the thrusting force required from each thruster is obtained. The calculated force is then converted into analog instructions that drive the thrusters to enable the AUV to move as expected. Meanwhile, the inspection module of the actuators and sensors transmits the information of the emergency module, acoustic visual module, optical visual module, and the thrusters to the memory where such data are stored.

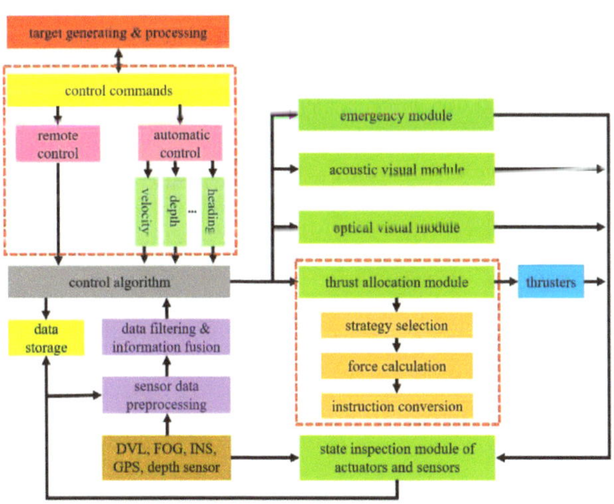

Figure 3. Information flow of the motion control sector.

3. Dynamics Modelling

3.1. Coordinate Systems

Based on the standard symbol system, the body coordinate system $G - xyz$ and the inertial coordinate system $E - \xi\eta\zeta$ are constructed, as shown in Figure 4 [36]. $G - xyz$ reflects the dynamic characteristics of the AUV while $E - \xi\eta\zeta$ describes the attitude angles.

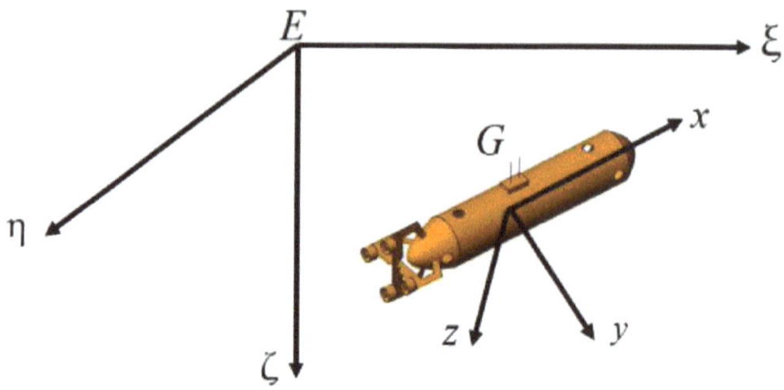

Figure 4. Body coordinate system and inertial coordinate system.

The two coordinate systems can be transformed as follows [37]:

$$\begin{bmatrix} \xi \\ \eta \\ \varsigma \end{bmatrix} = T^{-1} \begin{bmatrix} x \\ y \\ z \end{bmatrix} \tag{1}$$

$$T = \begin{bmatrix} c(\psi)c(\theta) & s(\psi)c(\theta) & -s(\theta) \\ c(\psi)s(\theta)s(\phi) - s(\psi)c(\phi) & s(\psi)s(\theta)s(\phi) + c(\psi)c(\phi) & c(\theta)s(\phi) \\ c(\psi)s(\theta)c(\phi) + s(\psi)s(\phi) & s(\psi)s(\theta)c(\phi) - c(\psi)s(\phi) & c(\theta)c(\phi) \end{bmatrix} \tag{2}$$

Where T is the transformation matrix, ϕ is the rolling angle, θ is the pitch angle, and ψ is the heading angle. s() stands for sin() and c() stands for cos().

3.2. Dynamical Model

The AUV motion equations are as follows [38]:

$$\begin{cases} X = m[(\dot{u} - vr + wq) - x_G(q^2+r^2) + y_G(pq - \dot{r}) + z_G(pr + \dot{q})] \\ Y = m\left[(\dot{v} - wp + ur) - y_G(r^2+p^2) + z_G(qr - \dot{p}) + x_G(qp + \dot{r})\right] \\ Z = m\left[(\dot{w} - uq + vp) - z_G(p^2+q^2) + x_G(rp - \dot{q}) + y_G(rq + \dot{p})\right] \\ K = I_x \dot{p} + (I_z - I_y)qr + m[y_G(\dot{w} + pv - qu) - z_G(\dot{v} + ru - pw)] \\ M = I_y \dot{q} + (I_x - I_z)rp + m[z_G(\dot{u} + wq - vr) - x_G(\dot{w} + pv - uq)] \\ N = I_z \dot{r} + (I_y - I_z)pq + m[x_G(\dot{v} + ur - pw) - y_G(\dot{u} + qw - vr)] \end{cases} \tag{3}$$

where X, Y, Z, K, M, N are the external forces and torques acting upon the AUV. m is the mass of the AUV. I_x, I_y, I_z are inertia moments. u, v, w stand for linear velocity and p, q, r are angular velocity. x_G, y_G, z_G mean the center of gravity.

Static loads such as buoyancy and gravity are variable since the AUV needs to be equipped with different facilities or sensors as required by different operation tasks. The research object in this paper is symmetrical in structure, so its movement model is established as follows with full consideration of the situational static loads and high speeds [39].

$$F_T - I(\nu)\nu - A(\nu)\nu - R_G = M\dot{\nu} \tag{4}$$

F_T is the force/moment vector provided by the thrusters, $I(\nu)$ is the inertia force, and $A(\nu)$ is the coefficient matrix of the damping force. R_G is the force/moment vector caused by the static loads in the body coordinate system. M is the mass matrix (including additional mass) and ν is the speed vector in the body coordinate system.

The AUV is presumed symmetrical in three planes, and thus:

$$I(\nu) = \begin{bmatrix} 0 & 0 & 0 & 0 & -Z_{\dot{w}}w & Y_{\dot{v}}v \\ 0 & 0 & 0 & Z_{\dot{w}}w & 0 & -X_{\dot{u}}u \\ 0 & 0 & 0 & -Y_{\dot{v}}v & X_{\dot{u}}u & 0 \\ 0 & -Z_{\dot{w}}w & Y_{\dot{v}}v & 0 & -N_{\dot{r}}r & M_{\dot{q}}q \\ Z_{\dot{w}}w & 0 & -X_{\dot{u}}u & N_{\dot{r}}r & 0 & -K_{\dot{p}}p \\ -Y_{\dot{v}}v & X_{\dot{u}}u & 0 & -M_{\dot{q}}q & K_{\dot{p}}p & 0 \end{bmatrix} \quad (5)$$

$$M = \begin{bmatrix} m - X_{\dot{u}} & 0 & 0 & 0 & 0 & 0 \\ 0 & m - Y_{\dot{v}} & 0 & 0 & 0 & 0 \\ 0 & 0 & m - Z_{\dot{w}} & 0 & 0 & 0 \\ 0 & 0 & 0 & I_x - K_{\dot{p}} & 0 & 0 \\ 0 & 0 & 0 & 0 & I_y - M_{\dot{q}} & 0 \\ 0 & 0 & 0 & 0 & 0 & I_z - N_{\dot{r}} \end{bmatrix} \quad (6)$$

Where $X_{\dot{u}}$, $Y_{\dot{v}}$, $Z_{\dot{w}}$ and $N_{\dot{r}}$ are the first-order hydrodynamic derivatives of the hull, and:

$$R_G = \begin{bmatrix} (W - B_F)s(\theta) \\ -(W - B_F)c(\theta)s(j) \\ -(W - B_F)c(\theta)c(j) \\ -(y_G W - y_B B_F)c(\theta)c(j) + (z_G W - z_B B_F)c(\theta)s(j) \\ (z_G W - z_B B_F)s(\theta) + (x_G W - x_B B_F)c(\theta)c(j) \\ -(x_G W - x_B B_F)c(\theta)s(j) - (y_G W - y_B B_F)s(\theta) \end{bmatrix} \quad (7)$$

where B_F is the buoyancy and W is the gravity of the AUV. x_B, y_B, z_B indicate the buoyancy center.

The AUV motion model shown in Equation (4) derives the following properties.

Property I. The mass matrix M is positive definite and symmetric. $M = M^T > 0$ and $0 < \lambda_{min}(M) \leq \|M\| < \lambda_{max}(M)$. $\lambda_{min}(M)$ and $\lambda_{max}(M)$ are the minimum and maximum of M. $\|M\|$ is the modulus of M.

Property II. $\dot{M} - 2I(\nu)$ is anti-symmetric. $x^T(\dot{M} - 2I(\nu))x = 0$.

Property III. $A(\nu)$ is a positive definite matrix. $A(\nu) > 0$.

Property IV. The uncertainty of the model is assumed to be determined with the upper bound based on a known function, with ^ standing for the estimation matrix, ~ standing for the estimation deviation matrix, and Δ standing for the upper limit of the modulus of the estimation deviation matrix; hence:

$$\begin{aligned} \|\tilde{M}\| &= \|M - \hat{M}\| \leq \Delta M < \infty \\ \|\tilde{I}(\nu)\| &= \|I(\nu) - \hat{I}(\nu)\| \leq \Delta I(\nu) < \infty \\ \|\tilde{A}(\nu)\| &= \|A(\nu) - \hat{A}(\nu)\| \leq \Delta A(\nu) < \infty \\ \|\tilde{R}_G\| &= \|R_G - \hat{R}_G\| \leq \Delta R_G < \infty \end{aligned}$$

4. Motion Controller

4.1. S-Plane Controller

Based on the combination of PD control structure with fuzzy control theory, the classic S-plane method is very effective with a simple structure and few parameters, which makes it rather practical in engineering applications. Its practicability in AUV motion control has been soundly verified in plenty of field trials [40].

The S-plane controller functions are based on the mathematical model [28]:

$$\begin{cases} O = \frac{2}{1+\exp(-k_e e - k_{\dot{e}}\dot{e})} - 1 \\ T_c = T_{max} O \end{cases} \tag{8}$$

O is the normalized output of control. exp () means the exponential function. e is the normalized deviation and $\dot{e}$ is its variation rate, where $e = (u_d - u\ y_d - y\ z_d - z\ \phi_d - \phi\ \theta_d - \theta\ \psi_d - \psi)^T$ and $\dot{e} = (a_u\ v\ w\ p\ q\ r)^T$. k_e and $k_{\dot{e}}$ are, respectively, the control parameters of e and $\dot{e}$. T_c is the expected thrusting force (or torque) calculated by the control algorithm. T_{max} is the maximum thrusting force (or torque) that the AUV can provide.

It can be seen that the static load of the AUV is not considered in the mathematical model. The existence of considerable static loads will surely influence the control quality. In addition, the damping force that varies with the increasing speed is also prone to cause difficulties in high-precision motion control.

4.2. Improved S-Plane Controller

With the reference to the sliding mode variable structure and S-plane function, the design of the improved S-plane control method derives from Equation (4), where the situational static load and high speed are initially considered. $K_2 v_t + K_1 e$ is used to indicate high-speed movement. The item $K_3 \mathrm{sgn}(S_E)$ reflects the concept of sliding mode surface. $K_{Es} \cdot f(S_E)$ means the reference to the S-plane function. $\hat{R}_G$ stands for the situational static load. Therefore, an improved S-plane control method is designed by bringing in the sliding mode variable structure and giving thought to the situational static load and high-speed movement.

$$\begin{cases} F_T = \hat{M}(K_2 \dot{v}_t + K_1 \dot{e}) + (\hat{I}(v) + \hat{A}(v))(K_2 v_t + K_1 e) + K_3 \mathrm{sgn}(S_E) + \hat{R}_G + K_{Es} \cdot f(S_E) \\ f(S_E) = \frac{2}{1+\exp(S_E)} - 1 \end{cases} \tag{9}$$

F_T is the control output matrix of the improved S-plane controller and $\hat{M}$ the estimated mass matrix. $\dot{v}_t$ indicates the derivative of the desired speed vector. K_1 and K_2 are positive definite diagonal gain matrices and K_3 is a positive definite diagonal matrix. sgn() means sign function. $\hat{I}(v)$ stands for the estimated matrix of inertia force coefficients, $\hat{A}(v)$ stands for the estimated matrix of damping force coefficients, and $\hat{R}_G$ stands for the estimated static load vector, including force and torque. K_{Es} means positive definite diagonal matrix. $f(S_E)$ is the chosen sigmoid function and S_E is similar to the sliding mode surface in sliding mode variable structure control.

In Equation (9), e and S_E are defined as:

$$\begin{cases} e = T\tilde{\xi} \\ S_E = K_1 e + K_2 \dot{e} = K_1 e + K_2 v_t - K_2 v \end{cases} \tag{10}$$

where $\xi = [\xi\ \eta\ \varsigma\ \phi\ \theta\ \psi]$ indicates the attitude angle and position matrix of the AUV in the geodetic coordinate system. $\tilde{\xi} = [\xi_t - \xi\ \eta_t - \eta\ \varsigma_t - \varsigma\ \phi_t - \phi\ \theta_t - \theta\ \psi_t - \psi]$ is the estimated deviation matrix of ξ.

To ensure the stability of the proposed control model, the Lyapunov function shown in Equation (11) is constructed:

$$V_L = \frac{1}{2} S_E{}^T M S_E \tag{11}$$

The derivation of V_L in Equation (11) leads to:

$$\dot{V}_L = \frac{1}{2} S_E{}^T \dot{M} S_E + S_E{}^T M \dot{S}_E \tag{12}$$

With respect to the bounded K_1 and K_2, the derivation of S_E in Equation (10) leads to:

$$\begin{aligned}\dot{S}_E &= \dot{K}_2 v_t + K_2\dot{v}_t - \dot{K}_2 v - K_2\dot{v} + \dot{K}_1 e + K_1\dot{e}\\ &= K_2\dot{v}_t - K_2\dot{v} + K_1\dot{e}\end{aligned} \tag{13}$$

When equivalently transformed, Equation (4) is changed into Equations (14)–(16).

$$-M\dot{v} = I(\nu)\nu + A(\nu)\nu + R_G - F_T \tag{14}$$

$$-M^{-1}M\dot{v} = M^{-1}[I(\nu)\nu + A(\nu)\nu + R_G - F_T] \tag{15}$$

$$-\dot{v} = M^{-1}[I(\nu)\nu + A(\nu)\nu + R_G - F_T] \tag{16}$$

When Equation (13) is substituted with Equation (16):

$$\dot{S}_E = K_2\dot{v}_t + K_1\dot{e} + K_2M^{-1}[(I(\nu) + A(\nu))\nu + R_G - F_T] \tag{17}$$

When Equation (12) is substituted with Equation (17):

$$\begin{aligned}\dot{V}_L &= \tfrac{1}{2}S_E^{\,T}\dot{M}S_E + S_E^{\,T}M\{K_2M^{-1}[(I(\nu) + A(\nu))\nu + R_G - F_T] + K_2\dot{v}_t + K_1\dot{e}\}\\ &= \tfrac{1}{2}S_E^{\,T}\dot{M}S_E + S_E^{\,T}\{K_2[(I(\nu) + A(\nu))\nu + R_G - F_T] + M(K_2\dot{v}_t + K_1\dot{e})\}\end{aligned} \tag{18}$$

When equivalently transformed, Equation (10) is changed into:

$$v = K_2^{-1}K_1e + v_t - K_2^{-1}S_E \tag{19}$$

When Equation (18) is substituted with Equation (19):

$$\begin{aligned}\dot{V}_L &= \tfrac{1}{2}S_E^{\,T}\dot{M}S_E + S_E^{\,T}\{K_2[(I(\nu) + A(\nu))(K_2^{-1}K_1e + v_t - K_2^{-1}S_E) + R_G - F_T] + M(K_2\dot{v}_t + K_1\dot{e})\}\\ &= \tfrac{1}{2}S_E^{\,T}\dot{M}S_E + S_E^{\,T}[K_2(I(\nu) + A(\nu))(K_2^{-1}K_1e + v_t) - K_2(I(\nu) + A(\nu))K_2^{-1}S_E + M(K_2\dot{v}_t + K_1\dot{e}) + R_G - F_T]\\ &= \tfrac{1}{2}S_E^{\,T}\dot{M}S_E + S_E^{\,T}[M(K_2\dot{v}_t + K_1\dot{e}) + K_2(I(\nu) + A(\nu))(v_t + K_2^{-1}K_1e) + R_G - F_T] - S_E^{\,T}K_2(I(\nu) + A(\nu))K_2^{-1}S_E\end{aligned} \tag{20}$$

According to Property II, Equation (20) can be simplified as:

$$\begin{aligned}\dot{V}_L &= [\tfrac{1}{2}S_E^{\,T}\dot{M}S_E - S_E^{\,T}K_2(I(\nu) + A(\nu))K_2^{-1}S_E] + S_E^{\,T}[M(K_2\dot{v}_t + K_1\dot{e}) + K_2(I(\nu) + A(\nu))(v_t + K_2^{-1}K_1e) + R_G - F_T]\\ &= \tfrac{1}{2}[S_E^{\,T}\dot{M}S_E - S_E^{\,T}K_2(2I(\nu))K_2^{-1}S_E - S_E^{\,T}K_2(2A(\nu))K_2^{-1}S_E] + S_E^{\,T}[M(K_2\dot{v}_t + K_1\dot{e}) + K_2(I(\nu) + A(\nu))(v_t + K_2^{-1}K_1e) + R_G - F_T]\\ &= \tfrac{1}{2}[S_E^{\,T}(\dot{M} - 2I(\nu))S_E] - S_E^{\,T}A(\nu)S_E + S_E^{\,T}[M(K_2\dot{v}_t + K_1\dot{e}) + K_2(I(\nu) + A(\nu))(v_t + K_2^{-1}K_1e) + R_G - F_T]\\ &= -S_E^{\,T}A(\nu)S_E + S_E^{\,T}[M(K_2\dot{v}_t + K_1\dot{e}) + K_2(I(\nu) + A(\nu))(v_t + K_2^{-1}K_1e) + R_G - F_T]\end{aligned} \tag{21}$$

When Equation (21) is substituted with Equation (9):

$$\begin{aligned}\dot{V}_L &= -S_E^{\,T}A(\nu)S_E + S_E^{\,T}[M(K_2\dot{v}_t + K_1\dot{e}) + (I(\nu) + A(\nu))(K_2v_t + K_1e) + R_G - \hat{M}(K_2\dot{v}_t + K_1\dot{e}) - (\hat{I}(\nu) + \hat{A}(\nu))(K_2v_t\\ &\quad + K_1e) - \hat{R}_G - K_{Es}\cdot f(S_E) - K_3\cdot \mathrm{sgn}(S_E)]\\ &= -S_E^{\,T}A(\nu)S_E - S_E^{\,T}K_{Es}\left[\frac{2}{1+\exp(S_E)} - 1\right] - S_E^{\,T}K_3\mathrm{sgn}(S_E) + S_E^{\,T}\left[\tilde{M}(K_2\dot{v}_t + K_1\dot{e}) + \left(\tilde{I}(\nu) + \tilde{A}(\nu)\right)(K_2v_t + K_1e) + \tilde{R}_G\right]\end{aligned} \tag{22}$$

According to Property I and III, and with the consideration of bounded K_{Es} in Equation (9), $\dot{V}_L$ in Equation (22) is bounded and:

$$\dot{V}_L \le -\lambda_{min}(A(\nu))\|S_E\|^2 - \lambda_{min}(K_{Es})\|S_E\| - \lambda_{min}(K_3)\|S_E\| + \left\|S_E^{\,T}\left[\tilde{M}(K_2\dot{v}_t + K_1\dot{e}) + \left(\tilde{I}(\nu) + \tilde{A}(\nu)\right)(K_2v_t + K_1e) + \tilde{R}_G\right]\right\| \tag{23}$$

where $\lambda_{min}(A(\nu))$, $\lambda_{min}(K_{Es})$, and $\lambda_{min}(K_3)$ are the minimum eigenvalues of their corresponding matrix. $\|S_E\|$ is the modulus of S_E.

According to Property IV and triangle inequality, Equation (23) can be changed into:

$$\begin{aligned}\dot{V}_L &\le -\lambda_{min}(A(\nu))\|S_E\|^2 - \lambda_{min}(K_{Es})\|S_E\| - \lambda_{min}(K_3)\|S_E\| + [\|\tilde{M}\|\|K_2\dot{v}_t + K_1\dot{e}\| + (\|\tilde{I}(\nu)\| + \|\tilde{A}(\nu)\|)\|K_2v_t + K_1e\| + \|\tilde{R}_G\|]\|S_E\|\\ &\le -\lambda_{min}(A(\nu))\|S_E\|^2 - \lambda_{min}(K_{Es})\|S_E\| - \lambda_{min}(K_3)\|S_E\| + [\Delta M\|K_2\dot{v}_t + K_1\dot{e}\| + (\Delta I(\nu) + \Delta A(\nu))\|K_2v_t + K_1e\| + \Delta R_G]\|S_E\|\end{aligned} \tag{24}$$

where ΔM, $\Delta I(\nu)$, $\Delta A(\nu)$, and ΔR_G are, respectively, the upper limits of the modulus of the estimated deviation matrix.

Based on the appropriate K_{Es} and K_3, when Equation (24) satisfies $\dot{V}_L < 0$, K_{Es} and K_3 satisfies:

$$[\lambda_{min}(K_{Es}) + \lambda_{min}(K_3)] \geq [\Delta M \|K_2\, \dot{v}_t + K_1\, \dot{e}\| + (\Delta I(\nu) + \Delta A(\nu))\|K_2 v_t + K_1 e\| + \Delta R_G] \tag{25}$$

The sliding mode surface of the improved S-plane controller is defined as:

$$S_E = K_1 e + K_2\, \dot{e} \tag{26}$$

where K_1 is the matrix of control parameters over deviation and K_2 is the matrix of control parameters over deviation variation rate.

At the time of high-speed movement, the AUV is controlled with speed in the surge direction while it is controlled with the position in the other degrees of freedom. For this reason, the control inputs in the surge direction are speed deviation and accelerated speed, while the inputs in the other degrees of freedom are the deviation and its variation rate of angle or position.

K_1 and K_2 are expressed as:

$$\begin{cases} K_1 = \mathrm{diag}(k_1, k_2, k_3, k_4, k_5, k_6) \\ K_2 = \mathrm{diag}(k_{21}, k_{22}, k_{23}, k_{24}, k_{25}, k_{26}) \end{cases} \tag{27}$$

The model of the improved S-plane controller is finalized as:

$$\begin{cases} F_{Ti} = \hat{M}_i(K_{2i}\, \dot{v}_{ti} + K_{1i}\, \dot{e}_i) + (\hat{I}(\nu)_i + \hat{A}(\nu)_i)(K_{2i} v_{ti} + K_{1i} e_i) + K_{3i} \cdot \mathrm{sgn}(S_{Ei}) + \hat{R}_{Gi} + K_{Esi} \cdot f(S_{Ei}) \\ f(S_{Ei}) = \frac{2}{1+\exp(S_{Ei})} - 1 \end{cases} \tag{28}$$

where F_{Ti}, $\dot{v}_{ti}$, v_{ti}, $\dot{e}_i$, e_i, $\hat{R}_{Gi}$, and S_{Ei} are the ith component of their corresponding vector. $\hat{M}_i$, K_{1i}, K_{2i}, K_{3i}, $\hat{I}(\nu)_i$, and $\hat{A}(\nu)_i$ are the element at the ith row and the ith column in their corresponding matrix.

4.3. Thrust Allocation Strategy

Due to the slender and symmetric outline of the research object, the movement in the roll direction is typically ignored [41]. In accordance with the configuration of the propelling system as illustrated in Figure 1, f_1 and f_2 are left thruster and right thruster, respectively, with l_1 and l_2 as the arm of force and α as the included angle to the center of the research object. f_3 and f_4 are the thrusting forces from side thrusters at the bow and the stern, with l_3 and l_4 being the arm of force to the center of the research object. f_5 and f_6 are the thrusting forces from the vertical thrusters at the bow and the stern, with l_5 and l_6 being the arm of force to the center of the research object. f_7 and f_8 are the thrusting forces from the upward thruster and downward thruster, with l_7 and l_8 being the arm of force and β is the included angle to the center of the research object. The forces required in the five degrees of freedom and the thrusting forces provided by the eight thrusters follow the relationship below.

$$\begin{bmatrix} F_x \\ F_y \\ F_z \\ M_y \\ M_z \end{bmatrix} = \begin{bmatrix} c(\alpha) & c(\alpha) & 0 & 0 & 0 & 0 & c(\beta) & c(\beta) \\ -s(\alpha) & s(\alpha) & 1 & 1 & 0 & 0 & 0 & 0 \\ 0 & 0 & 0 & 0 & 1 & 1 & -s(\beta) & s(\beta) \\ 0 & 0 & 0 & 0 & -l_5 & l_6 & -l_7 & l_8 \\ -l_1 & l_2 & l_3 & -l_4 & 0 & 0 & 0 & 0 \end{bmatrix} \begin{bmatrix} f_1 \\ f_2 \\ f_3 \\ f_4 \\ f_5 \\ f_6 \\ f_7 \\ f_8 \end{bmatrix} \tag{29}$$

It can be inferred from Equation (29) that the control over thrusters is redundant. For the purpose of simplified calculation, constraints are introduced for thrust allocation in accordance with practical task requirements.

In regard to the significant drag force in the sway direction, the heading angle is adjusted to achieve path tracking at the time of a high-speed cruise.

$$F_y = 0 \tag{30}$$

In the case of high-speed movement, since the movement in the heave direction produces great hydrodynamic force, the trimming angle is adjusted to achieve depth control. For this reason, there is no requirement for external force in the heave direction.

$$F_z = 0 \tag{31}$$

Considering the deduction of thrusting force, the four slot thrusters stop service at the time of high-speed movement; hence:

$$\begin{cases} f_3 \cdot l_3 - f_4 \cdot l_4 = 0 \\ f_5 \cdot l_5 - f_6 \cdot l_6 = 0 \end{cases} \tag{32}$$

The thrust allocation strategy for control in the surge direction is developed according to the respective maximum thrust provided by the upward thruster, downward thruster, left thruster, and right thruster, with τ as the ratio of the maximum force between the upward and downward thrusters, as well as the left and right thrusters.

$$\tau = \frac{(f_7 + f_8)c(\beta)}{(f_1 + f_2)c(\alpha)} \tag{33}$$

Equations (29)–(33) jointly lead to the force provided by each thruster at the time of high-speed navigation.

In position control, the AUV proceeds stably and potent forces can be provided by the slot thrusters. Therefore, the constraints are added as follows. The upward and downward thrusters are shut; thus, $f_7 = f_8 = 0$. The left and right thrusters provide force in the surge direction and steering moment. The side thrusters at the bow and the stern provide force in the sway direction. The vertical thrusters at the bow and the stern provide force in the heave direction. Accordingly, Equation (29) is simplified as:

$$\begin{bmatrix} F_x \\ F_y \\ F_z \\ M_z \end{bmatrix} = \begin{bmatrix} c(\alpha) & c(\alpha) & 0 & 0 & 0 & 0 \\ -s(\alpha) & s(\alpha) & 1 & 1 & 0 & 0 \\ 0 & 0 & 0 & 0 & 1 & 1 \\ -l_1 & l_2 & l_5 & -l_6 & 0 & 0 \end{bmatrix} \begin{bmatrix} f_1 \\ f_2 \\ f_5 \\ f_6 \\ f_7 \\ f_8 \end{bmatrix} \tag{34}$$

Equations (35) and (36) are the constraints.

$$f_3 \cdot l_3 - f_4 \cdot l_4 = 0 \tag{35}$$

$$f_5 \cdot l_5 - f_6 \cdot l_6 = 0 \tag{36}$$

Equations (34)–(36) jointly lead to the force provided by each thruster in the case of AUV position control.

5. Simulation Experiments and Analysis

The static loads of the AUV may vary with the sensors or facilities it carries as required in different operation tasks. Based on the experience of sea trials and analysis of the static loads in different situations, it is concluded that the possible static loads range between 20 N and 150 N. Given the applicability of the assumptions, simulation experiments were carried out with 20 N, 100 N, and 150 N selected as the static loads.

Contrastive simulation experiments were considered to be an effective way to highlight the superiority of the improved S-plane method over the classic S-plane method under the same static load. In the improved S-plane controller, $\hat{R}_{G3}$ was 50 N, with K_1 = diag (0.8,1.0,2.0,1.0,1.2,0.4) , K_2 = diag (0.4,0.5,1.0,0.5,0.6,0.2) , K_3 = diag (1.0,1.5,2.0,1.5,1.0,0.5), and K_3 = diag (120,120,150,100,100,500).

It is noteworthy that the static load of the AUV varies in different situations, it($\hat{R}_{G3} = 50$) was set differently from that in the simulator so as to verify the robustness of the improved controller. Moreover, different current conditions were also included to examine the robustness of the controller to external disturbances.

Simulation I. In the contrastive simulation experiment, the static load was set to 20 N, with no current. The AUV was stationary on the water's surface at the beginning. The desired velocity in the surge direction was 3.0 m/s and the desired depth was 5 m. The contrastive simulation results are shown in Figure 5.

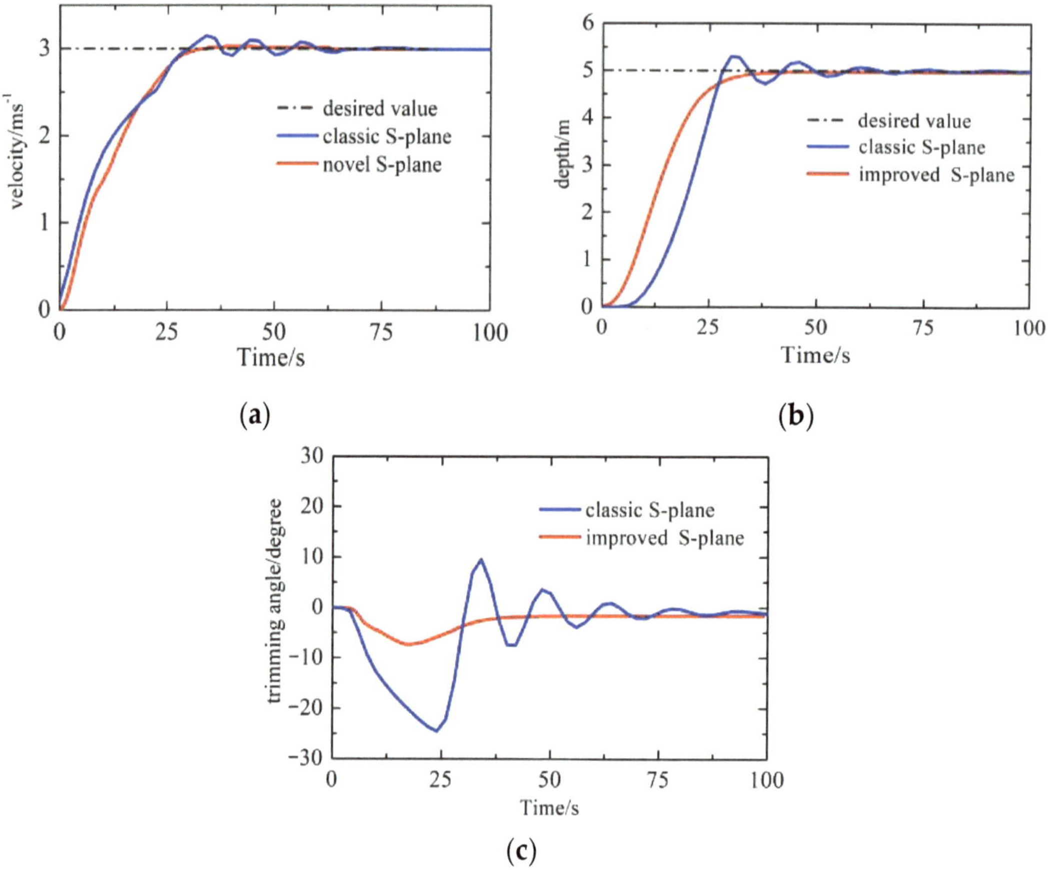

Figure 5. Contrastive control results with a static load of 20 N. (**a**) Results of velocity control; (**b**) results of depth control; (**c**) changes of trimming angle.

It can be seen from the results that both the classic and improved S-plane controllers reached the desired value with no steady-state error. With the classic S-plane controller, a maximum overshoot of 0.15 m/s and significant fluctuations can be seen in the velocity control in Figure 5a, and a maximum overshoot of 0.35 m together with oscillations from 30–70 s in depth control in Figure 5b. In contrast, based on the improved S-plane controller, there was barely any overshoot or oscillation in the velocity control and the depth control. Additionally, the system based on the improved S-plane method produced a much milder influence on the trimming angle in contrast with the drastic changes caused by the classic S-plane controller, as shown in Figure 5c.

Simulation II. The static load was set to 100 N, with a current velocity of 0.25 m/s and a direction of 0°. The rest of the conditions were the same as in Simulation I. The simulation results are shown in Figure 6.

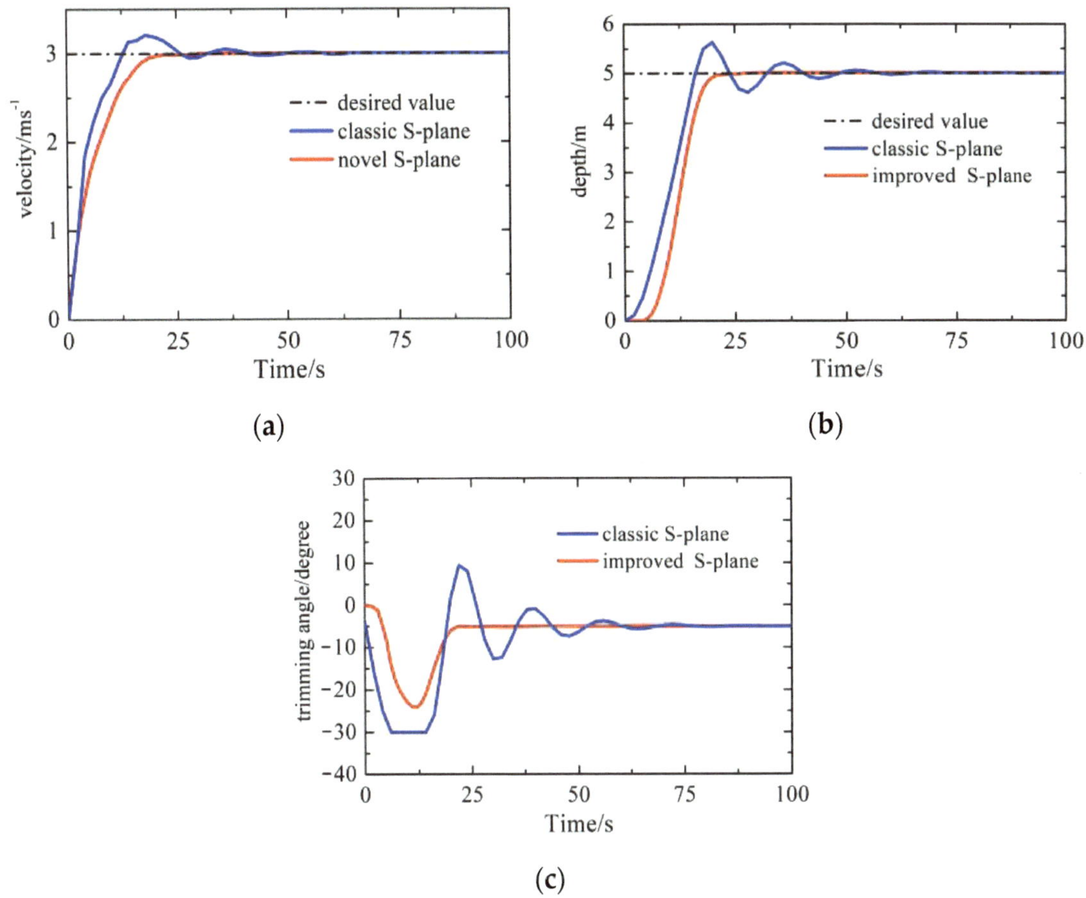

Figure 6. Contrastive control results with a static load of 100 N. (**a**) Results of velocity control; (**b**) results of depth control; (**c**) changes of trimming angle.

With a static load of 100 N, the classic S-plane controller showed a maximum overshoot of 0.20 m/s in the velocity control in Figure 6a and significant oscillations of over 0.5 m from 18–65 s in the depth control in Figure 6b. However, the improved S-plane controller provided admirable control results with almost no overshoot and reached the stability state in a smooth way. Although both controllers caused impacts on the trimming angle, the influence caused by the classic S-plane controller lasted almost 70 s during the control process, while that caused by the improved S-plane controller immediately cleared off within the first 25 s during the control period, as shown in Figure 6c.

Simulation III The static load was set to 150 N, with a current velocity of 0.25 m/s and a direction of 45°. The rest of the conditions were the same as in Simulation I. The simulation results are displayed in Figure 7.

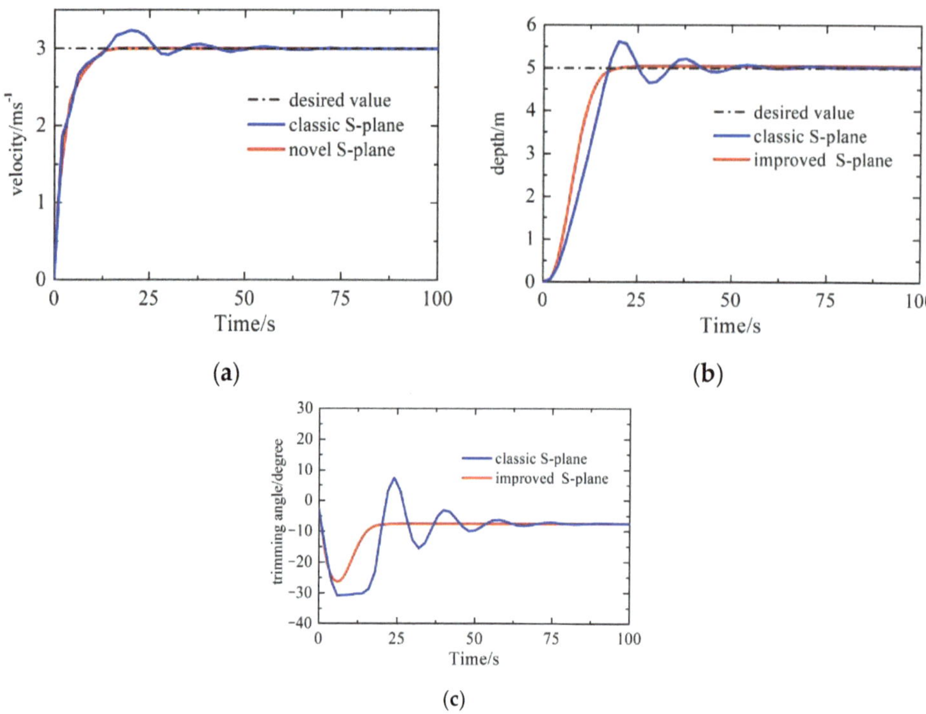

Figure 7. Contrastive control results with a static load of 150 N. (**a**) Results of velocity control; (**b**) results of depth control; (**c**) changes of trimming angle.

With a much greater static load this time, the classic S-plane controller produced a maximum overshoot of 0.25 m/s in the velocity control in Figure 7a and significant oscillations of over 0.6 m from 20–62 s in the depth control in Figure 7b. However, the improved S-plane controller provided marvelous results with a much shorter time reaching the stability state than that with a static load of 100 N. As shown in Figure 7c, the improved S-plane controller enabled the trimming angle to recover within a much shorter time than that of the classic S-plane controller. The contrastive simulation experiments have verified the robustness and superiority of the improved S-plane controller and justified the feasibility of the sliding mode variable structure.

The controller has no prior knowledge of the currents, so the currents are treated as external interferences in the controller. As the results show, the control quality presents barely any difference with and without the existence of currents. This means the controller is capable of dealing with external interference, which verifies its strong robustness. In addition, the smooth responses from the improved S-plane method benefit from the strategies developed to solve the chattering effect. Based on the strategies of slow start–smooth running–slow stop, the small input at the initial stage is provided to avoid abrupt state change. Moreover, the outputs of the symbolic function are also saturated. These measures jointly help to reduce the chattering effect of the sliding mode variable structure to a large extent.

6. Field Trials and Analysis

Field trials were conducted with the purpose of verifying the control performance, as well as the stability and reliability of the improved S-plane controller in continuous operation. Motion control is the foundation for all the other functions, so contrastive field trials of AUV motion control were carried out between the classic and the improved S-plane methods, and the long-range cruise and path point following were completed to examine the improved S-plane method.

The contrastive controls over velocity and heading angle in the horizontal surface were first carried out. Then, contrastive controls over velocity and depth in the vertical plane were conducted to verify the effectiveness of the proposed controller, followed by the long-range cruise and trial of path point following. The surroundings and environment of the sea trials are shown in Figure 8.

Figure 8. Environment of the sea trials.

6.1. Contrastive Trials of Velocity and Heading Control on a Horizontal Surface

On an equal basis, the AUV was stationary on the water surface at the beginning, with an initial heading angle of 0°. The desired velocity was 2.0 m/s in the surge direction and the desired heading angle was 60°. The results of velocity and heading control based on the classic S-plane method and the improved S-plane method are shown below, together with a detailed view of the control values from 200–250 s during the control process.

As shown in Figure 9a, the velocity control curve based on the classic S-plane method is satisfactory on the whole. It takes the system approximately 100 s to reach stability. In the detailed view shown in Figure 9b, oscillations go up and down around the desired value with the maximum offset over 0.112 m/s. In contrast, the system based on the improved S-plane method reaches stability within a much shorter time of about 60 s, and as shown in Figure 10b, the bias is smaller than that based on the classic S-plane method.

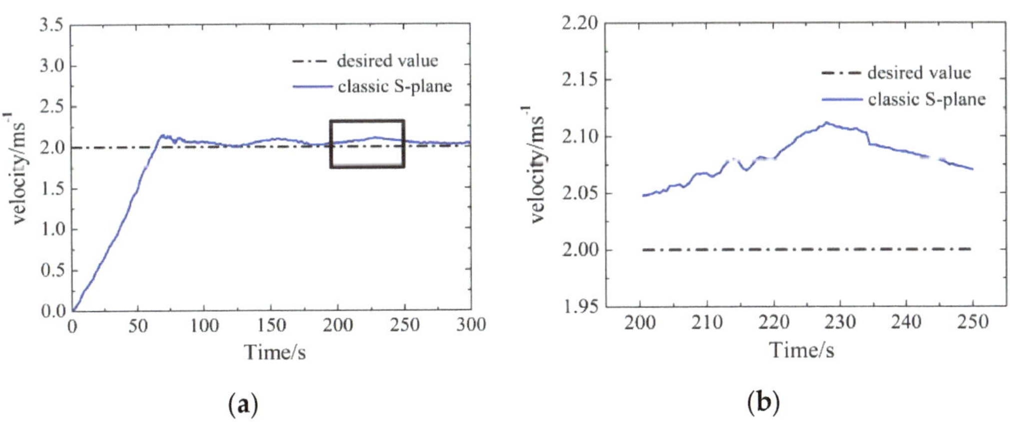

Figure 9. Results of velocity control in horizontal surface based on the classic S-plane method. (**a**) Results of velocity control; (**b**) detailed local values.

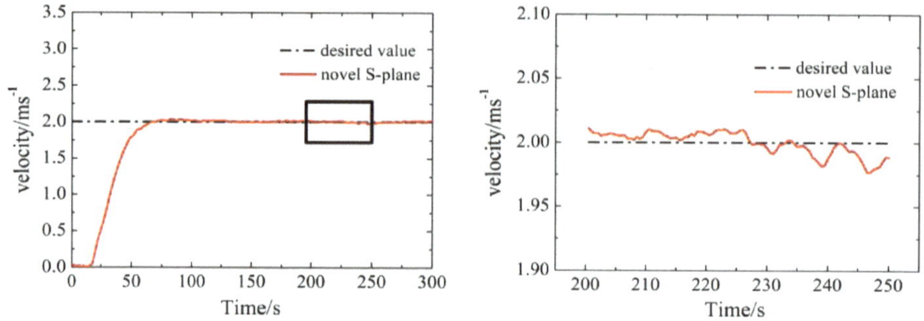

(a) (b)

Figure 10. Results of velocity control in horizontal surface based on the improved S-plane method. (**a**) Results of velocity control; (**b**) detailed local values.

As shown in Figures 11 and 12, the improved S-plane method proves its effectiveness and superiority again in the heading control with barely any fluctuations or overshoot. Although both methods have caused deviations, the improved S-plane method excels compared to the classic method with a smaller and milder offset.

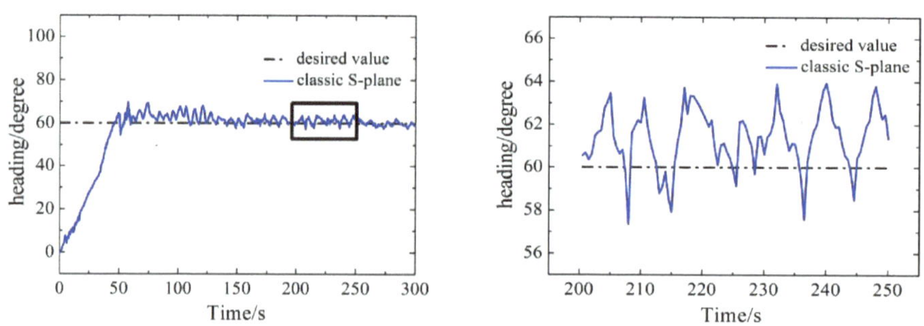

(a) (b)

Figure 11. Results of heading control based on the classic S-plane method. (**a**) Results of heading control; (**b**) detailed local values.

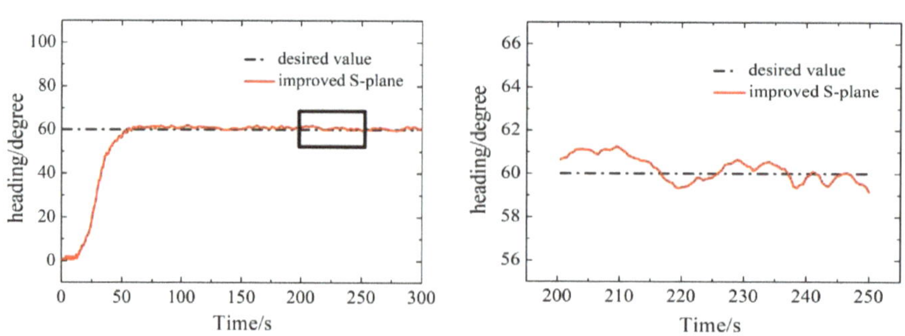

(a) (b)

Figure 12. Results of heading control based on the improved S-plane method. (**a**) Results of heading control; (**b**) detailed local values.

With maximum overshoot, standard deviation, and arithmetic mean value, the contrast of the velocity and heading control results based on the two methods ranging from 200–250 s during the control process are listed as shown in Table 1.

Table 1. The contrast of velocity and heading control results with two different methods (200–250 s).

Target	Methods	Maximum Overshoot	Standard Deviation	Arithmetic Mean Value
Velocity (2.0 m/s)	classic S-plane	0.112 m/s	0.017 m/s	2.081 m/s
	improved S-plane	0.011 m/s	0.008 m/s	2.000 m/s
Heading (60°)	classic S-plane	3.93°	1.39°	61.38°
	improved S-plane	1.24°	0.56°	60.23°

In the stability state, it can be seen from the parameters that the proposed improved S-plane controller provides admirable control quality and higher control precision in motion control on the horizontal surface, with greater robustness against the time-varying control system. Such superiority is especially noticeable in the heading control.

6.2. Contrastive Trials of Velocity and Depth Control on the Vertical Plane

Again, for the purpose of contrasting the two different methods, sea trials were carried out for velocity and depth control on the vertical plane.

On an equal basis, the velocity control proceeded with the AUV at the depth of 0.5 m for both methods. The velocity control began with an initial velocity of 0 m/s with a desired velocity of 1.5 m/s. In the depth control, the AUV was expected to dive from the water surface down to the desired depth of 1.8 m. The results of velocity and depth control based on the two methods are shown below, together with a detailed view of the control values from 200–250 s during the control process.

As shown in Figures 13 and 14, both methods present admirable performance in velocity control in the vertical plane. The improved S-plane method wins out with a control curve that almost perfectly fits the desired value as shown in Figure 14b.

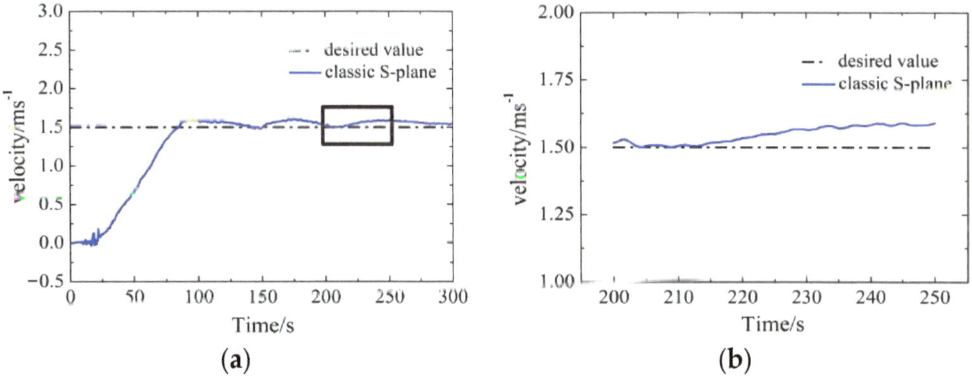

Figure 13. Results of velocity control in vertical plane based on the classic S-plane method. (**a**) Results of velocity control; (**b**) detailed local values.

As shown in Figures 15 and 16, in depth control, it takes both methods a little while to begin to go under the water's surface. When reaching the stability state, the classic S-plane method causes frequent oscillations, while the improved S-plane method enables the system to stabilize at the desired depth with mild oscillations and insignificant overshoot, as shown in Figure 16b.

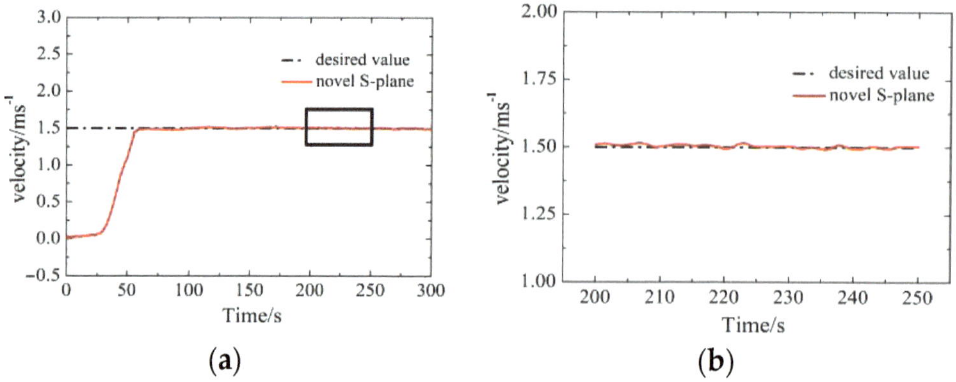

Figure 14. Results of velocity control in vertical plane based on the improved S-plane method. (**a**) Results of velocity control; (**b**) detailed local values.

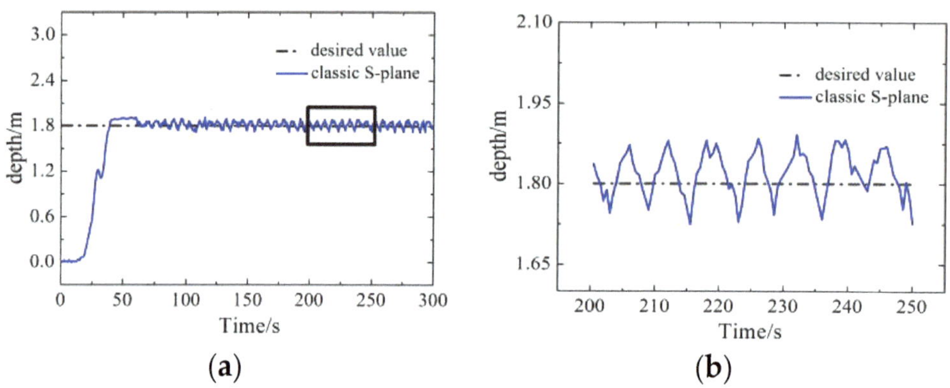

Figure 15. Results of depth control based on the classic S-plane method. (**a**) Results of depth control; (**b**) detailed local values.

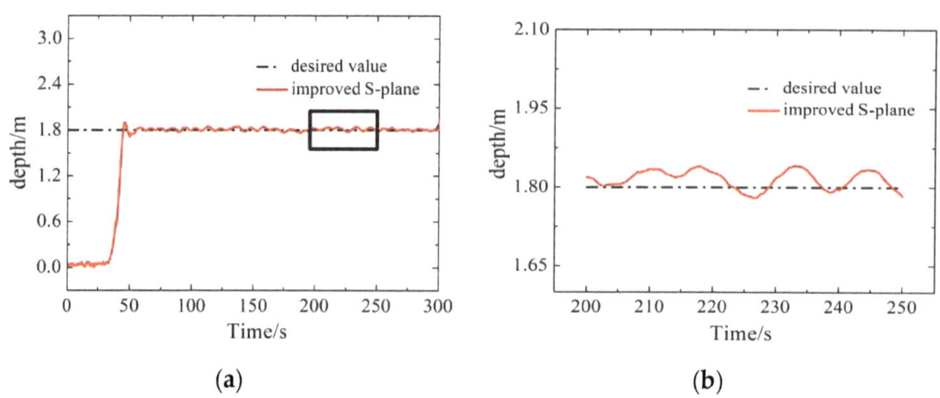

Figure 16. Results of depth control based on the improved S-plane method. (**a**) Results of depth control; (**b**) detailed local values.

With maximum overshoot, standard deviation, and arithmetic mean value, the contrast of the velocity and depth control results based on the two methods ranging from 200–250 s are listed as shown in Table 2.

Table 2. The contrast of velocity and depth control results with two different methods.

Target	Methods	Maximum Overshoot	Standard Deviation	Arithmetic Mean Value
Velocity (1.5 m/s)	classic S-plane	0.089 m/s	0.031 m/s	1.547 m/s
	improved S-plane	0.016 m/s	0.005 m/s	1.503 m/s
Depth (1.8 m)	classic S-plane	0.092 m	0.030 m	1.827 m
	improved S-plane	0.041 m	0.016 m	1.816 m

The improved S-plane method proves its superiority over the classic S-plane method again with a smaller maximum overshoot and standard deviation. Despite almost the same arithmetic mean value from both methods, the improved S-plane method showed much milder deviations in depth control, as shown in Figure 16b, which guarantees the stability of the AUV motion control system. Such superiority is of great significance to real operations with high requirements on control precision.

6.3. Long-Range Cruise

The long-range cruise was carried out to verify the performance, reliability, and stability of the improved S-plane controller. The AUV was expected to cruise along a quadrilateral route of 12 km × 4 km. The cruise lasted for approximately 4 h, with an average speed of 2 m/s and a maximum cruising speed of 2.5 m/s.

It can be seen from the long-range cruise results that the actual route fit the desired route. The improved S-plane controller proved its control competence, especially at the apex of the quadrangle. The detailed route data during the first part of the cruise are shown in Figure 17b. Although the torrential currents caused a few deviations, the offsets were within the permissible range of engineering requirements. The AUV control system functioned normally with no hardware or software failure during the long-range cruise, which verified the stability and reliability of the improved S-plane method.

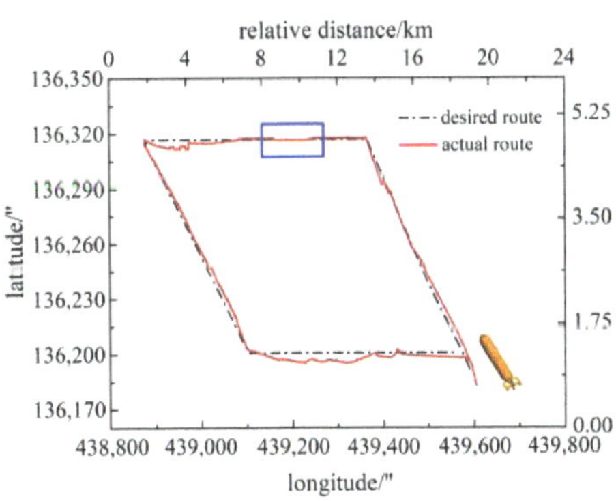

(a)

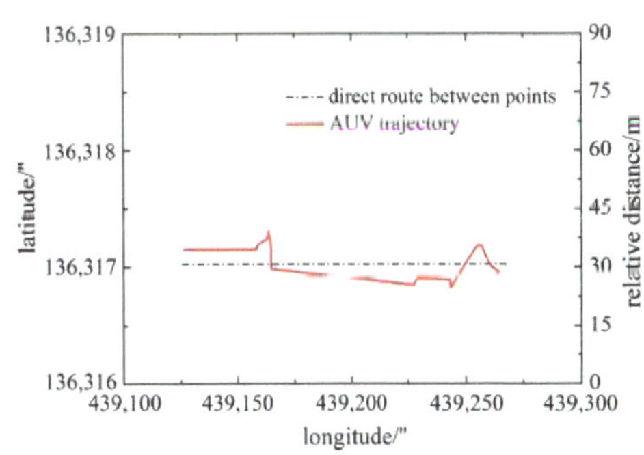

(b)

Figure 17. Results of the long-range cruise. (**a**) The curve of the long-range route; (**b**) detailed route values.

6.4. Path Point Following

The trial of the path point following was conducted to examine the AUV's competence in the search or scanning operations of underwater pipelines or cables based on the improved S-plane controller. The pectinate path was marked with eight points to be covered, as shown in Figure 18, with a long side of approximately 550 m and a short side of 200 m. To test the performance and robustness of the improved S-plane method, the eight points were selected to be non-orthogonal with the longitude or latitude. The AUV was expected to reach the eight points (A–H) in succession in the scanning. It ran along the path at a constant speed of 2.0 m/s. Considering the challenging water conditions, a circle with a radius of 10 m with each marked point as the center was defined as the acceptable range of reach.

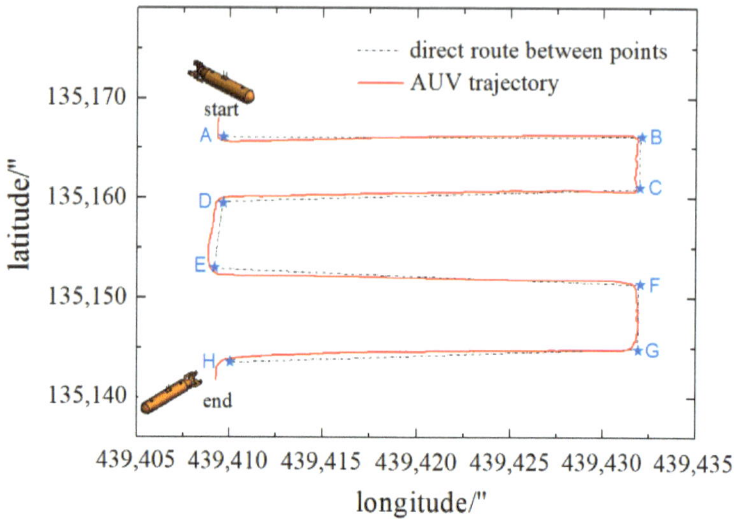

Figure 18. The trajectory of path points following.

Based on the detailed view of reaching the eight points, as shown in Figure 19, and the analysis of the deviations from the connection lines between the points, as shown in Table 3, the AUV was able to reach all eight acceptable circles in a stable way, with a maximum deviation of 9.44 m, which was within the permissible range in engineering. The following trial verified the control accuracy of the improved S-plane method, which could be of practical use in improving the efficiency of sweeping search operations.

Table 3. An analysis of path point following based on the improved S-plane method.

Position	Maximum Overshoot	Mean Deviation
A-B	8.65 m	4.95 m
B-C	7.56 m	5.15 m
C-D	6.29 m	3.63 m
D-E	9.44 m	8.53 m
E-F	7.29 m	3.85 m
F-G	6.97 m	3.86 m
G-H	8.61 m	4.73 m

The control trials in the horizontal plane, vertical plane, long-range cruise, and path point following have all witnessed the superiority and excellence of the improved S-pane method. This is because the novel controller is based on the sliding mode variable structure and takes the situational static load into account, which enhances the method's robustness to external disturbances. In addition, the control items introduced in the improved S-plane model could directly offset the adverse influences from the static load.

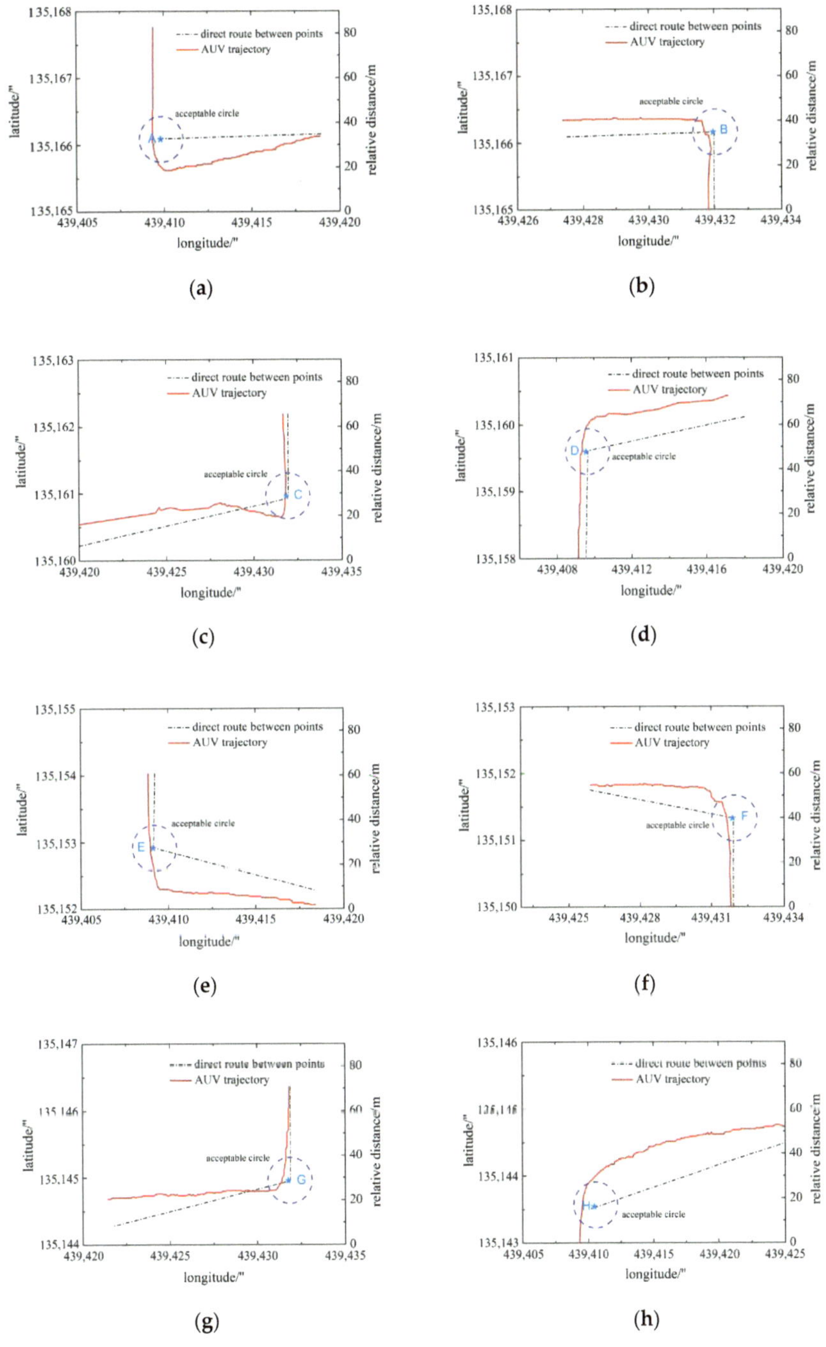

Figure 19. Detailed view of reaching the eight points. (**a**) Detailed view of reaching Point A; (**b**) detailed view of reaching Point B; (**c**) detailed view of reaching Point C; (**d**) detailed view of reaching Point D; (**e**) detailed view of reaching Point E; (**f**) detailed view of reaching Point F; (**g**) detailed view of reaching Point G; (**h**) detailed view of reaching Point H.

7. Conclusions

This paper focuses on an improved S-plane controller designed for high-speed AUV with different static loads, as required in multi-purpose operations. The existing control methods, including the classic S-plane method, can sufficiently deal with low- and medium-speed AUV motion control but they are limited in performance, with a lack of consideration of situational static load and hydrodynamic force as the result of high-speed movement.

On the basis of the classic S-plane control model and giving thoughts to the static load and hydrodynamic forces that change with AUV speed, an improved S-plane controller based on the sliding mode variable structure is developed by taking the static load and hydrodynamic forces as the control items. The stability of the proposed control method is testified by Lyapunov's stability theory.

In order to compare and analyze the influence of situational static loads on the control quality of the classic S-plane controller and the improved S-plane controller, simulation experiments were carried out. Given the commonly used sensors or facilities that the AUV may carry in operations, the static loads of minimum, maximum, and medium weights together with the current disturbances were set in the contrastive simulations. It is especially noteworthy that the situational static load $\hat{R}_G$ was intentionally set to be different from that of the simulator, so as to verify the robustness of the improved S-plane controller. The simulation results show that the system based on the classic S-plane method suffered from overshoot in the velocity control and oscillations in the depth control with drastic fluctuations in the trimming angle as the result of a failure to consider the static load. By contrast, the improved S-plane controller countered the static loads and guaranteed control quality. The contrastive simulation experiments have verified the robustness of the improved S-plane controller.

On the basis of the simulation experiments, field trials were also carried out to further test the performance of the improved S-plane controller. Since motion control is the foundation of all the other functions, it was firstly conducted in velocity, heading, and depth control in the horizontal plane and the vertical plane, respectively, followed by the long-range cruise and path point following to examine the stability and reliability of the improved S-plane method. In the horizontal plane, the improved S-plane controller presented a maximum overshoot and standard deviation of approximately 0.011 m/s and 0.008 m/s in the velocity control, as well as 1.24° and 0.56° in the heading control. In the vertical plane, the improved S-plane controller showed a maximum overshoot and standard deviation of approximately 0.016 m/s and 0.005 m/s, as well as 0.041 m and 0.016 m in the depth control. The precision in AUV high-speed motion control was verified. In addition, the stability of the improved S-plane method was proved during the approximately 4 h long-range cruise with no failures on the hardware and software architectures. The path point following was finally conducted to examine the AUV's competence in the search or scanning operations of underwater pipelines or cables based on the improved S-plane controller. The AUV was able to reach all eight path points, with a maximum deviation of 9.44 m, which was within the permissible range in engineering. The improved S-plane method can be of great use to field tasks that have high expectations on AUV motion control quality, especially in cases of high-speed movement or challenging working environments.

The research object in the study is driven by thrusters, but theoretically, the research findings are also of great value to AUVs with propellers and rudders. Later studies will concentrate on the effectiveness of the improved S-plane controller on AUVs with different propulsion structures or outer dimensions so as to lay the foundation for a wide application of the proposed technology.

Author Contributions: Conceptualization, C.J. and J.W.; methodology, C.J. and L.W.; software, C.J. and G.W.; validation, C.J., B.H. and G.W.; formal analysis, C.J.; investigation, L.W. and J.L.; resources, L.W. and C.J.; data curation, C.J. and J.W.; writing—original draft preparation, C.J.; writing—review and editing, L.W. and C.J.; visualization, B.H.; supervision, L.W. and C.J.; project administration, L.W. and C.J.; funding acquisition, L.W. and J.L. All authors have read and agreed to the published version of the manuscript.

Funding: This research was funded by the National Natural Science Foundation of China (grant numbers 51309148, 51779057, 51709061, and 61803119).

Institutional Review Board Statement: Not applicable.

Informed Consent Statement: Not applicable.

Data Availability Statement: Not applicable.

Conflicts of Interest: The authors declare no conflict of interest.

References

1. Kazantsev, Y.V.; Glazyrin, V.; Khalyasmaa, A.I.; Shayk, S.M.; Kuparev, M.A. Advanced Algorithms in Automatic Generation Control of Hydroelectric Power Plants. *Mathematics* **2022**, *10*, 4809. [CrossRef]
2. Kolsi, L.; Al Dahidi, S.; Kamel, S.; Aich, W.; Boubaker, S.; Nidhal, B.K. Prediction of Solar Energy Yield Based on Artificial Intelligence Techniques for the Ha'il Region, Saudi Arabia. *Sustainability* **2022**, *15*, 774. [CrossRef]
3. Sofos, F.; Papakonstantinou Christos, G.; Valasaki, M.; Karakasidis Theodoros, E. Fiber-Reinforced Polymer Confined Concrete: Data-Driven Predictions of Compressive Strength Utilizing Machine Learning Techniques. *Appl. Sci.* **2022**, *13*, 567. [CrossRef]
4. Karimi, H. Computer simulation and modeling the metal to insulating transition of liquid mercury via pair, empirical, and many-body potentials. *J. Mol. Model.* **2022**, *28*, 377. [CrossRef]
5. Gašparović, B.; Lerga, J.; Mauša, G.; Ivašić, K.M. Deep Learning Approach for Objects Detection in Underwater Pipeline Images. *Appl. Artif. Intell.* **2022**, *36*, 2146853. [CrossRef]
6. Cambridge, M. The application of inspection and monitoring in the reduction of risk for mine tailings dams. *Proc. Inst. Civ. Eng. Geotech. Eng.* **2022**, *175*, 00248. [CrossRef]
7. Casanova, E.; Knowles Timothy, D.J.; Bayliss, A.; Walton-Doyle, C.; Barclay, A.; Evershed, R.P. Compound-specific radiocarbon dating of lipid residues in pottery vessels: A new approach for detecting the exploitation of marine resources. *J. Archaeol. Sci.* **2022**, *137*, 105528. [CrossRef]
8. Al Shafatul Islam Subad, R.; Saikot, M.M.H.; Park, K. Soft Multi-Directional Force Sensor for Underwater Robotic Application. *Sensors* **2022**, *22*, 3850. [CrossRef]
9. Tholen, C.; El Mihoub Tarek, A.; Nolle, L.; Zielinski, O. Artificial Intelligence Search Strategies for Autonomous Underwater Vehicles Applied for Submarine Groundwater Discharge Site Investigation. *J. Mar. Sci. Eng.* **2022**, *10*, 7. [CrossRef]
10. Liwei, S.; Yao, H.; Shuxiang, S.; Shuxiang, G.; Huiming, X.; Xihuan, H.; Yu, L.; Zhan, C.; Zan, L.; Debin, X. A Fuzzy PID Algorithm for a Novel Miniature Spherical Robots with Three-dimensional Underwater Motion Control. *J. Bionic Eng.* **2020**, *17*, 959–969.
11. Hasan, M.W.; Abbas, N.H. Disturbance Rejection for Underwater robotic vehicle based on adaptive fuzzy with nonlinear PID controller. *ISA Trans.* **2022**, *130*, 360–376. [CrossRef] [PubMed]
12. Keymasi, K.A.; Bahrami, S. Finite-time sliding mode control of underwater vehicles in 3D space. *Trans. Inst. Meas. Control.* **2022**, *44*, 3215–3228. [CrossRef]
13. Zhilenkov, A.; Chernyi, S.; Firsov, A. Autonomous Underwater Robot Fuzzy Motion Control System with Parametric Uncertainties. *Designs* **2021**, *5*, 24. [CrossRef]
14. Cruz, U.C.; Terrile, S.; Barrientos, A. Soft Underwater Robot Actuated by Shape Memory Alloys "JellyRobcib" for Path Tracking through Fuzzy Visual Control. *Appl. Sci.* **2020**, *10*, 7160. [CrossRef]
15. Keymasi, K.A.; Haghjoo, M. Adaptive passivity-based control of an autonomous underwater vehicle. *Proc. Inst. Mech. Eng. Part C J. Mech. Eng. Sci.* **2022**, *236*, 10563–10572. [CrossRef]
16. Xue, G.; Liu, Y.; Shi, Z.; Guo, L.; Li, Z. Research on Trajectory Tracking Control of Underwater Vehicle Manipulator System Based on Model-Free Adaptive Control Method. *J. Mar. Sci. Eng.* **2022**, *10*, 652. [CrossRef]
17. Nikou, A.; Verginis Christos, K.; Heshmatialamdari, S.; Dimarogonas Dimos, V. A robust non-linear MPC framework for control of underwater vehicle manipulator systems under high-level tasks. *IET Control. Theory Appl.* **2020**, *15*, 323–337. [CrossRef]
18. Oliveira Éverton, L.; Orsino Renato, M.M.; Donha Décio, C. Disturbance-Observer-Based Model Predictive Control of Underwater Vehicle Manipulator Systems. *IFAC PapersOnLine* **2021**, *54*, 348–355. [CrossRef]
19. Shojaei, K. Neural network feedback linearization target tracking control of underactuated autonomous underwater vehicles with a guaranteed performance. *Ocean. Eng.* **2022**, *258*, 111827. [CrossRef]
20. Tony, S.; Xinbo, Y.; Wei, H.; Okyay, K. Adaptive Neural Network Control of Underwater Robotic Manipulators Tuned by a Genetic Algorithm. *J. Intell. Robot. Syst.* **2020**, *97*, 657–672.
21. Hadi, B.; Khosravi, A.; Sarhadi, P. Deep reinforcement learning for adaptive path planning and control of an autonomous underwater vehicle. *Appl. Ocean. Res.* **2022**, *129*, 103326. [CrossRef]
22. Elhaki, O.; Shojaei, K.; Mehrmohammadi, P. Reinforcement learning-based saturated adaptive robust neural-network control of underactuated autonomous underwater vehicles. *Expert Syst. Appl.* **2022**, *197*, 116714. [CrossRef]
23. Guo, L.; Liu, W.; Li, L.; Lou, Y.; Wang, X.; Liu, Z. Neural Network Non-Singular Terminal Sliding Mode Control for Target Tracking of Underactuated Underwater Robots with Prescribed Performance. *J. Mar. Sci. Eng.* **2022**, *10*, 252. [CrossRef]

24. Liang, J.; Huang, W.; Zhou, F.; Liang, J.; Lin, G.; Xiao, E.; Li, H.; Zhang, X. Double-Loop PID-Type Neural Network Sliding Mode Control of an Uncertain Autonomous Underwater Vehicle Model Based on a Nonlinear High-Order Observer with Unknown Disturbance. *Mathematics* **2022**, *10*, 3332. [CrossRef]
25. Karpenko, M.; Halverson Julie, K.; Besser, R. Waypoint following dynamics of a quaternion error feedback attitude control system. *Proc. Inst. Mech. Eng. Part G J. Aerosp. Eng.* **2022**, *236*, 281–293. [CrossRef]
26. Cortés, P.N.; Torres, M.L.A. A Mirror-Based Active Vision System for Underwater Robots: From the Design to Active Object Tracking Application. *Front. Robot. AI* **2021**, *8*, 542717. [CrossRef]
27. Shi, J.; Fang, J.; Zhang, Q.; Wu, Q.; Zhang, B.; Gao, F. Dynamic Target Tracking of Autonomous Underwater Vehicle Based on Deep Reinforcement Learning. *J. Mar. Sci. Eng.* **2022**, *10*, 1406. [CrossRef]
28. He, Y.; Xie, Y.; Pan, G.; Cao, Y.; Huang, Q.; Ma, S.; Zhang, D.; Cao, Y. Depth and Heading Control of a Manta Robot Based on S-Plane Control. *J. Mar. Sci. Eng.* **2022**, *10*, 1698. [CrossRef]
29. Koyippilly, S.K.; Bandana, M.; Varun, G.M. Into the World of Underwater Swarm Robotics: Architecture, Communication, Applications and Challenges. *Recent Adv. Comput. Sci. Commun.* **2020**, *13*, 110–119.
30. Neira, J.; Sequeiros, C.; Huamani, R.; Machaca, E.; Fonseca, P.; Nina, W. Review on Unmanned Underwater Robotics, Structure Designs, Materials, Sensors, Actuators, and Navigation Control. *J. Robot.* **2021**, *2021*, 5542920. [CrossRef]
31. Kot, R. Review of Obstacle Detection Systems for Collision Avoidance of Autonomous Underwater Vehicles Tested in a Real Environment. *Electronics* **2022**, *11*, 3615. [CrossRef]
32. Kaiser Carl, L.; Lindzey, L. Improving Navigation for Long Endurance Underwater Robots. *Oceanography* **2020**, *33*, 99.
33. Zuluaga Carlos, A.; Aristizábal Luis, M.; Rúa, S.; Franco Diego, A.; Osorio Dorie, A.; Vásquez Rafael, E. Development of a Modular Software Architecture for Underwater Vehicles Using Systems Engineering. *J. Mar. Sci. Eng.* **2022**, *10*, 464. [CrossRef]
34. Ernstsson, A.; Griebler, D. Kessler Christoph. Assessing Application Efficiency and Performance Portability in Single-Source Programming for Heterogeneous Parallel Systems. *Int. J. Parallel Program.* **2022**, *51*, 61–82. [CrossRef]
35. Amandeep, K.; Ashutosh, D.; Grover, P.S. A Framework for Evaluating Extensibility in an Aspect-oriented Software System and its Validation. *Recent Pat. Eng.* **2020**, *14*, 557–566.
36. Manimaran, R. Hydrodynamic investigations on the performance of an underwater remote operated vehicle under the wave using OpenFOAM. *Ships Offshore Struct.* **2022**, *17*, 2186–2202. [CrossRef]
37. Bobkov, V.; Kudryashov, A.; Inzartsev, A. A Technique to Navigate Autonomous Underwater Vehicles Using a Virtual Coordinate Reference Network during Inspection of Industrial Subsea Structures. *Remote Sens.* **2022**, *14*, 5123. [CrossRef]
38. McFarland Christopher, J.; Whitcomb Louis, L. Stable adaptive identification of fully-coupled second-order 6 degree-of-freedom nonlinear plant models for underwater vehicles: Theory and experimental evaluation. *Int. J. Adapt. Control. Signal Process.* **2021**, *35*, 786–810. [CrossRef]
39. Tanveer, A.; Ahmad Sarvat, M. High fidelity modelling and GA optimized control of yaw dynamics of a custom built remotely operated unmanned underwater vehicle. *Ocean. Eng.* **2022**, *266*, 112836. [CrossRef]
40. Zhang, G.; Liu, J.; Sun, Y.; Ran, X.; Chai, P. Research on AUV Energy Saving 3D Path Planning with Mobility Constraints. *J. Mar. Sci. Eng.* **2022**, *10*, 821. [CrossRef]
41. Borovik, A.I.; Rybakova, E.I.; Galkin, S.V.; Mikhailov, D.N.; Konoplin, A.Y. Experience of Using the Autonomous Underwater Vehicle MMT-3000 for Research on Benthic Communities in Antartica. *Oceanology* **2022**, *62*, 709–720. [CrossRef]

MDPI AG
Grosspeteranlage 5
4052 Basel
Switzerland
Tel.: +41 61 683 77 34

Journal of Marine Science and Engineering Editorial Office
E-mail: jmse@mdpi.com
www.mdpi.com/journal/jmse